T0260764

GREEN CHEMISTRY AND ENGINEERING

GREEN CHEMISTRY AND ENGINEERING

A Pathway to Sustainability

Anne E. Marteel-Parrish
Department of Chemistry
Washington College

Martin A. Abraham
College of Science, Technology, Engineering,
and Mathematics
Youngstown State University

AIChE

WILEY

Cover design: John Wiley & Sons, Inc.
Cover images: © Martin A. Abraham and Elizabeth C. Abraham

Copyright © 2014 by the American Institute of Chemical Engineers, Inc.

Published by John Wiley & Sons, Inc., Hoboken, New Jersey. All rights reserved

Published simultaneously in Canada

No part of this publication may be reproduced, stored in a retrieval system, or transmitted in any form
or by any means, electronic, mechanical, photocopying, recording, scanning, or otherwise, except as
permitted under Section 107 or 108 of the 1976 United States Copyright Act, without either the prior
written permission of the Publisher, or authorization through payment of the appropriate per-copy fee
to the Copyright Clearance Center, Inc., 222 Rosewood Drive, Danvers, MA 01923, (978) 750–8400,
fax (978) 750–4470, or on the web at www.copyright.com. Requests to the Publisher for permission
should be addressed to the Permissions Department, John Wiley & Sons, Inc., 111 River Street, Hoboken,
NJ 07030, (201) 748–6011, fax (201) 748–6008, or online at http://www.wiley.com/go/permission.

Limit of Liability/Disclaimer of Warranty: While the publisher and author have used their best efforts
in preparing this book, they make no representations or warranties with respect to the accuracy
or completeness of the contents of this book and specifically disclaim any implied warranties of
merchantability or fitness for a particular purpose. No warranty may be created or extended by sales
representatives or written sales materials. The advice and strategies contained herein may not be suitable
for your situation. You should consult with a professional where appropriate. Neither the publisher nor
author shall be liable for any loss of profit or any other commercial damages, including but not limited to
special, incidental, consequential, or other damages.

For general information on our other products and services or for technical support, please contact our
Customer Care Department within the United States at (800) 762–2974, outside the United States at
(317) 572–3993 or fax (317) 572–4002.

Wiley also publishes its books in a variety of electronic formats. Some content that appears in print may
not be available in electronic formats. For more information about Wiley products, visit our web site at
www.wiley.com.

Library of Congress Cataloging-in-Publication Data:

Marteel-Parrish, Anne E.
   Green chemistry and engineering : a pathway to sustainability / Anne E. Marteel-Parrish,
Department of Chemistry, Washington College; Martin A. Abraham, College of Science, Technology,
Engineering, and Mathematics, Youngstown State University.
      pages   cm
   Includes bibliographical references and index.
   ISBN 978-0-470-41326-5 (hardback)
1. Environmental chemistry–Industrial applications.   I. Martin A. Abraham.   II. Title.
   TP155.2.E58A27 2014
   660.6′3–dc23

                                                                                    2013011104

Printed in the United States of America

10 9 8 7 6 5 4 3 2

*I dedicate this book to Damon Parrish, Ph.D., my endlessly supportive
and patient husband, and to Martin and Marie, my two children
and best accomplishments in life. Without them I would not be
the person I am today.*

*I thank Sara Martin, Washington College Chemistry major 2014,
and Damon Parrish for their assistance with the figures.*
Anne E. Marteel-Parrish, Chestertown, MD, January 2013

*I dedicate this book to my parents, Sam and Barbara Abraham, who have
always been supportive of my goals and aspirations. They have encouraged
me to dream big and reassured me that I could achieve anything that I
wanted. I further dedicate this book to my family, my wife Nancy, and my
children Elizabeth and Josh, who have put up with my long hours in the lab
and the office. Without their support, none of this would have been possible.*
Martin A. Abraham, Youngstown, OH, January 2013

CONTENTS

PREFACE

When green chemistry was first described in 1998 through the publication of *Green Chemistry: Theory and Practice* by Paul Anastas and John Warner, nobody could have predicted the role that it would play today in the world's politics, economics, and education.

The success of green chemistry has been driven by academia, industry, and governmental agencies. It is a central theme within the American Chemical Society and the American Institute of Chemical Engineers, the professional societies for chemists and chemical engineers, and leading organizations that will determine the future of our professions.

The importance of education in driving the future of our profession cannot be understated. The future generations of scientists and engineers, our students of today, who learn chemistry from a green chemistry point of view, will be able to make connections between real-world issues and the challenges that chemistry presents to the environment, and to understand environmentally preferable solutions that overcome these challenges.

This book provides the springboard for students to be exposed to green chemistry and green engineering, the understanding of which will lead to greater sustainability. As Paul Anastas mentioned: "Green chemistry is one of the most fundamental and powerful tools to use on the path to sustainability. In fact, without green chemistry and green engineering, I don't know of a path to sustainability."

This book is aimed at students who want to learn about chemistry and engineering from an environmentally friendly point of view. This book can be used in the first undergraduate course in general chemistry and would be appropriate for a two-semester sequence to allow a more complete understanding of the role of chemistry in society. Portions of this text would be suitable as the basis for a one-semester introductory course on the principles of science and engineering for nontechnical majors, as well.

This book gives students a new outlook on chemistry and engineering in general. While covering the essential concepts offered in a typical introductory course for science and engineering majors, it also incorporates the more fascinating applications derived from green chemistry. This book spans the breadth of general, organic, inorganic, analytical, and biochemistry with applications to environmental and materials science. A novel component is the integration of introductory engineering concepts, allowing the reader to move from the fundamental science included in a typical course into the application areas. As much as the excitement of green chemistry and green engineering occurs at the interface between science and engineering, it is that interface at which we aimed our attention.

This book is divided in three main areas: the first three chapters introduce the birth of green chemistry (Chapter 1), fundamental principles of green chemistry and green engineering (Chapter 2), and the role of chemistry as an underlying force in ecosystem interactions (Chapter 3). After having been provided the foundation of green chemistry and engineering, readers will realize how applications of green chemistry and engineering are relevant while acquiring knowledge about matter, the atomic structure, different types of compounds, and an introduction to chemical reactions (Chapter 4). Readers will also discover the different types of reactions and the quantitative aspect of chemistry in reactions and processes (Chapter 5), while learning about the role of kinetics and catalysis in chemical processes (Chapter 6) and the role of thermodynamics and equilibrium in multiphase systems and processes (Chapter 7). The last four chapters look into novel applications of green chemistry and engineering through the use of renewable materials (Chapter 8) and through the current and future state of energy production and consumption (Chapter 9), while unveiling the relationship between green chemistry and economics (Chapter 10). The importance of toxicology to green chemistry, and the identification of hazards and risks from chemicals to ecological, wildlife, and human health targets conclude this book (Chapter 11).

We hope that this book will enlighten students' perception about chemistry and engineering and will demonstrate the benefits of pursuing a career in the chemical sciences, while contributing to their knowledge of sustainability for our planet and its well-being for our future generations.

ANNE E. MARTEEL-PARRISH
MARTIN A. ABRAHAM

1

UNDERSTANDING THE ISSUES

1.1   A BRIEF HISTORY OF CHEMISTRY

Chemistry (from Egyptian *kēme* (chem), meaning "earth"[1]) is the science concerned with the composition, structure, and properties of matter, as well as the changes it undergoes during chemical reactions.

Chemists and chemical engineers have the tools to design essential molecules, and impart particular properties to these molecules so they play their expected role in an efficient and standalone manner. Chemicals are used throughout industry, research laboratories, and also in our own homes. Discoveries and development of fundamental chemical transformations contribute to longer, healthier, and happier lives. We need chemistry and chemicals to live.

However, chemophobia and the unnatural perception that all chemicals are bad have origins in the remote past, but are still in people's minds today. The following historical background sheds some light on the evolution of the environmental movement.

Green Chemistry and Engineering: A Pathway to Sustainability,
Anne E. Marteel-Parrish and Martin A. Abraham.
© 2014 American Institute of Chemical Engineers, Inc. Published 2014 by John Wiley & Sons, Inc.

1.1.1 Fermentation: An Ancient Chemical Process

Fermentation, an original chemical process that was discovered in ancient times, led to the production of wine and beer. With relatively crude techniques, a simple enzyme contained in yeast was found to catalyze the conversion of sugar into alcohol. Control of the ingredients in the fermentation broth would impact the flavor of the alcohol, and the effectiveness of the conversion was controlled by the length of time the fermentation was allowed to proceed and the temperature of the reaction.

Today, ethyl alcohol, acetic acid, and penicillin are produced through fermentation processes. Separation of the product (which is usually a dilute species in an aqueous solvent) and recycle of the enzyme is required to make these processes operate economically.

1.1.2 The Advent of Modern Chemistry

In the 19th century chemistry was viewed as the "central discipline" around which physics and biology gravitated. The medical revolution with the synthesis of drugs and antibiotics coupled with the development of chemicals protecting crops and the expansion of organic chemistry in every aspect of life increased the life expectancy from 47 years in 1900 to 75 years in the 1990s and to over 80 years in 2007.

Chemistry has contributed greatly to improve the quality of human life. For many years, manufacturers took the approach that the world is big and chemical production is relatively small, so chemicals could be absorbed by the environment without effect. The high value of the chemicals produced created an atmosphere in which the manufacturers believed that successful production was the only concern, and control of their waste stream was irrelevant to success. Eventually, the public developed concerns about the impact of chemicals on health and the environment.

1.1.3 Chemistry in the 20th Century: The Growth of Modern Processes

In the 20th century the growth of chemical and allied industries was unprecedented and represented the major source of exports in the most powerful nations in the world. Among some of the major exports were chemicals derived from the petrochemical, agricultural, and pharmaceutical industries.

1.1.3.1 Petrochemical Processes

In the 19th century, oil was discovered. Originally extracted and refined to produce paraffin for lamps and heating, oil was rapidly adopted as a source of energy in motor cars. Eventually, techniques were developed that allowed oil to be converted to chemicals, and its availability and financial accessibility allowed the petrochemical industry to grow at a tremendous rate. Developments in the modern plastics, rubbers, and fibers industries led to significant demand growth for synthetic materials.

TABLE 1.1. End Products Made from Common Hydrocarbons

Hydrocarbons	Trade Names	Consumer Products
Ethylene (C_2H_4)	Polyethylene (Polythene)	Plastic bags, wire and cable, packaging containers, plastic kitchen items, toys
Propylene (C_3H_6)	Polypropylene (Vectra, Herculon)	Carpets, yogurt pots, household cleaners' bottles, electrical appliances, rope
Butadiene (C_4H_6)	Copolymers with butadiene named Nipol, Kyrnac, Europrene	Synthetic rubber for automobile tires, footwear, golf balls
Benzene (C_6H_6)	Polystyrene	Insulation, cups, packaging for carry-out foods
Toluene (C_7H_8)	Polyurethanes	Furniture, bedding, footwear, varnishes, adhesives
Paraxylene (C_8H_{10})	Polyesters	Clothes, tapes, water and soft drink bottles

Fossil resources, which include oil, natural gas, and coal, are the major sources of chemical products impacting our modern lives. Hydrocarbons, the principal components of fossil resources, can be transformed through a number of refining processes to more valuable products. One of these processes is called cracking, in which the long carbon chains are cracked (broken down) into smaller and more useful fractions. After these fractions are sorted out, they become the building blocks of the petrochemical industry such as olefins (ethylene, propylene, and butadiene) and aromatics (benzene, toluene, and xylenes). These new hydrocarbon products are then transformed into the final consumer products. Table 1.1 gives examples of some end products made from hydrocarbons.

More than 10 million metric tons of oil is used in the world every day. The increasing world population (expected to reach 10 billion people in a few decades) puts increasing pressure on this nonrenewable resource to provide the raw material for a growing consumer demand. Fossil resources also produce 85% of the world's energy supply, and the growing population and increasing energy consumption puts even greater demand on their use. Because society is increasing its consumption of this non-renewable resource, identification of alternative, renewable sources of energy and raw materials for chemicals is emerging as one of the biggest challenges for the 21st century.

1.1.3.2 Agriculture and Pesticides

As the rate of population grew in the 20th century, the demand for food increased dramatically. Production kept up with demand through the use of new technologies such as the synthesis of fertilizers, pesticides, new crop varieties, and extensive irrigation [2]. To provide the necessary cropland, forests were destroyed and prairies and similar types of rangelands were converted.

As new lands were made available for farming, it was discovered that most soils lacked sufficient nitrogen to permit maximum plant growth. Through the nitrogen cycle, bacteria convert atmospheric nitrogen to ammonia and nitrates, which are then absorbed by the plants through their roots. In a natural environment, nitrogen-containing compounds are eventually returned to the soil when plants die and decompose. A natural balance is achieved between the amount of nitrogen removed from the soil through plant growth and the amount returned to the soil through decay. In order to boost the amount of nitrogen required for plant growth, synthetic inorganic fertilizers containing ammonia and nitrates were often applied by farmers. The excessive addition of fertilizers led to runoff of the extra nitrogen-containing compounds in the rivers and lakes and damage to the environment.

More damage to the environment and human health resulted from the development of pesticides to control the impact of insects and other pests. Health issues associated with pesticides were substantial, especially in less developed countries where farmers and employees of the pesticide industries did not take adequate precautions when spraying pesticides. The worst insecticide accident happened in 1984 in Bhopal, India (see Highlight 1.4). One well-known pesticide based on inorganic arsenic salts was used extensively to destroy rodents, insects, and fungi. However, arsenic was recognized as a carcinogen, increasing the risk of bladder cancer. Pesticides based on organophosphates (organic compounds containing phosphorus) were also developed but are especially toxic to human health. A further problem arose when some pests and insects developed resistance to pesticides following repeated uses. In order to overcome the resistance, a more potent pesticide would be applied until resistance was gained, and the cycle repeats. The farmers found themselves on a "pesticide treadmill" [3, p. 451].

A third factor contributing to the increase of grain production was the development of new varieties of crop plants. To produce high-yielding crops, selective cross-breeding was introduced into India, South America, Africa, and other developing countries. Genetically engineered crops started to appear on grocery store shelves in the late 20th century. Through enzymatic transformations, the structure of DNA in living organisms can be modified. Molecular biologists are able to incorporate wanted genes into the DNA of living organisms. For example, in 1994, the first genetically engineered tomato was marketed. Tomatoes are known to be sensitive to frost. To postpone the ripening process, scientists incorporated the "antifreeze" gene of a flounder into a tomato. However, the sales were not profitable so the first genetically engineered tomato was removed from the market. Today, the U.S. Food and Drug Administration (FDA) approves the sale of genetically modified canola, corn, flax, cotton, soybeans, squash, and sugar beet, just to name a few.

Likewise, irrigation systems have been put in place all over the world to make use of arid lands. In hot and humid climates and in the absence of rain, this practice created an accumulation of salts on the soil surface due to the high evaporation rate of water from the soil. The only way to remove excess salts on the surface is to irrigate more. The increase in the salinity of the irrigation water, often recycled through many irrigation cycles, led to a decrease in the productivity of crops, especially beans, carrots, and onions [3, p. 236].

Meeting the food demand of the 21st century is an increasingly difficult challenge, since these new technologies have already been exploited to their maximum potentials, especially in developed countries. Food shortages are expected due to grain productivity decline and growth in the world demand for food.

1.1.3.3 Pharmaceuticals

The modern pharmaceutical industry was born in the 20th century with the mass production of new medicines. The fast growing field of biotechnology and bio-catalysis provided the ability to explore new technological applications through a vital drug discovery process. Among the highlights of the pharmaceutical sector in the 20th century were the discovery and development of insulin, new antibiotics to fight a greater range of diseases, and the development of new drugs for cancer treatment.

The discovery of insulin, a hormone that regulates blood sugar, changed the lives of diabetic patients whose malfunctioning pancreas leads to an inability to produce the required hormone. In 1921, Canadian physician Frederick Banting first isolated the hormone. In the laboratories of Eli Lilly, now the 10th largest pharmaceutical company in the world, the process was developed to extract, purify, and mass produce insulin. Insulin was introduced commercially in 1923.

The second famous discovery happened in 1928 when Dr. Alexander Fleming, a bac-teriologist at London's Saint Mary's Hospital, found that a "magic mold" resisted the action of bacteria. He named the mold penicillin. It was not until 1940 that penicillin was developed into a therapeutic agent by Oxford University scientists Howard Florey and Ernest Chain. Unfortunately, an insufficient supply of penicillin existed until the beginning of World War II, when several U.S.-based companies purified and mass produced penicillin to treat the wounds of U.S. soldiers on the battlefield. A long series of new antibiotics followed in the 1950s, known as the "decade of antibiotics."

Substantial progress in the fight against cancer also occurred during the 20th century. Named *karkinos* by Hippocrates, a Greek physician and the father of medicine, cancer found its origin as early as 1500 BCE. Although typically grouped together, there are a wide variety of cancerous diseases. When cells in our organs continue to multiply without any need for them, a mass or growth called tumor appears. These masses of cells can either be benign (noncancerous, not life threatening, and easily removed) or malignant (cancerous, spread to tissue and organs). Malignant cells can be identified by magnetic resonance imaging (MRI) used in radiology to distinguish pathologic tissue such as a brain tumor from normal tissue. The fight against cancer was pursued with assiduity in the 20th century when chemotherapy and radiation therapy were discovered. The first chemotherapy agent for cancer was actually mustard gas used in World War I. However, the gas killed both healthy and cancerous cells. Since then, many antimetabolites ("any substance that interferes with growth of an organism by competing with or substituting for an essential nutrient in an enzymatic process" [4]) have been developed and deaths from all cancers combined declined.

John E. Niederhuber, M.D., the 13th director of the National Cancer Institute, opined on the growth of biotechnology and its impact on human health. "The continued decline in overall cancer rates documents the success we have had with our aggressive efforts to reduce risk in large populations, to provide for early detection, and to develop new therapies that have been successfully applied in this past decade. … Yet we cannot be content with this steady reduction in incidence and mortality. We must, in fact, accelerate our efforts to get individualized diagnoses and treatments to all Americans and our belief is that our research efforts and our vision are moving us rapidly in that direction" [5].

The contribution of the pharmaceutical sector to health and welfare, the importance of this sector to the economy, and the springboard it provided for research in the medical field have been unprecedented. The challenge of the pharmaceutical industry in the 21st century is to ensure the safety and efficacy of drugs on the market. Unexpected side effects lead to greater numbers of recalls, even after being approved by the FDA. In 2004, a nonsteroidal anti-inflammatory drug named Vioxx, marketed by Merck and prescribed for osteoarthritis, menstruation, and adult pain, was recalled from the U.S. market after it was discovered that the drug caused an increased risk of heart attacks and strokes. The challenge is to maximize the therapeutic benefits of the drug while eliminating or reducing the toxic side effects.

1.1.4   Risks of Chemicals in the Environment

Industrialization and materialization came with a price, sometimes easily recognized but often more obscure. Over time, we have come to realize that the development and use of new chemicals is not without risk, and the associated risk of chemicals in the environment must be managed carefully. In 1962, Rachel Carson, in her well-known book *Silent Spring*, pointed out that "chemicals are the sinister and little-recognized partners of radiation in changing the very nature of the world—the very nature of life." Our planet has been despoiled, and the environment in which we live today is one of a fear of chemicals, and a lack of recognition of their importance in our lives.

There are numerous examples of instances in which advances in chemistry and the introduction of new chemicals did not fully take into account their impacts on our lives. At times the negative side effects were covered over for many years after they were known, but more frequently this was simply a lack of knowledge.

1.1.4.1 Lead Paint

Lead is a toxic metal found mostly in paint, dust, drinking water, and soil. According to the U.S. Environmental Protection Agency the walls of houses built before 1978 are likely to have lead-based paint. Lead from paint chips and lead dust from old painted toys and furniture are particularly dangerous to children, since children are more likely to put hands covered with lead dust in their mouths or eat paint chips containing lead. The growing body of a child absorbs lead rapidly, making a child more sensitive to lead's destructive effects. Lead causes damage to the brain and nervous system, slowed growth, hearing problems, and behavior and learning problems. In late 1991, the Secretary of the Department of Health and Human Services, Louis W. Sullivan, called lead the "number one environmental threat to the health of children in the United States." In 1996 requirements for sales and leases of older housing became effective under the "Residential Lead-Based Paint Disclosure Program Section 1018 of Title X." In 2001 hazard standards for paint, dust, and soil were established by the EPA for most pre-1978 housing and child-occupied facilities.

1.1.4.2 Thalidomide

In the late 1940s and into the following decade, biologists and chemists determined that thalidomide could be used by pregnant women to combat morning sickness and help them sleep. This was a remarkable advance in human health care, as it alleviated a major discomfort. However, all of the biological impacts of the drug within the body were not understood, especially as concerned the relationship with the growing fetus in the womb. From 1956 to 1962, approximately 10,000 children were born with malformations. Scientists had not understood that the use of the chemical could cause birth defects in children, outweighing all of the parental benefits from the use of the drug. This undesirable outcome caused outrage in the general public about the unintended effect of drugs and led to implementation of new governmental regulations for testing new drugs. In 1962, the use of thalidomide during pregnancy was discontinued (Highlight 1.1).

Highlight 1.1 Use of Thalidomide Drug and Pregnancy—Irreversible Effect

During the early 1960s thalidomide was prescribed to pregnant women in Europe and Canada to treat morning sickness. This drug was not approved by the FDA due to insufficient proof of the drug's safety in humans. However, according to the March of Dimes, "more than 10,000 children around the world were born with major malformations, many missing arms and legs, because their mothers had taken the drug during early pregnancy. Mothers who had taken the drug when arms and legs were beginning to form had babies with a widely varying but recognizable pattern of limb deformities. The most well-known pattern, absence of most of the arm with the hands extending flipper-like from the shoulders, is called phocomelia. Another frequent arm malformation called radial aplasia was absence of the thumb and the adjoining bone in the lower arm. Similar limb malformations occurred in the lower extremities. The affected babies almost always had both sides affected and often had both the arms and the legs malformed. In addition to the limbs, the drug caused malformations of the eyes and ears, heart, genitals, kidneys, digestive tract (including the lips and mouth), and nervous system. Thalidomide was recognized as a powerful human teratogen (a drug or other agent that causes abnormal development in the embryo or fetus). Taking even a single dose of thalidomide during early pregnancy may cause major birth defects."

New therapeutic uses are being found for thalidomide. In 1998 the FDA approved the use of thalidomide to treat leprosy and studies are currently looking at the effectiveness of this drug to relieve symptoms associated with AIDS, inflammatory bowel syndrome, macular degeneration, and some cancers.

1.1.4.3 Toxic Chemicals in the Environment

Limited understanding of the role of pharmaceuticals in contact with humans was paralleled by a limited understanding of the impact of chemicals in the environment. Examples of poor management of chemical waste abound.

On June 22, 1969, the Cuyahoga River in Cleveland, Ohio, caught on fire, when oil-soaked debris was ignited by the spark from a passing train car. Although only a brief river fire, this incident brought national attention to the poor state of the nation's urban rivers.

The Love Canal in Niagara Falls, New York, was used as a waste disposal site by Hooker Chemical and the City of Niagara Falls from the 1930s to 1950s (Highlight 1.2).

Highlight 1.2   Niagara Falls and the Love Canal—Not a Love Affair After All [6]

> If you get there before I do
> Tell 'em I'm a comin' too
> To see the things so wondrous true
> At Love's new Model City
>   —From a turn-of-the-century advertising jingle promoting
>   the development of Love Canal

Love Canal, named after William T. Love, was supposed to be a dream community. Love's vision was to dig a canal between the upper and lower Niagara Rivers to generate cheap electricity to the soon-to-be Model City. However, the dream shattered when economic strain and discovery of the alternating current to transmit electricity over long distances came into play. In the 1920s the partial ditch was turned into a municipal and chemical dumpsite by Hooker Chemical Company, owners and operators of the property at the time. They used the site as an industrial dumpsite until 1953 when, after covering the canal with soil, they sold it to the city for one dollar. About 100 homes and a school were built on the ticking time bomb until it exploded in 1978. After a record amount of rainfall, corroded waste-disposal drums started to leach their contents into the backyards and basements of the homes and school built on the banks of the canal. The air was filled with a choking smell and children had burns on their hands and faces from playing in the neighborhood. Birth defects and a high rate of miscarriages started to surface.

Residents were evacuated and relocated after New York Governor Hugh Carey announced on August 7, 1978 that the state would purchase their homes. On the same day the first emergency financial aid fund was approved by President Carter for something other than "natural disaster."

> Give me Liberty. I've Already Got Death.
>   —From a sign displayed by a Love Canal resident, 1978

The site was later sold to the city for construction of a school, with Hooker disclosing that the site had been used as a waste repository. The school was built nearby in 1955.

In Times Beach, Missouri, the roads were sprayed with waste oil to reduce dust formation. Unfortunately, the contractor combined waste oil with other hazardous chemicals, including dioxin, one of the main components of Agent Orange. As a result of the contractor's actions, the entire town of Times Beach was determined to be contaminated with dioxin, the town was quarantined, and the inhabitants were relocated by the government.

General Electric (GE) Corporation produced polychlorinated biphenyls (PCBs) at its plants in Fort Edward and Hudson Falls, New York, for use as dielectrics and coolant fluids in transformers, capacitors, and electric motors. From 1947 through 1977, they discharged the runoff from this process into the Hudson River. In 1983, the U.S. Environmental Protection Agency declared 200 miles of the Hudson River a superfund site, and sought to develop a cleanup and remediation plan to remove the PCBs that contaminated the sediment at the bottom of the river. Phase 1 cleanup was completed in 2009, at a cost to GE of $460,000,000. A projected Phase 2 effort will be even larger and more expensive (Highlight 1.3).

Highlight 1.3   Impact of Industry on Local Environment— Example of the Hudson River and GE

The Hudson River is not only famous for being the site of the successful ditch of the U.S. Airways Flight 1549 on January 15, 2009 by Captain Chesley "Sully" Sullenberger, it was also the waste disposal site of approximately 1.3 million pounds of polychlorinated biphenyls (PCBs) by General Electric (GE) Corporation between 1947 and 1977. Polychlorinated biphenyls are long lived and semivolatile and do not dissolve in water; therefore they can travel a long distance. They are also fat soluble and concentrate very rapidly in animal tissues and go up in the food chain. Experts have reported that PCBs are proved to cause cancer in animals and are probable human carcinogens.

Two GE capacitor plants located in Fort Edward and Hudson Falls, New York, discharged PCBs now found in water, sediment, fish, and the whole Hudson River ecosystem. GE agreed to perform Phase 1 of the cleanup process, which started in May 2009. The dredging of the upper Hudson River was set for about six months to remove approximately 10% of the PCBs. GE has not committed to the removal of the full scope of the contaminants, which is the goal of Phase 2. The issue in this story is not the cost (the cost of the EPA's proposal to GE was $460 million), but rather if the cleanup will work. GE does not believe that dredging is the solution to the problem and has invested "$200 million on a groundwater pump to reduce the flow of PCBs from the bedrock below its Hudson Falls facility from 5 pounds to 3 ounces a day." GE officials have pointed out that the level of PCBs in fish is down 90% since 1977. The Hudson River is only one site out of 77 other sites where GE is responsible for the cleanup.

1.1.4.4 Bhopal

The industrial disaster of 1984 in Bhopal, India, was caused by the release of 40 tons of methyl isocyanate gas by a Union Carbide pesticide plant, resulting from a series of worker errors and safety issues that had not been properly addressed. The official government report documents 3787 deaths as a result of this leakage, although reports of as many as 20,000 deaths are widely accepted. Today, more than 100,000 people still suffer from painful symptoms, most of which doctors are not sure how to treat. Furthermore, most of the waste left behind is in evaporation ponds outside the factory walls and this poses a danger for the health of nearby residents who get their drinking water from hand pumps and wells. The plant is still not dismantled (Highlight 1.4), and legal wrangling over responsibility for cleanup of the site continues today.

Highlight 1.4   Tragic Wake-up in Bhopal, India, in 1984

In the late 1960s Union Carbide built a chemical plant supplying pesticides to protect Indian agricultural crops. Methyl isocyanate was used in the production of a carbamate insecticide called Sevin. Initially, methyl isocyanate was shipped from the United States but in the late 1970s a plant was specifically built on the outskirts of Bhopal for the manufacturing of methyl isocyanate. On December 3, 1984 at approximately 12:30 in the morning an explosion releasing a cloud of poisonous gas killed between 2500 and 5000 people and injured up to 200,000 people. Approximately 100,000 people lived within a 1-kilometer radius of the plant at the time of the tragedy [7].

The source of the explosion is believed to be the reaction of methyl isocyanate with water, which created an exothermic reaction accompanied by the formation of carbon dioxide, methylamine gases, and nitrogenous gases. The wind was blowing at the time of the accident and 27 tons of toxic gas traveled over the city, contaminating water and food supplies. Little was known about the acute toxicity and long-term effects of exposure to methyl isocyanate at the time of the pesticide manufacture. By 3 a.m. the first deaths were reported and tens of thousands of people were seen in hospitals within the first 24 hours. The inhalation of the toxic gas resulted in chronic respiratory illnesses among Bhopal residents and deaths due to bronchial necrosis and pulmonary edema. Other toxic effects such as acute ophthalmic effects and maternal–fetal, gynecological, and genetic effects were also accounted for.

This tragedy raised many issues, such as addressing the close proximity of heavily populated settlements to chemical plants, assessing the risk of toxic compounds being used or produced, and developing a plan to maintain a safe operation of chemical industries and to protect workers and nearby residents in case of disaster.

Union Carbide (now owned by Dow Chemical Company) agreed to pay US$470 million in damages.

1.1.5   Regulations: Controlling Chemical Processes

With the growing environmental awareness throughout the 1960s and into the early 1970s, the United States initiated a series of legislative initiatives that controlled the release of toxic materials into the environment, and set standards for clean air and clean water. A brief and noncomprehensive timeline includes the following breakthrough actions:

- The Clean Air Act of 1970 addresses and regulates emissions of hazardous air pollutants. One of the main goals of this act was to reduce the formation of ground-level ozone, an ingredient of smog.
- The Clean Water Act of 1972 regulates discharges of pollutants into waters of the United States.
- The Resource Conservation and Recovery Act (RCRA) of 1976 allows the U.S. EPA to control hazardous waste from a cradle-to-grave perspective.
- The Toxic Substances Control Act of 1976 (TSCA) gave the U.S. EPA "the ability to track the 75,000 industrial chemicals currently produced or imported into the United States."
- The Comprehensive Environmental Response, Compensation and Liability Act (CERCLA) of 1980 was established "to clean up such sites and to compel responsible parties to perform cleanups or reimburse the government for EPA-led cleanups" [8]. Included within CERCLA legislation was the Superfund authorization, which allowed the EPA to address and compel private industry to address abandoned hazardous waste sites.
- The Emergency Planning and Community Right-to-Know Act (EPCRA) of 1986 was established to "help local communities protect public health, safety, and the environment from chemical hazards" [9]. Most notable in this legislation was the development of the Toxic Release Inventory (TRI), which is a database containing detailed information on about 650 chemicals and chemical categories. In 2006, there were 179 known or suspected carcinogens on the TRI list, of which lead and lead compounds accounted for 54% and arsenic and arsenic compounds for 14%.

The Pollution Prevention Act (also called P2 Act) of 1990 designated the EPA to embark on a mission of source reduction, rather than monitoring and cleanup (Highlight 1.5). "Congress declared it to be the national policy of the United States that pollution should be prevented or reduced at the source whenever feasible; pollution that cannot be prevented should be recycled in an environmentally safe manner, whenever feasible; pollution that cannot be prevented or recycled should be treated in an environmentally safe manner whenever feasible; and disposal or other release into the environment should be employed only as a last resort and should be conducted in an environmentally safe manner" [10].

At the same time the EPA was working on the Clean Air Act and the Clean Water Act, the first United Nations Conference on the Human Environment (UNCHE) was

> ### *Highlight 1.5   The Pollution Prevention Act of 1990* [11]
>
> The Pollution Prevention Act of 1990, passed by Congress, authorized the U.S. Environmental Protection Agency to develop cost-effective approaches and control of pollution from dispersed or nonpoint sources of pollution. Pollution prevention, also called "source reduction," is the first step to reduce risks to human health and the environment.
>
> Dealing with pollutants at "the end of the pipe" or after disposal was not cost effective in terms of pollution control and treatment costs. This act states that "pollution should be prevented or reduced at the source whenever feasible; pollution that cannot be prevented should be recycled in an environmentally safe manner, whenever feasible; pollution that cannot be prevented or recycled should be treated in an environmentally safe manner whenever feasible; and disposal or other release into the environment should be employed only as a last resort and should be conducted in an environmentally safe manner." The Office of Pollution Prevention was established following passage of the act.
>
> To encourage source reduction and recycling, owners and operators of industrial facilities must report on their releases of toxic chemicals to the environment under the EPCRA of 1986.
>
> In January 2003, The National Pollution Prevention Roundtable published "An Ounce of Pollution Prevention Is Worth Over 167 Billion Pounds of Cure: A Decade of Pollution Prevention Results 1990–2000." The 167 billion pounds of pollution prevented included data from air, water, waste, and electricity. More than 4 billion gallons of water were also conserved. The main implementation barriers to the pollution prevention (P2) program were lack of capital, high rate of staff changes, and lack of management commitment.

held in Stockholm, Sweden, in 1972. This conference acknowledged the need to reduce the impact of human activities on the environment, the specificity of the environmental issues to developing countries versus developed countries, as well as the need for international collaboration to work on these global problems. The United Nations Environmental Program (UNEP) whose mission is "to provide leadership and encourage partnership in caring for the environment by inspiring, informing, and enabling nations and peoples to improve their quality of life without compromising that of future generations" was launched as a result of this conference. A step forward defining sustainability was accomplished.

Recognizing that pollution does not respect the boundaries between countries, it became clear that international agreements would be required to control the more onerous of environmental issues. One of the most successful examples of such international cooperation has been the Montreal Protocol, signed by 100 countries in September 1987, and made effective in 1989. This agreement led to a ban on ozone-depleting chemicals, such as chlorofluorocarbons (CFCs).

1.2 TWENTY-FIRST CENTURY CHEMISTRY, aka GREEN CHEMISTRY

1.2.1 Green Chemistry and Pollution Prevention

Throughout their history, chemists have discovered some revolutionary molecules and synthetic pathways that bring new products and technologies to society. Production techniques have often neglected the impact of these materials and processes on the environment. Today, with the increased environmental awareness, it is crucial to discover new ways of producing the same or similar molecules with desirable properties but with zero waste and zero pollution. New materials that are inherently nontoxic and have functionality that replaces hazardous chemicals, and processes to make these materials without the use of toxic intermediates or release to the environment, need to be developed.

1.2.1.1 Understanding Risk

As mentioned in the fact sheet published in December 2005 by the State of Ohio Environmental Protection Agency [12]: "Each day we are exposed to risks. Some risks are the result of our own behavior: choices we make such as diet, smoking, speeding on the freeway or playing a contact sport. Other risks come from factors we don't directly control: hazardous weather conditions, environmental pollutants, even our own genetic history."

One accepts risk on a daily basis. When you drive in your car, there is a real potential that you will have a wreck, and that wreck might even lead to death. The insurance company can determine how likely you are to have an accident and, from that calculation, will assign you to a specific risk category, from which they determine the amount of the premium. The insurance company assumes a portion of your risk, and you pay a price for that.

We can also consider risk from a chemical standpoint. Consider two relatively similar chemicals. Benzene is a known carcinogen, is a likely mutagen, and is a known nervous system toxin. Based on its impact on humans, benzene is a particularly nasty chemical. Toluene, however, is an irritant without any known cancer-causing activity and is listed as possibly causing damage to the central nervous system. In short, benzene is known to be a particularly harmful chemical, whereas toluene is simply a chemical deserving of concern but without any special toxicity issues. You wouldn't want to risk coming into contact with benzene without special protective equipment, but you might be willing to work with toluene.

Risk assessment is the process by which we evaluate the potential adverse health effects associated with people coming in contact with an environmental hazard. It consists of two specific activities:

- Determining the extent to which a group of people may be exposed to a particular chemical hazard, and
- Determining the intrinsic hazard associated with exposure to the chemical through a particular exposure route.

Once these two components are determined, then the total risk may be determined through a combination of these two elements:

$$\text{Risk} = \text{hazard} \times \text{exposure}$$

Traditional engineering approaches focus on risk management; understand the hazard and take steps to minimize the likelihood of exposure. This has resulted in the use of personal protective equipment, such as the safety goggles you wear in the laboratory. Safety protocols are in place throughout the manufacturing sector, and existed in Bhopal. Because these risk management initiatives are imperfect, hazardous chemicals continue to find their way into the environment.

1.2.1.2 Benign by Design

Green chemistry, on the other hand, is more attuned to reducing the hazard through inherently safer design (ISD). If the chemical of concern is replaced by a less hazardous material, then that chemical cannot possibly be harmful, since it is removed from the process. An inherently safe process is one in which nothing bad can happen, even if something does go wrong. Inherently safer design reduces cost, since less expense is incurred in ensuring that workers and the public are not exposed to an unsafe material. As described by Berkeley (Buzz) Cue, Founder and President of BWC Pharma Consulting, LLC, inherently safer design is "just a question of changing your mind set."

ISD requires the chemist and engineer to go back to the drawing board and think about alternative feedstocks, solvents, and synthetic pathways when developing a new process. It is about redesigning products. As an example, paint used to be made from oil-based materials, and its application created volatile organic compounds in the ambient air. Today, all paints are water-based (latex). They work the same as previous paints, but they are inherently less hazardous. Latex paint is a green chemistry product.

Green chemistry is not a new type of chemistry, but a new philosophy of chemistry, one focused on the reduction of risks and inherent safety. It is about new benign by design alternatives. It is not more complex than traditional chemistry but it is a distinctive approach based on the evaluation of toxicity of materials and their by-products when designing a safer, cleaner, and cost-efficient process.

1.2.2 Sustainability

In Stockholm, Sweden, in 1972, the First International Conference on the Human Environment focused the attention of the world on the transnational issues associated with pollution. Twenty years later, the United Nations Conference on Environment and Development (UNCED), also known as the Earth Summit, occurred in Rio de Janeiro, Brazil, in June 1992. This summit featured 178 countries discussing global problems such as poverty, war, and sustainable development. For the first time, the protection of the welfare of the planet was seen as the key driver to promote long-term economic and social progress. This was the Earth Summit's most important achievement. Some notable results of the UNCED included:

- The Rio Declaration, a set of 27 principles aimed at ensuring environmental protection by participating governments,
- Agenda 21, which is the "blueprint for sustainability" and defines an international plan of action for achieving sustainable development, and
- A Statement of principles for the Sustainable Management of Forests intended to value reforestation and forest conservation.

Ten years after Rio's Earth Summit, the United Nations World Summit on Sustainable Development (WWSD), also known as Earth Summit II or Rio +10, took place in Johannesburg, South Africa, in 2002. Even if some have argued that the agenda of this summit may have been too ambitious, it acknowledged that sustainable development addresses not just environmentally related issues but also economic and sociopolitical matters. Key points of discussion included water and sanitation, energy, human health, agricultural productivity, biodiversity, and ecosystem management. The main outcome of the summit was the Johannesburg Declaration, which reinforced agreements made at the Rio de Janeiro and Stockholm summits. Specific goals outlined in the declaration pointed at:

- Reducing poverty through the establishment of a solidarity fund.
- Cutting in half by 2015 the number of people living on less than a dollar per day and the number of people who lack clean water and basic sanitation and who suffer from hunger (these goals were already mentioned as Millennium Development Goals derived from the United Nations Millennium Declaration in 2000).
- Increasing sources of renewable energy.
- Restoring depleted fish stocks by 2015.

Heads of the richest industrialized countries (France, Germany, Italy, Japan, the United Kingdom, the United States, Canada, Russia), the President of the European Commission, and representatives from five developing countries, including Brazil, China, India, Mexico, and South Africa, referred to as the Outreach Five, attend annual G8 summits. At the 2008 G8 summit in Japan, world priorities such as the current global food crisis, African development, climate change, intellectual property rights, and some political issues were discussed. The eight richest nations in the world pledged to cut global greenhouse emissions by 50% by 2050. However, a baseline year was lacking for those cuts. Concerns and ideas were shared, challenges and long-term impacts were described, but the implementation of ideas did not happen.

Four years later, at the G8 summit in Camp David, Frederick County, Maryland, on May 18–19, 2012, the leaders of Britain, Canada, France, Germany, Italy, Japan, Russia, and the United States met to address major global economic, political, and security challenges. Most importantly, it was once again recognized that development and access to environmentally safe and sustainable sources of energy is essential to global economic growth. The issue of food security was addressed, and a goal of providing a new alliance with African leaders to allow 50 million people to live above the poverty level by 2022.

1.2.2.1 Energy, Food, and Water

Energy, food, and water—resources fundamentally linked to one another—are becoming scarce in the 21st century. Called the "perfect storm" of food, energy, and water shortages by the 2009 U.K. government chief scientist Professor John Beddington, scarcity for these resources will happen at an international level. Professor Beddington predicts that as the world population will get to 8.3 billion by 2030, "demand for food and energy will increase by 50% and for fresh water by 30%" [13].

If these issues are not addressed now, widespread shortages will be followed by increased prices, that could create international conflicts. Security of these resources is now a high priority on the political agenda of the heads of many countries as the prices of oil and goods continue to rise.

One way to tackle the problem, according to Professor Beddington, is to improve agricultural productivity since "30 to 40% of all crops are lost due to pest and disease before they are harvested." More disease- and pest-resistant crops as well as plants resistant to drought and salinity coupled with more efficient irrigation and harvesting practices are needed in the 21st century.

1.2.2.2 Global Climate Change

Many concerns have already emerged in the developed countries, but the developing countries face similar sustainability challenges. In his book, *Natural Capitalism: Creating the Next Industrial Revolution* with co-authors Amory Lovins and L. Hunter Lovins, Paul Hawken laid out some potential disasters coming up, such as decline in grain production and in fish per capita, decline in jobs (1 billion people looking for work), loss of rain forests, disappearing wetlands, nuclear waste cleanup, greenhouse gases, global warming, stratospheric ozone depletion, and rising global temperatures. Global warming and climate change are widely recognized among scientists, although controversy still rages among the political community. By the end of the 21st century temperatures may rise between 1.8 and 4 degrees Celsius (3.2 to 7.2 degrees Fahrenheit). Greater impact is observed in the polar regions, and Greenland has warmed by 4 °C since 1991. Impacts are seen on every continent and every ocean [14, pp. 26–27].

The abundance of coal and its low cost make it the world's choice to provide the growing electricity demand. However, "worldwide coal-plants are responsible for 20% of human-caused greenhouse gas emissions" [14, p. 10]. The combustion of coal, natural gas, and biomass (according to the American Heritage Dictionary of the English language biomass is defined as "the total mass of living matter within a given unit of environmental area or as plant material, vegetation, or agricultural waste used as a fuel or energy source"), the rapidly growing population, and increasing development in Asia and Africa are adding to the challenge. Extreme dry weather has contributed to famine and the spread of disease, especially in African countries such as Ethiopia, which are heavily dependent on rain-fed agriculture and cannot afford adequate watering technology.

1.2.2.3 Global Carbon Footprint

One way to measure the impact of an individual on the environment is through the carbon footprint, which varies substantially from region to region. As published in the magazine *National Geographic*, Special Report on "Changing Climate," early in 2008, families from Botswana, the United States of America, and India have very different carbon footprints. In Botswana, the traditional way of cooking uses wood and biomass for fuel, which leads to annual CO_2 emissions of 2.2 metric tons per person, twice the annual emissions per person in India. Since the population of Botswana is only 1.8 million, the global impact is small. In the United States, the total CO_2 emissions in 2005 reached 5957 million metric tons, or approximately 18.5 metric tons per person. While most people in the United States are using energy-efficient appliances and taking other actions to reduce their personal carbon foot-print, the average emissions include their contributions through the transportation, manufacturing, and residential sectors.

The United States is making strides in reducing its carbon footprint, with per capita emissions in 2011 falling to the lowest levels since 1992. The State of California is, in many respects, leading the way. In California, energy efficiency and conservation, sustainability, green building, and green purchasing are put into prac-tice through Executive Order S-20-04 known as the "Green Building Initiative" and the Green Building Action Plan. Governor Arnold Schwarzenegger created the Green Action Team, leading the call for public buildings to be 20% more energy efficient by 2015. California, often referred to as "the Green State," is devoted to creating healthier environments for its citizens in which to work, live, and learn (Highlight 1.6).

Highlight 1.6   Welcome to Green California! [15]

The State of California is "working to reduce energy and resource consumption in state buildings, while lowering greenhouse gas emissions, and creating healthier environments in which to work, live, and learn."

Some of the goals of Governor Schwarzenegger's Green Building Executive Order were to attain a LEED-EB (Leadership in Energy and Environmental Design for Existing Buildings) standard for all state occupied buildings over 50,000 square feet, to seek ENERGY STAR leases over 5000 sq. ft beginning in 2006 (for new leases) and 2008 (for existing leases), to develop solar photovoltaic projects and biomass fueled energy generation systems, to acquire goods and services that have a lesser or reduced impact on the environment and health, to include alternative fuel use and fuel cell vehicle utilization as well as purchase additional hybrid vehicles.

The launch of new education and outreach efforts by The California Building Standards Commission in 2009 provides a new way to increase awareness among the Californian population.

Development and population growth add to the global carbon footprint. In India, the growing population, strong economic development, and even a small increase in the living standard are having a tremendous impact on the average carbon footprint. Economists foresee that by 2050 more than 600 million automobiles will be on India's roads, an increase that has never been seen in any other country in the world. The footprint of China, which represents a geographical area similar to that of the 48 contiguous states and has a population five times that of the United States, has grown to 5.3 metric tons per person and continues to increase. Growth in the developing countries is a central question, as voiced by Maurice Strong, the cofounder and chairman emeritus of the Earth Council: "The fate of the Earth will be decided in the developing world." One asks, "Should the developing countries follow the model provided by the developed countries?" [16]. The Agenda 21/Rio summit states clearly that the developing world must have the right to strive for the same rights, chances, and justice as the developed world, but at what cost? Is there a path to prosperity for these regions that does not result in a similar level of emissions?

1.3   LAYOUT OF THE BOOK

The current image of chemical industries is associated with pollution makers and environment destructors, and the image of chemicals is associated with hazards and toxicity. But the chemical industry has also been responsible for great advances in medicine and science, and the development of new materials such as plastics. Chemistry is a part of society, whether we want/like it or not. Chemistry is essential to maintain and improve the quality of life, but chemistry must work with other disciplines to develop the new technologies needed for the advancement of society. It goes hand-in-hand with disciplines such as ecology, biology, toxicology, and chemical engineering. Together with these additional disciplines, the chemical industry will be transformed into one that creates the materials and products demanded by society without the environmental consequences that society is no longer willing to accept.

Getting from "here to there" takes patience, collaboration across disciplines, and a sense of achievement and influence. This book will help to show that there are environmentally benign alternatives to current syntheses and chemicals. Understanding these inherently safer design alternatives will enable us to reduce our local and global impacts on the environment, health, society, and economy. As mentioned by L. Hunter Lovins: "We need to invent whole new institutions, new ways of doing business, and new ways of governing." Where do we want to be 50 years from now? What do we want our planet to look like? How do we get out of our comfort zone and change our way of thinking? If you are interested in having the answers to these questions and if you are ready to pursue science in a creative, innovative, and responsible manner, then this book is for you.

Green Chemistry and Engineering: A Pathway to Sustainability is divided into 11 chapters. The first three chapters provide background on the practices and principles of green chemistry and engineering, the impacts of chemistry in natural systems, and some basic tenets of life cycle analysis and pollution prevention. The following four chapters of the book enable the readers to learn about matter, the heart of chemistry, to understand chemical reactions and processes commonly used and their greener alternatives, concepts about kinetics, catalysis, and reaction engineering from an environmentally benign point of view, and the role of thermodynamics and equilibrium. The last four chapters focus on applications of renewable materials and current and future states of energy sources, as well as the relationship between green chemistry and economics and green chemistry and toxicology.

Green chemistry and green engineering are not new scientific disciplines. There is even a belief that the word green will disappear because "green" will be the only way to do chemistry. This new way of thinking, doing, teaching, and learning is part of the solution to our global problems. It is a new philosophy of life beneficial to our own health, environment, business, and sense of community. Our own drive may be different from our neighbor's but the end product will be the same. Let's begin by looking at how it all started.

REFERENCES

1. http://en.wikipedia.org/wiki/Chemistry#cite_note-0.
2. http://www.icistrainingsite.com.
3. Girard, J. *Principles of Environmental Chemistry*, 2nd ed., Jones and Bartlett Publishers, Sudbury, MA, 2010.
4. http://dictionary.reference.com/browse/anti+ metabolites?s=t.
5. http://www.cancer.gov/newscenter/pressreleases/Report Nation2009Release.
6. http://www.epa.gov/history/topics/lovecanal/01.htm.
7. http://www.tropmed.org/rreh/vol1_10.htm.
8. http://www.epa.gov/superfund/.
9. http://www.epa.gov/oem/content/epcra/index.htm.
10. http://www.epa.gov/p2/pubs/basic.htm.
11. http://www.eoearth.org/article/Pollution_Prevention_ Act_of_1990,_United_States.
12. http://www.sbcapcd.org/airtoxics/OhioEPA-Understanding RiskAssessment.pdf.
13. http://www.guardian.co.uk/science/2009/mar/18/ perfect-storm-john-beddington-energy-food-climate.
14. *National Geographic* Special Report June 2008.
15. http://www.green.ca.gov/default.htm.
16. Anderson, R. C. *Mid-course Correction: Towards a Sustainable Enterprise: The Interface Model*, Chelsea Green Publishing Company, White River Junction, VT, 1998.

2

PRINCIPLES OF GREEN CHEMISTRY AND GREEN ENGINEERING

2.1 INTRODUCTION

Green is the color of chlorophyll; it is also the color of money. For some countries it is associated with a political party, for others it may be everything that is beneficial to the environment. Green chemistry is not a new branch of chemistry or about the chemistry of molds, leaves, or plants. It is apolitical and is rather the new color of science. It is cleaner, smarter, and cheaper chemistry.

Green chemistry strives to minimize waste production, to promote the use of renewable and recycled resources, and to achieve the highest possible energy efficiency. It is based on the design or redesign of chemicals at the basic molecular level. When we asked ourselves in the first chapter: "Will future generations be able to live adequately with sufficient food supplies, energy resources, and means of transportation?" we realized that for the answer to be yes, our way of thinking had to change. "We've got to get over the idea that all of this is here for us," said Janine Benyus, founder of the Biomimicry Guild and author of *Biomimicry: Innovation Inspired by Nature* [1]. "What we have to do is learn from what's out there." Nature

Green Chemistry and Engineering: A Pathway to Sustainability,
Anne E. Marteel-Parrish and Martin A. Abraham.
© 2014 American Institute of Chemical Engineers, Inc. Published 2014 by John Wiley & Sons, Inc.

- Sustainability
  - Ecosystems
  - Human Health
- Green Engineering
  - Lifecycle
  - Systems
  - Metrics
  - Energy efficiency
- Green Chemistry
  - Feedstocks and reagents
  - Reactions and catalysts
  - Solvents
  - Thermodynamics
  - Toxicology

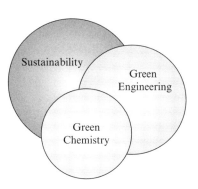

Figure 2.1. Sustainability, green engineering, and green chemistry.

should be our mentor. Effectively, "What use is a house if you haven't got a tolerable planet to put it on?" asked Henry David Thoreau, American author and naturalist (1817–1862).

Green chemistry addresses the issue of "doing well by doing good" [2]. We all know that it is more efficient to prevent cancer than to try to cure it after being diagnosed. The same concept applies to pollution. Prevention of pollution rather than treating pollution and disposing of chemicals is a sustainable solution. We all have heard "An ounce of prevention is worth a pound of cure" by Benjamin Franklin, one of the Founding Fathers of the United States of America and renowned inventor (1706–1790). Green chemistry is not applying bandage technology anymore. As Paul Anastas, former Director of the Center for Green Chemistry and Green Engineering and Professor in the Practice of Green Chemistry in the School of Forestry and Environmental Studies at Yale University mentioned, "Green chemistry is powerful because it starts at the molecular level and ultimately delivers more environmentally benign products and processes" [3].

Green chemistry is distinct from green engineering, which is itself distinct from sustainability. Although the terms are often used interchangeably, each of these concepts embodies slightly different ideas and encompasses different scopes of activity. They are related terms with some overlap, but as indicated in Figure 2.1, they have unique foci that create distinct flavors in the application and use of these concepts.

Within the remainder of the chapter, we provide greater details on the definition of each of these terms. For now, it is worthwhile to consider the concepts in their unique flavor. Green chemistry focuses on the design of chemical products to minimize their inherent hazard, using fundamental principles of chemistry. Green engineering is more oriented toward the process and, as such, to identify ways in which manufactured products (including those that are not inherently chemical in nature) are produced and used. On the other hand, sustainability is focused on the system, not only the nature of the material and how it's produced, but also on why it is needed in the first place. Because of its breadth and scope, issues of sustainability require more than just scientists and engineers; sustainability science requires contributions from social scientists, economists, health professionals, and more as it seeks to transform the fundamental value system of the human population.

2.2 GREEN CHEMISTRY

2.2.1 Definition

So what exactly is green chemistry? Green chemistry applies fundamental chemical principles to produce chemical products that are inherently less toxic, either to humans or to the ecosystem, than currently existing chemical products. Green chemistry may be applied to any of the various elements of the chemical product life cycle, from manufacture, to use, and ultimately to disposal. Thus, green chemistry may be applied to the production of a particular chemical to minimize the hazard associated with its use, or it may focus on the manufacture of the chemical to minimize the environmental consequences of the by-products or the synthesis, or it may equally well look into the development of more environmentally friendly alternatives to a specific chemical. Regardless, green chemistry seeks to reduce the hazard associated with chemical species.

Green chemistry is distinct from environmental chemistry. The latter focuses on understanding the natural environment, the interactions of chemicals in the eco-system, and, ultimately, the nature and fate of pollutants within the environment. Environmental chemistry leads to an understanding of the interactions of chemical species in the environment; green chemistry emphasizes minimizing the addition of undesired species into the ecosystem.

Green chemistry was first defined by Paul Anastas and John Warner. In their revolutionary book *Green Chemistry: Theory and Practice* published in 1998, green chemistry is defined as "the utilization of a set of principles that reduces or eliminates the use or generation of hazardous substances in the design, manufacture and application of chemical products" [4, p. 11]. One of the distinct features of green chemistry is the idea of "benign by design," that is, the deliberate action to make a chemical product or process inherently less hazardous. This chemical philosophy was described in the previous chapter as inherently safer design and is also referred to as "environmentally benign chemical synthesis" or "alternative synthetic pathways for pollution prevention."

Green chemistry is an outgrowth of the 1990 Pollution Prevention Act (PPA) and draws from all chemical disciplines including organic, inorganic, analytical, biochemistry, and even physical chemistry. Prior to 1990, the U.S. government focused on command and control strategies that prescribed mechanisms to limit the impact of the chemical enterprise on the environment. Pollution was considered acceptable, as long as it was kept below some acceptable value. If a manufacturer exceeded the permitted emission limit, then a penalty would be imposed. Thus, the business could make an economic decision based on the cost of installing recovery or treatment equipment relative to the cost of violating a permissible emission limit.

The PPA changed the economic equation, promoting an opportunity for the manufacturer to seek new technologies or chemical products that would be inherently benign, thereby making the regulation irrelevant. This is the inherent value of green chemistry; no longer would a company be forced to install certain types of treatment technology, but within the green chemistry paradigm the company is able to meet its environmental targets through innovation. Green chemistry is "an opportunity rather than a limitation" [4, p. 12].

Green chemistry is not just a way to protect the environment by preventing pollution before its creation; it is also a way to increase efficiency and reduce costs of production, an opportunity for businesses to lighten their environmental burdens and make money. To demonstrate how green chemistry can be applied, Anastas and Warner have provided some directions detailed in the 12 principles of green chemistry [4, pp. 29–55].

2.2.2   Principles of Green Chemistry and Examples

The 12 principles are summarized below and examples of each principle are provided.

1. *Prevention: It is better to prevent waste than to treat or clean up waste after it has been created.* Prevention starts by avoiding the use or generation of hazardous substances. If hazardous materials are not produced, then treatment and disposal are not required. Moreover, extraordinary safety measures needed for the manufacture of hazardous materials are not required, making the green product less costly to produce and easier to use. Why waste time, money, and effort to deal with these avoidable consequences? Remember the quote from Benjamin Franklin: "An ounce of prevention is worth a pound of cure."

2. *Atom economy: Synthetic methods should be designed to maximize the incorporation of all materials used in the process into the final product.* The concept of atom economy was invented by Professor Barry Trost from Stanford University [5]. Atom economy is a measure of how much of the reactants are actually incorporated into the products. This concept differs completely from the concept of yield, which is still taught in traditional general chemistry courses. With yield, one is only focused on the molar quantity of product formed relative to the molar amount of reactant. Atom economy looks more closely at what happens to the side or waste products. One can easily imagine the case where a product is formed in 100% yield through an elimination reaction, but because the moiety being eliminated is large, the reaction may only have 50% or 75% atom economy. Atom economy is an essential tool in measuring how much waste is being produced and thus provides a valuable metric for reducing the use of nonrenewable resources, minimizing the amount of waste, and reducing the number of steps used to synthesize chemicals (Highlight 2.1). The ultimate measure of "reaction efficiency" is calculated as being the product of chemical yield by experimental atom economy.

Highlight 2.1 Definition and Calculations Dealing with the Application of Atom Economy

Chemists are interested in assessing the reaction efficiency and, most of the time, use the percent yield of a reaction calculated as the ratio of the actual yield of a specific product divided by the theoretical yield (based on the limiting reactant). This tool is used to quantify the efficiency of a chemical reaction and to compare the expected product quantity to the actual one.

However, just looking at a percent yield does not tell the complete story. A percent yield calculation does not measure how efficiently the reactants have been used in generating the desired product and therefore does not take into account how much waste (toxic or benign) is created.

To express how efficiently reactant atoms are used in the products, Barry Trost of Stanford University developed the concept of atom economy, for which he received the Presidential Green Chemistry Challenge Award in 1998. The percent atom economy is calculated as:

$$\left(\text{Mass of atoms in desired product / mass of atoms in reactants}\right) \times 100\%$$

This tool does not take into account the amount of reactants often used in excess as well as the yield of the reaction. However, it helps in making the final decision about the efficiency of a synthetic pathway versus another.

Let's work out an example. Propene (C_3H_6) can be synthesized via two routes:

Route 1: Propene is generated from the reaction of trimethylpropylamine ion $[C_6H_{16}N]^+$ with a strong base $[OH^-]$ under heat conditions. The by-products of this reaction are trimethylamine $[C_3H_9N]$ and water.

Route 2: Propene can also be generated from the decomposition of 1-propanol $[C_3H_8O]$ under heat and acidic conditions. In this case the only by-product is water.

When calculating the molecular mass of each reactant and product we need to add the atomic weight of all elements present in the molecule (see Chapter 5). The mass of propene is $42\,g/mol$ ($3 \times 12\,g/mol + 6 \times 1\,g/mol = 42\,g/mol$); the mass of trimethylpropylamine and OH^- combined is $119\,g/mol$ whereas the mass of 1-propanol is $60\,g/mol$.

Route 1:

$$\text{Percent atom economy} = 42\,g / \left(102\,g + 17\,g\right) \times 100\% = 35\%$$

Route 2:

$$\text{Percent atom economy} = 42\,g / 60\,g \times 100\% = 70\%$$

We conclude that route 2 offers the highest atom economy (twice the one obtained for route 1) but the only by-product in route 2 is water; so by comparing these two synthetic pathways we can determine that route 2 is the most environmentally friendly way to produce propene.

We need to exercise caution when looking at atom economy. A synthetic pathway with the highest percent atom economy is not always the less toxic one. Assessing the toxicity and hazards associated with the by-products (in route 2 the only by-product was a benign compound, water) is essential to make the final call.

3. *Less hazardous chemical syntheses: Wherever practicable, synthetic methods should be designed to use and generate substances that possess little or no toxicity to human health and the environment.* The use of hazardous chemicals is linked to risk of harm. Risk can be written as the product of exposure and hazard. A quick segue into nuclear power illustrates this concept perfectly. The hazard associated with a nuclear reactor meltdown is catastrophic. Because of the severity of the hazard, we demand that huge costs be expended to minimize the exposure. Ultimately, the combination of these two features makes nuclear power acceptable to a large segment of the population, despite the huge environmental risk. Put mathematically, if we multiply infinity (the hazard) by zero (the exposure), we get none. Since we can never reduce the exposure to zero, there is always a finite risk with any potentially hazardous material. On the other hand, chemists and scientists have the knowledge and skills to develop safer reactions and chemicals. Hazard is an intrinsic characteristic: it will not change. We can minimize the risk by reducing the hazard, and thus reduce the costs associated with minimizing the exposure, creating a new dynamic for determining an acceptable risk (Highlight 2.2).

4. *Designing safer chemicals: Chemical products should be designed to achieve their desired function while minimizing their toxicity.* Many chemicals are multifunctional, with a specific portion of the molecule providing the desired function while another portion of the molecule might impart some undesirable toxicity characteristics. Careful analysis of a selected chemical can elicit which elements of the molecule provide the desired and undesired functions. If the mechanism of action of a chemical is known, then it is possible to modify its structure and reduce its toxicity. Even if the mechanism of action is unknown, chemical structure analysis of the functional groups can be used to distinguish between the functional elements and the potentially hazardous moieties. By redesigning the molecule, the functionality related to the toxic effect can then be avoided, minimized, or totally suppressed, while the functionality providing the desired activity is retained (Highlight 2.3).

5. *Safer solvents and auxiliaries: The use of auxiliary substances (e.g., solvents and separation agents) should be made unnecessary wherever possible and innocuous when used.* Anecdotal information suggests that many reactions are performed in a selected solvent for expediency, and not for chemical or environmental reasons. In other words, a solvent that was once identified as "good" is simply chosen for all subsequent versions of the chemistry, without

Highlight 2.2   Risk, Hazard, and Exposure?

How can we assess Risk $=f$ (hazard, exposure)?

Risk is often expressed as the probability of adverse effects associated with a particular activity. For example, the fatality risk associated with smoking 10 cigarettes per day is 1 in 1000 compared to the risk associated with being hit by automobiles which is 4 in 100,000. Some risks are voluntary risks (such as smoking) while others are not (e.g., natural disasters).

The risk assessment process always starts with identification of hazards (e.g., what chemicals are potentially harmful based on toxicity score). The toxicity score is a function of the maximum concentration of a chemical and its acceptable daily intake. For example, when comparing toxicity score values for chlorobenzene (previously used in the manufacture of pesticides such as DDT) and BEHP (bis-2-ethylhexyl phthalate used as solvent in glowsticks and in the manufacture of PVC), the toxicity score for BEHP is 11,500 compared to 320 for chlorobenzene. The evaluation of other properties of the contaminant such as bioaccumulation and persistence in the environment as well as treatability is also important to assess. Most importantly, the second step in assessing risk is to look into the dose–response relationship. In other words, quantifying the dose in relation to its adverse effects will enable a safe dose to be established. There is normally a proportional relationship between an increase in the dose administered or received and the incidence of an adverse health effect and its severity for an exposed population. A dose is often quantified as mg chemical/kg of body weight. The third step is to assess means of exposure (occupational exposure, i.e., workplace and community exposure) and routes of exposure (dermal exposure through the skin, lung exposure through inhalation, and ingestion). Intake rate calculations can help identify chronic daily inhalation intake as a function of the concentration of contaminant, contact rate, frequency, exposure duration, and body weight. The final step in the risk assessment process is the integration of hazard identification, dose response, and exposure assessment. This ultimately leads to the characterization of a contaminant as carcinogenic or noncarcinogenic. Since data is often missing, educated guesses and assumptions are often necessary, which leads to uncertainty. Computational analysis is a very helpful tool to make more informed decisions regarding risk management (see Chapter 11 for general information on toxicity, and Section 11.5 for more on computational methods).

regard to whether it is "best." But these solvents may be released into the environment, and if they don't impart specific chemical benefit, then their value must be questioned. The impacts of solvents on the environment, health, and ecosystem are well known. Whenever possible, reactions should be developed with the environmental impacts of the solvent in mind, and benign solvents such as supercritical CO_2 or water, immobilized solvents, or solventless systems should be considered. An example of a conversion to a solventless process is detailed in Highlight 2.4.

Highlight 2.3   Designing Molecules While Retaining Functionality and Suppressing Toxicity

In 2011 the Sherwin-Williams company reinvented oil-based paints by developing water-based acrylic alkyd paints with low levels of volatile organic compounds (VOCs). These new paints are made from recycled soda bottle plastic (polyethylene terephthalate or PET), acrylics, and soybean oil.

Oil-based "alkyd" paints generate high levels of VOCs as the paint dries. Acrylic-based paints were developed in the past with production of low VOCs but their performance was not satisfactory. Instead, Sherwin-Williams designed a new technology taking advantage of the combination of a PET-based polymer dispersion (for rigidity, hardness, and hydrolytic resistance) with acrylic (for improved dry times and durability) and soya functionality (for gloss formation and flexibility). This new technology combines the benefits of alkyd and acrylic paints with lower VOCs, without surfactants and with minimum odor. By redesigning the molecular content of this paint product, Sherwin-Williams was able to eliminate 800,000 pounds of VOC solvents and petroleum-based solvents since 2010. Sherwin-Williams won the 2011 Presidential Green Chemistry Challenge award in the designing greener chemicals category.

Highlight 2.4   Solvents: Are They Necessary?

The cosmetics and personal care manufacturers are heavy consumers of esters, usually synthesized using strong acids and hazardous organic solvents at high temperatures. In 2009, Eastman Chemical Company managed to replace strong acids and organic solvents and eliminated the production of undesirable by-products using immobilized enzymes (such as lipase involved in the formation or breaking of lipids) in the esters synthesis. These enzymes serve as biocatalysts in the production of esters and can easily be removed by filtration. They also allow the synthesis to be performed under mild processing conditions and therefore save energy and increase yield. This enzymatic route saves over 10 liters of organic solvent per kilogram of product. Overall, the quality of the product is superior to the one obtained via conventional esterification, while the cost and environmental footprint are improved. Eastman Chemical Company received the 2009 Presidential Green Chemistry Challenge Award in the greener synthetic pathways award category.

6. *Design for energy efficiency: Energy requirements of chemical processes should be recognized for their environmental and economic impacts and should be minimized.* Synthetic methods that are conducted at ambient temperature and atmospheric pressure consume less energy than those that are done at high temperature and pressure. Recovery of energy resources and reuse within the

Highlight 2.5   Heat Integration for Sustainable Energy

Heat integration analysis (or process integration) is a way to save energy by examining the "potential of exchanging heat between heat sources and heat sinks via the use of heat exchangers and reducing the amount of external heating and cooling required"[6].

Several case studies have been published regarding hospitals as important thermal energy users [7, 8]. Herrera and co-workers [6] looked into a hospital complex including an institute, a general hospital, a regional laundry center, a sports center, and other public buildings. Input of high-priced diesel fuel in boilers to produce steam usually met the heat demand. However, several sinks of heat were identified such as the hot streams coming from the laundry and condensed steam not recovered in the condensation network. A heat integration analysis suggested the addition of four heat exchangers to use energy that would otherwise be wasted (two in the laundry, one in the machinery room to help heat the boiler feed water, and the last one in the condensation tank area that heats the sanitary water) [7]. Kemp furthered this analysis by suggesting various hospital operating procedures to further reduce energy consumption, such as changing the periods of high utility demands since energy consumption seemed to vary considerably between day and night, or recovering exhaust gases to supply a space heating system, or to invest in a standby diesel generator to supply emergency power rather than expand overall generation capacity [8]. New designs based on heat integration led to energy savings of over 30% in major producing companies. Heat integration can be applied equally well to small or large buildings and offices and is a way to promote ecobuilding technologies.

process can minimize the need for external energy resources. Analysis of the heating requirements of a reaction for maximum heat integration minimizes the overall need for fossil resources. An example of heat integration analysis is presented in Highlight 2.5.

7. *Use of renewable feedstocks: A raw material or feedstock should be renewable rather than depleting whenever technically and economically practicable.* Depleting or nonrenewable fossil resources are subject to the criterion of time and cannot be replenished nearly as rapidly as they are consumed. Renewable feedstocks are biological and plant-based starting materials that can be replaced through their natural processes. Because there are limited supplies of depleting resources, we must take care to use these judiciously and look for opportunities in which renewable materials may be substituted economically. Opportunities for renewable feedstocks as the basis for chemical production require novel chemical techniques and new processing methods, but should be considered as an important method to reduce the reliance on nonrenewable fossil resources. The reuse and recycle of metals and other nonrenewable natural resources must also be considered as an opportunity to minimize the consumption of these depleting materials.

8. *Reduce derivatives: Unnecessary derivatization (use of blocking groups, protection/deprotection, temporary modification of physical/chemical processes) should be minimized or avoided if possible, because such steps require additional reagents and can generate waste.* During the manufacture of pharmaceuticals, pesticides, and dyes, it is sometimes necessary to derivatize certain functional groups or parts of the molecule to protect them from reaction with reagents that facilitate a necessary transformation. Conceptually, protection/deprotection requires two extra steps to accomplish a simple addition or a reaction step to a particular functional group. Extra reaction steps mean extra processing steps and extra waste, all of which reduces the atom economy of the overall process. The unnecessary derivatization generates additional waste upon deprotection.

9. *Catalysis: Catalytic reagents (as selective as possible) are superior to stoichiometric reagents.* Catalysts increase the rates of chemical reactions but are not consumed in the reactions. As such, a catalyst may decrease the temperature at which a reaction can be performed economically, thus saving energy costs. Because a catalyst will accelerate the rates of certain reactions more than others, a catalyst can also be used to effect the selectivity in a sequence of reactions. In some reactions, stoichiometric reagents are used to promote a particular pathway or to influence a specific reaction step. Substitution of a catalytic material, particularly one that can easily be recovered, enhances the performance of the reaction in terms of speed and decreases the consumption of materials. Chapter 6 covers some of the basic catalytic tools used in typical chemical reactions and in industry.

10. *Design for degradation: Chemical products should be designed so that at the end of their function they break down into innocuous degradation products and do not persist in the environment.* Some chemicals such as plastics made from petroleum intermediates or pesticides persist in the environment. For example, there are billions of plastic bags manufactured every year. The usable average life of a plastic bag is only 20 minutes but it can take up to 1000 years for the bag to degrade under ambient conditions. New technologies are being developed to insert photoactive materials into plastic bags that would cause them to degrade in the presence of sunlight, or for them to be more easily digested by microorganisms in the environment. Today, only about 10% of plastic bags are biodegradable (Highlight 2.6).

11. *Real-time analysis for pollution prevention: Analytical methodologies need to be further developed to allow for real-time, in-process monitoring and control prior to the formation of hazardous substances.* The monitoring of a chemical process is beneficial for several reasons: the formation of toxic by-products can be detected early on and parameters can be adjusted to reduce or eliminate formation of these substances. Real-time monitoring ensures that the reaction is running smoothly, and reduces the amount of

Highlight 2.6   Plastic Bags and Microbes?

Most of the traditional methods used to decompose plastic require high temperatures and harsh chemicals. In 2009, a 16-year-old high school student at Waterloo Collegiate Institute in Ottawa, Canada, won a science fair for finding an answer to the famous question: "Are there microorganisms able to decompose plastic bags?" Daniel Burd managed to find a way to breed microorganisms able to degrade plastic in a very short amount of time [9]. He immersed ground plastic in a yeast solution and was able to isolate the effective strains and maintain their growth so that 43% of the plastic degraded in 6 weeks. Unfortunately, the experiments were run under highly controlled environments that would need to be in place for the biodegradation to happen. Several bacteria-based solutions have been found since 2009 and microbial degradation is one of the most effective and environmentally friendly ways to incorporate green chemistry in a new design pathway.

off-specification product formed. Control systems can be designed to autocorrect as process conditions change, ensuring optimal performance of the system.

12. *Inherently safer chemistry for accident prevention: Substances and the form of a substance used in a chemical process should be chosen to minimize the potential for chemical accidents, including releases, explosions, and fires.* Toxicity, explosivity, and flammability must be part of the design of chemical products and processes. Pollution prevention relies on accident prevention to minimize the likelihood of chemicals leaking into the environment. However, if one uses inherently safe materials, then even in the event of a leak, there is minimal hazard. For example, when water is used as a reaction solvent, the loss of solvent is of no concern, because the water is nonhazardous. On the other hand, elaborate precautions must be taken with the use of organic solvents, since volatile organic compounds in the atmosphere are hazardous to the environment and human health. Other techniques to minimize the presence of hazardous materials include their generation as part of the reaction scheme, followed by rapid consumption of these materials. By controlling the process conditions, the amount of the hazardous substance is kept to a minimum at all times, which minimizes the impact of that substance in the event of an accident or spill.

2.2.3   Presidential Green Chemistry Challenge Awards

The Presidential Green Chemistry Challenge Program was created in 1996 to promote innovative research in and uses of green chemistry for pollution prevention. This partnership program is led by the EPA's Office of Chemical Safety and Pollution Prevention in cooperation with the American Chemical Society, other EPA offices, other federal agencies, and others. Although the program has had a varying scope over the years, including a research component, the program is currently limited to the administration of the Presidential Green Chemistry Challenge Awards.

Every year, the Presidential Green Chemistry Challenge Awards Program seeks nominations from individuals, groups, and both nonprofit and for profit organizations, including academia, government, and industry to compete for awards in recognition of innovative technology incorporating the principles of green chemistry. Nominations are judged by an independent panel of experts. Three areas of focus describe these awards.

2.2.3.1 Use of Greener Synthetic Pathways

The use of greener synthetic pathways involves either implementing a novel, green pathway for a new chemical product or redesigning the synthesis of an existing chemical product. According to the EPA, "examples include synthetic pathways that:

- Use greener feedstocks that are innocuous or renewable (e.g., biomass, natural oils).
- Use novel reagents or catalysts, including biocatalysts and microorganisms (e.g., protein enzymes).
- Are natural processes, such as fermentation or biomimetic synthesis (replication of reactions occurring in nature using *in vitro* system).
- Are atom-economical (this is the topic of principle 2 of green chemistry).
- Are convergent syntheses. Convergent syntheses aim at improving the yield of multistep chemical reactions by reordering the reaction steps, which serves to increase the final yield of the desired product. Convergent synthesis can be highly effective when applied to the production of complex molecules such as proteins made up of 300 amino acids."

A recent use of greener synthetic pathways was exemplified in the production of 1,4-butanediol using a microbial synthetic pathway and genetic engineering, for which the company Genomatica won the 2011 Presidential Green Chemistry Challenge Award. Genomatica is involved in the production of basic and intermediate chemicals made from renewable feedstocks such as sugars and biomass. Genomatica created a successful bio-based fermentation process of sugars for production of 1,4-butanediol [$HO(CH_2)_4OH$] as feedstock for polymers and commodity chemicals that are used in products such as spandex, automative plastics, and running shoes. Several outcomes were observed: reduction of energy by 60% compared to the acetylene-based process; reduction of carbon dioxide emissions by 70%; no organic solvent was used in the fermentation process and the water used was recycled; and a safer working environment. This bio-based process is expected to be economically competitive with traditional processes based on current oil and natural gas prices.

2.2.3.2 Use of Greener Reaction Conditions

The use of greener reaction conditions includes:

- Replacing hazardous solvents with solvents that reduce the impact on human health and the environment.
- Operating under solventless reaction conditions and performing solid-state reactions.
- Novel processing methods.
- Eliminating energy- or material-intensive separation and purification steps.
- Improving energy efficiency, including running reactions closer to ambient conditions.

As an example, biocatalysis has found its way into greener manufacturing of drugs. This is the case for the second generation production of sitagliptin, the active ingredient in Januvia™, used in the treatment of type-2 diabetes. Prominent companies in this field are Merck, a leading health-care and pharmaceutical proider, and Codexis, a leading provider of biocatalysts. Merck and Codexis collaborated to design a greener manufacture of sitagliptin using biocatalysts while reducing waste, improving yield and safety, and removing the need for a metal-based catalyst. This new route relies on the use of evolved transaminases (important enzymes in the production of amino acids). The traditional pathway involved a high-pressure catalytic hydrogenation step with a rhodium catalyst, ultimately leading to the production of undesired enantiomers. Major benefits include a 56% improvement in productivity, a 10–13% increase in yield, and a 19% reduction in waste generation mostly due to the elimination of the purification steps. Both Merck and Codexis won awards individually in 2006 and co-received the 2010 Presidential Green Challenge Award in the greener reaction conditions category.

2.2.3.3 Design of Greener Chemicals

Designing greener chemicals involves the development of new materials that are less hazardous than chemicals used in traditional technologies. Greener chemicals would have the following characteristics:

- Less toxic than currently available chemical products.
- Inherently safer with regard to accident potential.
- More easily recycled or biodegradable after use.
- Reduction in atmospheric impacts (e.g., nonozone depleting or smog forming).

In this electronic world, the dependence on paper products has decreased, but the paper and packaging industry still plays an important role in the economy of the United States, employing about 400,000 people and posting sales up to $115 billion per year. The traditional pathway to increase strength and quality of paper from cellulose in wood relies on either significant mechanical treatments and therefore intense energy requirements, or the use of nonrenewable chemicals. Buckman International, Inc. based in Memphis, Tennessee, developed Maximize® enzymes to increase the number of fibrils in cellulose, thereby improving the strength and quality of paper without additional mechanical or chemical treatment. Maximize enzymes are a combination of cellulose enzymes and enzymes from natural sources produced from fermentation. While these enzymes use less energy and electricity for refining, the overall treatment is also less toxic, safer to handle than the current treatments, and also completely recyclable. The commercialization of this biotechnology took place in 2010. One company using Maximize technology saved over $1 million per year and another one claimed that the machine speed increased by 20 feet per minute for a 2% increase in production. It is expected that this new technology will be applied in over 50 paper mills in the United States and beyond.

Five awards are typically given to an academic investigator for a green chemistry technology in any of the three focus areas mentioned above, to a small business* for a green chemistry technology in any of the three focus areas, to an industry sponsor for the use of greener synthetic pathways, to a second industry sponsor for the use of greener reaction conditions, and to a third industry sponsor for the design of greener chemicals.

A complete list of the award winners can be accessed on the EPA website: `http://www.epa.gov/greenchemistry/pubs/pgcc/past.html`.

2.3 GREEN ENGINEERING

2.3.1 Definition

The Environmental Protection Agency (EPA) defines green engineering as "the design, commercialization, and use of processes and products that are feasible and economical while minimizing (1) generation of pollution at the source and (2) risk to human health and the environment" [9]. The goal of green engineering is to minimize the impact of chemical processes on human health and the environment. The impact of environmentally conscious strategies is to reduce risks to ecosystems, workers, consumers, and the general population while manufacturers and suppliers are cutting their cost of production.

Engineers apply risk assessment to pollution prevention in their strategies. Risk is expressed as a mathematical function of hazards and exposures. Risk assessment methods help quantify the degree of environmental impact for individual chemicals. Engineers apply technologies to control the risks as an element of the design processes and products, taking into account the likelihood that certain actions will occur. Thus,

*According to the EPA: "A small business is defined here as one with annual sales of less than $40 million, including all domestic and foreign sales by the company, its subsidiaries, and its parent company."

a fundamental difference between green engineering and green chemistry is the assumption of risk as an acceptable element to be controlled, rather than attempting to design the risk out of the particular chemical species.

As mentioned by the EPA: "By applying risk assessment concepts to processes and products, the engineer can accomplish the following:

- Estimate the health and environmental impacts of specific chemicals on people and ecosystems.
- Prioritize those chemicals that should be minimized or eliminated.
- Optimize the process design to avoid or reduce health and environmental impacts.
- Assess feed and recycle streams based on risk, rather than volume, when evaluating the impact of a chemical process.
- Design greener products and processes."

Green engineering also involves systems analysis, creating an opportunity to consider the life cycle impacts of a particular product or process. In a life cycle analysis, one considers the environmental impacts at various stages of the process. For example, one may consider that the electric vehicle is "zero" emissions. However, when one recognizes that the electricity that is required to maintain charge on the battery is produced from a power plant (often a coal-fired power plant), then it is appropriate to assign a portion of the environmental impacts of that power plant to the operation of the electric vehicle, and it is no longer truly "zero" emissions. Similar types of analyses allow one to properly assign all of the impacts of a product and to thus choose the product that has the least environmental impact over its entire life cycle. Failure to look at the full life cycle results in the selection of a process improvement that enhances the environmental performance in a particular aspect while creating greater environmental challenges in the overall consumption of the product.

2.3.2 Principles of Green Engineering

The 12 principles of green engineering were proposed by Paul Anastas and Julie Zimmerman and published in *Environmental Science and Technology* on March 1, 2003 [10]. According to Anastas and Zimmerman, there are two fundamental concepts that designers should integrate in their design: "life cycle considerations and the first principle of green engineering, inherency."

1. *Inherent rather than circumstantial: Designers need to strive to ensure that all material and energy inputs and outputs are as inherently nonhazardous as possible.* Materials and energy sources that are inherently benign minimize the possibility of building hazardous substances into the design of the product or process. Thus, it is inherently preferred to use water as a solvent than an organic solvent, since water is inherently benign, whereas the vapors of the organic solvent need to be controlled and the solvent itself

needs to be recovered and recycled. The use of inherently benign materials reduces the need for constant monitoring and control throughout the hazard's lifetime.

2. *Prevention instead of treatment: It is better to prevent waste than to treat or clean up waste after it is formed.* Waste involves a resource that is not used. Any material that is not used, even if it is inherently innocuous, needs to be recovered and recycled. Recovery requires additional processing steps, the use of additional materials, and the consumption of increased energy resources. If the material cannot be recovered, it must be disposed of in some way, and thus it creates a burden on the environment. From a business standpoint, the company pays once for the material it does not need, and then a second time to dispose of the waste. Clearly, reducing the amount of waste enhances the process, both environmentally and economically.

3. *Design for separation: Separation and purification operations should be designed to minimize energy consumption and materials use.* Separation and purification steps consume significant amounts of energy and may also be materials intensive, using large amounts of hazardous solvents. The separation process is often one of the most energy-intensive steps in the manufacturing process. Thus, the incorporation of self-separated reaction products prevents the need to spend energy and materials to recover the output product.

4. *Products, processes, and systems should be designed to maximize mass, energy, space, and time efficiency.* The concept of efficiency focuses on proper consumption of a resource. Processes that occur in small spaces and over short times can contain the hazard, minimizing the need for elaborate safety and risk management methods. Mass efficiency again refers to the incorporation of the maximum mass into the process. The second law of thermodynamics, which states that the amount of energy embodied in a product can never be fully recovered, suggests that it is best to maximize the energy intensity in the process. That is, small periods of intense energy use are preferred to using low energy levels over extended periods.

5. *Output-pulled versus input-pushed: Products, processes, and systems should be output-pulled rather than input-pushed through the use of energy and materials.* This approach is based on Le Châtelier's principle, which states that "when a stress is applied to a system at equilibrium, the system readjusts to relieve or offset the applied stress." Stress is defined as temperature, pressure, or concentration gradient. It is possible to increase the productivity of a process by overwhelming the feed with a specific reactant. That leads to incredible excess of the reactant in the product stream, which requires separation, recovery, and recycle. On the other hand, if you can remove a product from the process stream as it is produced, then the reaction will drive to greater quantities of product, without any need for use of excess reactants.

6. *Conserve complexity: Embedded entropy and complexity must be viewed as an investment when making design choices on recycle, reuse, or beneficial disposition.* Highly complex materials can be broken down into simple molecules, which can then be used in the production of different complex materials. However, this is often more energy and material intensive than using the natural complexity of the existing material. In a similar way, complex end products that can be reused are easier to recycle than those that need to be broken down into individual subcomponents.

7. *Durability rather than immortality: Targeted durability, not immortality, should be a design goal.* Immortality of a product is associated with environmental issues such as persistence and bioaccumulation, and ultimately the need for secure waste disposal. Products and processes should be designed to accomplish their task through their anticipated operating conditions and lifetimes, but then be recyclable and reusable afterwards. The "greener" design should take into account effective and efficient maintenance and repair without the addition of extra material and energy. By designing a durable and biodegradable product, long-term environmental impacts are significantly reduced.

8. *Meet need, minimize excess: Design for unnecessary capacity or capability (e.g., "one size fits all" solutions) should be considered a design flaw.* Unnecessary resources are often spent to "overdesign" a product or process with the intention of designing for worst-case scenarios regardless of local time, space, and operating conditions. This tendency often results in a lot of excess materials and energy wasted throughout the process. Greener designs need to incorporate the specific needs and functions of materials without going overboard.

9. *Minimize material diversity: Material diversity in multicomponent products should be minimized to promote disassembly and value retention.* Earlier principles discuss the desire for materials to be recovered and recycled. Minimizing the number of unique substances included within any product simplifies the product and makes recovery easier. A single material such as a polymer with tailoring backbones specifically engineered to accomplish desired properties is easier to recycle and reuse than multicomponent materials.

10. *Integrate local material and energy flows: Design of products, processes, and systems must include integration and interconnectivity with available energy and materials flows.* Utilizing existing energy and material flows will increase efficiency. Reusing wasted heat or existing materials from adjacent processes reduces the consumption of raw materials and improves the life cycle efficiency of the process and the sustainability of the product. The waste product from one facility may be a raw material for another facility. Energy can be recovered from the wastewater stream of an industrial plant. Integrating processes provides the maximum opportunity for recovery of energy and material resources.

11. *Design for commercial "afterlife": Products, processes, and systems should be designed for performance after their intended function has been completed.* Afterlife should be part of the design strategy of a product. Incorporating components whose function and value can be recovered for reuse and reconfiguration after a premature end of life or a change in style is necessary. For example, pesticides will remain in the environment after the crop has been harvested. A pesticide that will decompose to innocuous materials in the natural environment in a reasonable time frame minimizes its long-term impact.

12. *Renewable rather than depleting: Material and energy inputs should be from renewable resources to the maximum extent possible.* Clearly, material and energy resources that are consumed at a rate that exceeds the ability of the natural environment to replenish that resource will eventually be fully depleted and not available for future use. Thus, processes and products should be designed to consume raw materials only at the rate at which they can be replenished. Because materials can be recycled, net consumption is the important parameter to be considered. In other words, aluminum that is obtained from recycled material does not deplete the raw material in the ground and thus may be considered renewable if there is sufficient recycled aluminum to fully meet the process needs.

2.4   SUSTAINABILITY

The idea of sustainability goes beyond the engineering and science associated with green chemistry and engineering, to incorporate social/health and economic factors into the discussion of the most appropriate technologies to implement. At the beginning of the chapter, we indicated that sustainability includes the concepts of ecosystems and human health. Incorporating these issues into engineering design creates new challenges in terms of valuation; one must consider the value of the rain forest relative to the value of clean drinking water or sufficient food supply. Clearly, sustainability involves analysis of the system, and thinking on a global scale.

According to the Brundtland Commission (formerly known as the World Commission on Environment and Development), sustainable development is generally defined as "providing for human needs without compromising the ability of future generations to meet their needs" [11].

The U.S. Environmental Protection Agency describes sustainability from two perspectives [12]: A public policy perspective would define sustainability as the satisfaction of basic economic, social, and security needs now and in the future without undermining the natural resource base and environmental quality on which life depends. From a business perspective, the goal of sustainability is to increase long-term shareholder and social value, while decreasing industry's use of materials and reducing negative impacts on the environment.

An alternative business perspective can also be used to describe the concepts of sustainability in terms of the triple bottom line [13]:

Economic viability: the business aspects of a project.

Social concerns: human health and social welfare.

Natural or ecological issues: environmental stewardship.

Regardless of the perspective one takes, sustainability involves the intersection of environment, economics, and social welfare and requires analysis of the trade-offs that weigh the value of each of these goals. In January 2004, a group of 65 engineers, scientists, and others met on the shores of San Destin, Florida, to develop a set of principles for sustainable engineering. These principles, with further analysis, follow [14].

Sustainable engineering transforms existing engineering disciplines and practices to those that promote sustainability. Sustainable engineering incorporates development and implementation of technologically and economically viable products, processes, and systems that promote human welfare while protecting human health and elevating the protection of the biosphere as a criterion in engineering solutions.

To fully implement sustainable engineering solutions, engineers use the following principles:

- Engineer processes and products holistically, use systems analysis, and integrate environmental impact assessment tools.

One must consider all of the impacts and all of the opportunities available to deliver a desirable service for society. The choice of how this outcome can be delivered results from a combination of discrete decisions that lead to the development of a system for the delivery of that service. Transportation provides a good opportunity to consider systems analysis. The use of the internal combustion engine as the power source for the automobile requires the presence of fueling stations throughout the country. In Europe, where the diesel engine is more commonly used, fueling stations provide diesel fuel. Trade-offs are made through the choice of power source: diesel engines produce greater particulates than Otto-cycle engines, but they are also more efficient. Which is preferred? Thinking beyond the choice of the fuel source, the same activity could be achieved without the use of a personal vehicle, if a robust mass transit system was in place. Again, decisions are made that impact delivery of the service. In Europe, a significant rail presence exists and many people travel by train on a regular basis. In the United States, the car is king, and trains are rarely a viable alternative.

- Conserve and improve natural ecosystems while protecting human health and well-being.

What is an ecosystem? There are several different definitions, but all focus on the interactions of plants, animals, and microorganisms within their environment.

Natural ecosystems are those that exist without input from humans. The preservation of natural ecosystems relates to the concept of using renewable resources, but goes further to reveal that even the use of renewable materials will disturb the natural cycles in nature. Maintaining these cycles reduces the impact of humanity on the planet and allows historical interactions to continue unabated.

- Use life cycle thinking in all engineering activities.

Life cycle is a key concept of green engineering and simply requires that one consider all aspects of the development of a product or process. While the use of a particular product may not lead to great hazard, the manufacture of that product may be particularly problematic and thus should be developed with great care. For example, a solar panel has very limited environmental impact while in use, but because of its need for hazardous and rare metals, it could have significant impacts during manufacturing or in disposal. Decisions must be made to weigh the impact of the product throughout its life cycle, to determine the true costs of the product against a sustainable outcome.

- Ensure that all material and energy inputs and outputs are as inherently safe and benign as possible.

This concept returns to familiar topics from the principles of green chemistry and green engineering. The use of benign materials is preferred, since these materials do not require elaborate engineering schemes to be used safely. And, if a material is truly benign, then disposal can usually be achieved without risk to the natural environment, since it will be assimilated into the environment through natural processes.

- Minimize depletion of natural resources.

Whereas the second principle discusses preservation of ecosystems, this principle focuses on conservation of natural resources. Recognizing that natural resources can be depleted, their use should be minimized. Attempts to recycle and reuse resources can minimize their depletion and leave them available for future generations.

- Strive to prevent waste.

Although seemingly straightforward, this concept is deceptively complex. A waste is a resource that is not available for another purpose. But when one brings systems together, it is possible that the waste stream from one process can become the feed for a second process. The wastewater of a chemical plant can be fed into a digester to produce methane gas that can become the fuel to operate a fuel cell, providing electricity to power the lights of the neighboring village.

- Develop and apply engineering solutions, while being cognizant of local geography, aspirations, and cultures.

The best solution for one area in the world may not be a viable solution for another. In the developed world, electricity is produced in large power plants and transported by power lines to locations in which it is needed. But in a world in which electricity is not reliable, local sources of power generation become critical, and the use of small-scale solar arrays that produce only sufficient energy for one family is more practical.

- Create engineering solutions beyond current or dominant technologies; improve, innovate and invent (technologies) to achieve sustainability.

One must recognize that the technologies that have been used to create the world in which we currently live will not be sufficient to provide a sustainable world for our children. It is critical to look beyond what is currently known, to seek out bold new ideas that deliver services in a way which is not yet conceived.

- Actively engage communities and stakeholders in development of engineering solutions.

Because sustainability is fundamentally a systems analysis, in which social aspects are equally important to engineering solutions, it is critical that the stakeholder community become involved in the selection of choices to deliver a desired service. If one needs a transportation solution in a community without cars, delivering a fueling station will provide no value. Delivering workforce training via the Internet to an inner city neighborhood in which people do not have access to a computer cannot be effective. One must understand the resources of the stakeholders, and their cultural limitations, in order to provide a solution of value.

- There is a duty to inform society of the practice of sustainable engineering.

The participants in the workshop were not satisfied simply stating a series of principles, and hoping that society would accept these concepts as self-evident. Because sustainability is built on the aggregation of billions of individual choices, it is vitally important that all of society understands the trade-offs associated with these choices. When individuals seek out the best choice for society as a whole, rather than the easy choice for themselves, the future of the world's ecosystems will be sustained.

REFERENCES

1. Benyus, J. M. In: *Biomimicry: Innovation Inspired by Nature*, First Quill Edition, 1998.
2. Anderson, R. C. In: *Mid-course Correction: Toward a Sustainable Enterprise: The Interface Model*, Chelsea Green Publishing Company, White River Junction, VT, 1998, p. 63.

3. American Chemical Society website—*Green Chemistry Educational Resources: Bleaching With Green Oxidation Chemistry, Chemistry and Compost Chemistry and Compost, Cleaning Up With Atom Economy, Fuel Cells: Energy From Gases Instead of Gasoline, Green Chemistry: A Greener Clean, Phytochemistry Activity, and Simple Green Cleans with Green Chemistry.* Available at `http://portal.acs.org/portal/acs/corg/content?_nfpb=true&_pageLabel=PP_SUPERARTICLE&node_id=1444&use_sec=false&sec_url_var=region1`.

4. Anastas, P. T.; Warner, J. C. *Green Chemistry: Theory and Practice*, Oxford University Press, New York, 1998.

5. `http://www.epa.gov/greenchemistry/pubs/pgcc/winners/aa98a.html`.

6. Herrerra, A.; Islas, J.; Arriola, A. *Applied Thermal Engineering*, **2003**, *23*, 127–139.

7. Kemp, I. *Pinch Analysis and Process Integration. A User Guide on Process Integration for Efficient Use of Energy*, Butterworth-Heinemann Elsevier, New York, 2007.

8. `http://www.mnn.com/green-tech/research-innovations/blogs/boy-discovers-microbe-that-eats-plastic`.

9. `http://www.epa.gov/opptintr/greenengineering/`.

10. Anastas, P. T.; Zimmerman, J. B. *Environ. Sci. Technol.*, **2003**, *37*(5), 94A–101A.

11. `http://www.un-documents.net/wced-ocf.htm`.

12. `http://www.epa.gov/sustainability/basicinfo.htm`.

13. Barrera-Roldan, A.; Saldivar-Valdes, A. *Ecol. Indic.*, **2002**, *2*, 251–256.

14. Abraham, M. A., *Sustainability Science and Engineering, Volume 1: Defining Principles*, Elsevier Science, New York, 2006.

3

CHEMISTRY AS AN UNDERLYING FORCE IN ECOSYSTEM INTERACTIONS

The previous chapters provided the foundations of green chemistry, green engineering, and sustainability. We discussed these concepts in high-level terms, describing the principles of practice and goals to which one should strive in order to achieve a healthier environment. Here, we start to put the concepts of chemistry to work in understanding the interactions between various ecosystems and how specific actions propagate through the environment.

Within this chapter, we emphasize the importance of chemistry within the following topics: nature and the environment, energy and its production from chemical sources, waste and pollution prevention, ecotoxicology, and green living. An understanding of the chemistry surrounding each of these areas will provide the background one needs to understand the challenges associated with protecting the resources, and the alternatives that can be applied and are aligned with green chemistry and engineering. Some of these concerns have already been raised in previous chapters but are expanded to greater details here.

Green Chemistry and Engineering: A Pathway to Sustainability,
Anne E. Marteel-Parrish and Martin A. Abraham.
© 2014 American Institute of Chemical Engineers, Inc. Published 2014 by John Wiley & Sons, Inc.

3.1   NATURE AND THE ENVIRONMENT

3.1.1   Air and the Atmosphere (Outdoor and Indoor Pollution)

3.1.1.1 The Makeup of the Atmosphere

The quality of the air that we breathe directly impacts our health. Pure dry air at ground level contains nitrogen (78.08% by volume), oxygen (20.94% by volume), argon (0.93% by volume), carbon dioxide (0.04% by volume), and other gases such as neon, helium, krypton, and methane (0.01% by volume). Additional components such as particulates and ozone also play a vital role in the makeup of the atmosphere. Issues of air pollution have been a challenge for more than 30 years and remain an area of great concern today.

In the Clean Air Act of 1970, described previously in Chapter 1, the EPA established two types of national ambient air quality standards (NAAQS): one aimed at the protection of public health (including "sensitive" populations such as asthmatic people, children, and the elderly) and the second one targeting protection against decreased visibility and damage to vegetation, crops, and buildings. The EPA articulated NAAQS for six common pollutants: carbon monoxide (CO), ground-level ozone (O_3), lead (Pb), nitrogen dioxide (NO_2), particulate matter (PM) known as particle pollution, and sulfur dioxide (SO_2). Since the establishment of the Clean Air Act in 1970, the gross domestic product increased by 207%, the number of vehicle miles traveled in the United States increased by 179%, and the U.S. population grew by 47% while the energy consumption was intensified by 49%. During the same period, the total emissions of the six principal pollutants decreased by 57%. Figure 3.1 compares growth areas and emissions in the period 1970–2007.

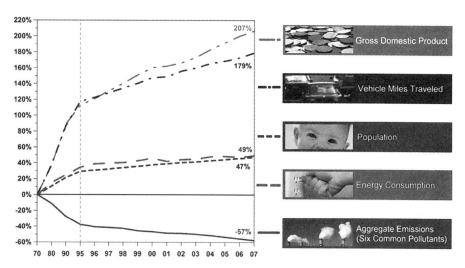

Figure 3.1. Comparison of growth areas and emissions from 1970 to 2007 [1].

As Figure 3.1 shows, for the past 37 years, the levels of the six principal pollutants have been declining, even as Americans drive more miles, consume more energy, and increase in population. The EPA has emphasized regulations and controls, and monitors air quality through programs and rules at the federal, state, local, and regional levels, as well as through voluntary partnerships. However, despite this overall progress in combined levels of the six principal pollutants, ground-level ozone and fine particle pollution continue to be higher than national air quality standards in some parts of the United States. Contrary to the overall trend, the annual emissions for ozone and particle pollution are subject to the impact of weather changes, which contribute to their formation in the atmosphere.

Some pollutants, such as carbon monoxide, sulfur dioxide, and lead, are produced in specific locations and then move through the environment. However, the pollution pathways for other components, such as ozone and nitrogen dioxide, are dependent on chemical transformations that occur in the atmosphere. They are impacted by temperature and the presence of sunlight and are therefore more difficult to assess. Figure 3.2 presents the most common air pollution pathways.

DISCUSSION OF OZONE.   Ozone exists both near the ground (in the troposphere) and in the stratosphere. We need the ozone layer in the stratosphere (10 to 30 miles above the earth) to provide protection from the harmful UV rays from the sun. It is believed that 95–99% of the sun's ultraviolet radiation is captured by the ozone layer before reaching earth's surface, through a complex series of chemical reactions.

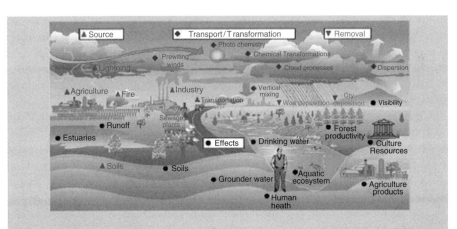

The interrelationship among pollutants, sources, transport and transformation pathways, and environmental effects are complex. For example,
- Emissions from various sources contribute to ozone, particle pollution, and acid rain formation in the atmosphere.
- The photochemistry involved to from thes pollutants is enhanced by sunlight.
- Fires contribute to the build-up of particle pollution.
- Winds disperse and transport pollution over large distances.
- Rain washes particles out of the atmosphere into streams and lakes.

The processes and interrelationships create many pathways and feedback systems through which human health and ecosystems are affected.

Figure 3.2. Common air pollution pathways [2].

The formation of ozone in the upper levels of the atmosphere is catalyzed by ultraviolet radiation that breaks apart oxygen to form two oxygen radicals as products. This process is called photodissociation. Then, the oxygen radical reacts with a second oxygen molecule to produce ozone, O_3, which is relatively stable at atmospheric conditions.

Formation of ozone:

$$O_2(g) \xrightarrow{\text{UV radiation}} \cdot O \cdot (g) + \cdot O \cdot (g)$$
$$\cdot O \cdot (g) + O_2(g) \rightarrow O_3(g)$$

Under normal conditions the rate at which ozone is formed is balanced by the rate at which it is destroyed.

Destruction of ozone:

$$\cdot O \cdot (g) + O_3(g) \rightarrow 2O_2(g)$$

Therefore, the concentration of ozone in the stratosphere should stay relatively constant. However, the presence of ozone-depleting substances, such as chlorofluorocarbons (CFCs), interferes with the ozone cycle and impacts the concentration of ozone in the upper atmosphere. These materials can break down in the stratosphere to form Cl or Br radicals, which can interact with the ozone molecules to produce other radicals and oxygen (Highlight 3.1).

While ozone in the stratosphere performs an important function for life on earth, ozone near the ground is considered a toxin for both humans and the environment. Ozone is known as a pollutant especially during the summer when emissions of nitrogen oxides react with volatile organic compounds in the presence of sunlight. Ground-level ozone is considered a greenhouse gas and is known to play a role in climate change (Figure 3.3).

Ozone can be transported throughout the atmosphere by a variety of chemical and physical properties, as described in Figure 3.4.

Changes in weather patterns contribute to differences in ozone concentrations from year to year and from region to region. The concentration also varies by time of year, with greater ozone concentrations reported in the summer when atmospheric temperatures are higher. The presence of the ozone hole over the Antarctic results from the combination of normal ozone cycling and depleted ozone layers. Since the inception of the Montreal Protocol in 1989 and the elimination of the use of CFCs in the late 1970s, the ozone concentration in the stratosphere has rebounded somewhat (Figure 3.5).

3.1.1.2 Air Pollutants

Carbon monoxide (CO) is a colorless and odorless gas found mainly as a product of incomplete combustion of carbon in fuel. Carbon monoxide emissions are higher in heavy traffic areas, but also near metal processing industries, residential wood burning, and when forest fires are occurring. Burning carbon-containing substances in oxygen is called combustion and is presented in Highlight 3.2.

Highlight 3.1   Chlorofluorocarbons

Chlorofluorocarbons, CFCs, contain elements such as carbon, fluorine, and chlorine. Examples of CFCs are $CFCl_3$ and CF_2Cl_2. CFCs were developed in the early 1930s and have been used in a variety of applications such as coolants for commercial and home refrigeration units, aerosol propellants, industrial solvents, and blowing agents. Ironically, CFCs are relatively inert, nonflammable, and nontoxic. Their useful properties are also the cause of their destructive effect on stratospheric ozone. Because they are not chemically reactive, they live for about 100 years on average in the troposphere (from the surface to an altitude of about 10km). When rising to the stratosphere (just above the troposphere, from about 12 to 50km above earth's surface), CFCs are decomposed by high-energy solar radiation. The decomposition of CFCs results in a lower concentration of ozone in the stratosphere and ultimately participates in the destruction of the ozone layer.

Let's use dichlorofluoromethane (CF_2Cl_2 sold under the brand-name Freon-12) to illustrate the sequence of ozone-destroying reactions:

Step 1: Decomposition in the presence of sunlight

$$CF_2Cl_2(g) \xrightarrow{hv} CF_2Cl \cdot (g) + Cl \cdot (g)$$

When a chlorine atom, a free radical, reacts with a molecule of ozone, the ozone is destroyed through conversion to oxygen.

Step 2:

$$Cl \cdot (g) + O_3(g) \rightarrow ClO \cdot (g) + O_2(g)$$

Two free radicals of ClO can react together, eventually leading to the formation of a new Cl radical, allowing for the destruction of ozone in a continuous cycle in which the chlorine radical survives.

Step 3:

$$ClO \cdot (g) + ClO \cdot (g) \rightarrow ClOOCl$$

$$ClOOCl \xrightarrow{hv} \cdot ClOO + \cdot Cl$$

$$\cdot ClOO \xrightarrow{hv} \cdot Cl + O_2$$

The net reaction is

$$2O_3(g) \rightarrow 3O_2(g)$$

These chain reactions disrupt the balance of ozone in the troposphere, since the presence of the chlorine radicals causes ozone to be destroyed faster than it is produced. Ultimately, the concentration of stratospheric ozone decreases.

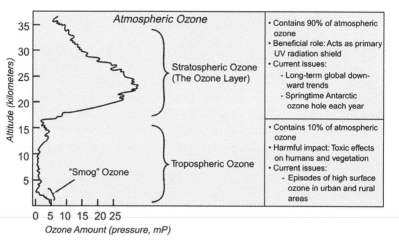

Figure 3.3. Comparison between tropospheric and stratospheric ozone [3].

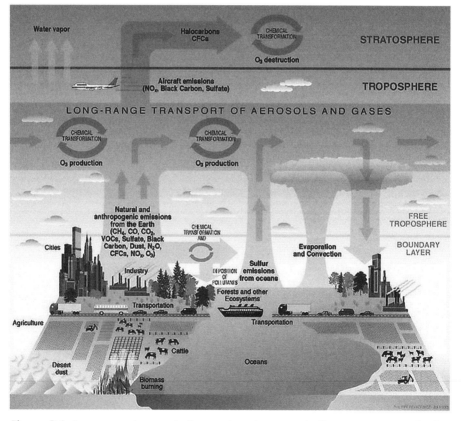

Figure 3.4. Long-range transport of aerosols and gases and effects on ozone production and destruction [4].

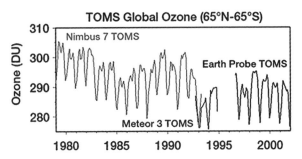

Figure 3.5. Changing ozone concentrations through annual and periodic cycling [5].

Highlight 3.2   Combustion of Carbon-Containing Substances

Combustion reactions involve burning a fuel with an oxidant and are accompanied with production of heat. When a hydrocarbon substance containing C and H only (C_xH_y) burns completely in oxygen, only two products are formed: carbon dioxide (CO_2) and water (H_2O).

A typical combustion reaction involving propane is the following:

$$C_3H_8(l) + 5O_2(g) \rightarrow 3CO_2(g) + 4H_2O(l)$$

Under many conditions, the hydrocarbon is not completely oxidized, leading to products of incomplete combustion. Carbon in the fuel may produce carbon monoxide, CO, and other partial oxidation products may also be formed. For example, the partial oxidation of ethanol (CH_3CH_2OH) produces acetaldehyde (CH_3CHO), one of the most important naturally occurring aldehydes. It is this partial oxidation product that is commonly believed to be a cause of hangovers.

When fuels such as diesel oil or coal are not completely burned, pyrolysis, which is the decomposition of organic compounds at elevated temperatures in the absence of oxygen, can occur. Pyrolysis can lead to the formation of soot, which is the basis for particulate matter in air. High molecular weight hydrocarbons, including polyaromatic hydrocarbons or PAHs can be produced. These are widely believed to be cancer-causing chemicals, in sufficient quantity.

Because of the potential for incomplete combustion, and the highly toxic nature of some of the reaction products, correct air flow and ventilation are crucial in maintaining a safe and efficient fireplace. Between 1999 and 2004, an average of 439 persons died annually from unintentional, non-fire-related CO poisoning (range: 400 in 1999 to 473 in 2003) [6].

Sulfur dioxide (SO_2) is a gas also formed from the combustion of sulfur-containing fuels such as coal or oil. Almost 90% of sulfur dioxide present in the air is due exclusively to fuel combustion. Sulfur dioxide is also a major contributor to the formation of sulfuric acid, which then causes acid rain when mixed with moisture in the atmosphere. In the presence of oxygen and water, sulfuric acid is produced through a two-step reaction process:

Step 1: Sulfur dioxide reacts with oxygen to produce sulfur trioxide (SO_3)

$$2SO_2(g) + O_2(g) \rightarrow 2SO_3(g)$$

Step 2: Sulfur trioxide can further react with water to produce sulfuric acid (H_2SO_4)

$$SO_3(g) + H_2O(l) \rightarrow H_2SO_4(aq)$$

Large amounts of acidic precipitation can cause serious environmental damage by changing the pH of the soil (effect on pH of the soil is discussed in Section 3.1.3, Chemistry of the Land). Sulfuric acid also reacts with limestone, which is one of the principal components of many of the ancient statues and historical buildings, damaging the preserved artifacts of the past.

Nitrogen oxides, NO_x, originate from combustion reactions, especially reactions that use air as the source of oxygen. The nitrogen and oxygen present in the fuel react to form nitrogen monoxide (NO), which can subsequently be converted to nitrogen dioxide (NO_2) by reaction with oxygen at high temperatures:

$$2NO(g) + O_2(g) \rightarrow 2NO_2(g)$$

Major sources of nitrogen oxide emissions are automobiles, power plants and any industry where high temperature combustion is involved.

As mentioned in Highlight 3.2, combustion of carbon-based substances such as coal, petroleum, and natural gas form CO_2 as the oxidation end product. These products are released into the atmosphere, leading to billions of metric tons of anthropogenic (human-made) carbon dioxide in the atmosphere.

Volatile organic compounds or VOCs are both human-made and naturally occurring. The biologically derived VOCs are from plants, especially the leaves and stomata. Conifers are known to emit terpenes, major components of resin and a major class of VOCs. Principal human-made sources include solvents used in paints and coatings, cleaning products and refrigerants based on CFCs, now banned or highly regulated, the toxic and volatile organic compound formaldehyde or methanal, used as a precursor for polymers, and methyl *tert*-butyl ether or MTBE, a persistant flammable and colorless liquid used in engine gasoline as antiknocking agent in the 1970s and 1980s.

3.1.1.3 Particulates

Particle pollution is a general term for a mixture of solid particles, which can be seen as dust or dirt or which are too small to be detected with our naked eyes (fine particles).

Larger solid particles are called particulates. When pollutant particles are inserted into water droplets or when finely divided solids are dispersed in a gas, they form a type of colloid called an aerosol. Fogs and smoke are common aerosols.

3.1.1.4 Indoor Air

Indoor air pollution is becoming more and more of an issue. Asthma mainly affects young children under five years old. According to the EPA, about 25.7 million people in the United States were reported to have asthma in 2012, including 7.1 million children. There were nearly 2 million emergency room visits due to asthma. Asthma is ranked third in the cause of hospitalizations for children under 15, with a total estimated cost of almost $56 billion, and approximately 10.5 million school days missed each year [7]. Poor ventilation in schools, molds, and radon are the most common sources of indoor air pollution and are the source of thousands of deaths every year. Radon, a natural radioactive gas that can't be seen, smelled, or tasted, is the second leading cause of lung cancer in the United States today [8]. See Highlight 3.3.

There are many other indoor air pollutants that contribute further to the degradation of human health. Formaldehyde resins are often used in carpet and other plastic

Highlight 3.3 Radon and Radon Action Week in October Annually

Radon, named after the element radium, was discovered in 1900 by Friedrich Ernst Dorn in Germany. Radon is a radioactive gas found in high concentrations where soil and rocks contain uranium, graphite, shale, or phosphate or where soils may have been contaminated with by-products of uranium or phosphate mining. Some homes were built using sand-like uranium pulverized rock as construction material.

Radon is the product of the natural breakdown or radioactive decay of uranium commonly used in nuclear weapons and nuclear reactors. Normally, radon gas is naturally diluted in open air; it becomes an issue in confined spaces such as homes. Radon gas can seep through dirt and solid floors, cracks in concrete walls, sumps, joints, and basement drains and can also permeate the water supply when trapped in well water.

When radon decays, it emits particles known as alpha particles, which, upon being inhaled, can radiate and penetrate the bronchial and lung tissues.

Even if there is no warning sign for the presence of radon, there are several ways to detect radon using detectors available at hardware stores. Reduction of radon exposure starts with sealing cracks and openings in basement floors and increasing subfloor ventilation as well as ventilation within the home.

Since 1990, when a congressional resolution to draw attention to radon as the silent killer took place, a special week is dedicated every October to motivate Americans to protect themselves from radon health effects.

products. These materials continuously degas solvent trapped in the product during the manufacturing process. In the confined space of a home, these toxic gases can accumulate, creating a hazardous condition in the home.

3.1.2   Water (Water Pollutants, Issues Associated with Nonpotable Drinking Water)

As a counterpart to the Clean Air Act, the Clean Water Act of 1972 promoted regulations for discharge of pollutants into U.S. waters and for quality standards for surface waters.

Water covers 71% of the earth's surface. Ocean water contains about 3.5% salt. The salt is primarily sodium chloride (NaCl) but other ionic species in sea salt include magnesium (Mg^{2+}), sulfate (SO_4^{2-}), calcium (Ca^{2+}), and potassium (K^+) ions. The minerals present in seawater are produced when rain deposits on the land and degrades the soils and rocks to produce dissolved minerals, which ultimately accumulate in the oceans. The concentration of salt in the water is termed salinity and describes the amount of dissolved salt.

The oceans contain 97.2% of the earth's water, making sources of fresh water scarce but hugely important. Fresh water also contains salt, although at concentrations typically less than about 1000 ppm (or 0.1% by weight).

Water contains dissolved gases as well. Oxygen dissolved in water is a vital element that allows sea life to thrive, since oxygen is essential for all living things. Fish can extract the oxygen from the seawater through their gills. Nitrogen can also be found in the water, although in this case it is likely from runoff of nitrogen applied to the land as a crop fertilizer. Nitrogen works equally well as a fertilizer on the land or in the ocean. The presence of high levels of nitrogen in the water can lead to eutrophication, a situation that occurs when too much plant matter consumes all of the available oxygen. Under these situations, there is insufficient oxygen in the water for fish and other marine animals.

Carbon dioxide is another important dissolved gas, as the dissolution of CO_2 into the ocean impacts the amount of carbon dioxide present in the atmosphere. Carbon dioxide can also react with water to produce carbonates and carbonic acid, which changes the pH of the oceans.

The reversible reaction of carbon dioxide with water produces carbonic acid or hydrogen carbonate:

$$CO_2(g) + H_2O(l) \Leftrightarrow H_2CO_3(aq)$$

As more CO_2 accumulates, the equilibrium shifts to increase the amount of acid formed, lowering the pH. The pH, which is a measure of the acidity or basicity of a solution, varies from 0 to 14 on a logarithmic scale. Acidic solutions have a pH between 0 and 7 whereas basic or alkaline solutions have a higher pH between 7 and 14. Discussion on acids and bases and pH of solutions is the topic of Chapter 5. The change in ocean acidity significantly affects coral in the ocean, since coral growth requires carbonate, the concentration of which is reduced by increasing acidity.

There are many types of water pollutants coming either from point sources such as sewage and industrial waste or from nonpoint sources such as agricultural runoff and storm water drainage. Water impairment includes excess levels of nutrients, metals (primarily mercury), sediment and organic enrichment due to agricultural activities, industrial and municipal discharges, atmospheric deposition, and unknown specific sources. More recently, pharmaceuticals and personal care products have been detected in U.S. waters at low concentrations.

Availability of potable water for food irrigation, industry, and domestic uses is likely to decline due to climate change, population increase, urban encroachment, and increasing levels of pollutants in water supplies. Preservation of water to sustain ecosystems is a topic of interest to environmentalists, ecologists, and biologists. During the 20th century, more than half of the world's wetlands have been lost through increased irrigated land and intensive dam construction. Increased demand, through growth in industrial and energy sectors, increases the challenge. Numbers are alarming: "40–50% of the globally available freshwater is used by humans, from which 70% is used for agriculture, 22% for industry, and 8% for domestic purposes" [9]. It is predicted that the total amount of water needs for human consumption will increase five times relative to that at the beginning of the 20th century.

The lack of potable water is even more severe in developing countries, as increased urbanization threatens existing water supplies. Water security (and thus food security) remains among the most important issues in these places. Acute contamination of drinking water by arsenic and 29 other metals such as lead, cadmium, and mercury in the rural population of Bangladesh and India became a major public health problem in the mid-1990s. Tens of millions of people were drinking untreated well water from naturally occurring aquifers in the flood plains. Thousands of people are already affected by arsenic poisoning, which can lead to several types of cancers. The arsenic poisoning in Bangladesh is the largest mass poisoning of a population in history according to the World Health Organization (WHO) [10].

3.1.3 Chemistry of the Land

The earth's surface contains rocks and minerals, as well as clays and soils. The minerals may take many forms, including primary aluminosilicate minerals and other complex species that makeup the hard surface, or crust, of the planet. Aluminosilicates are made from aluminum, silicon, and oxygen and account for approximately 82% of the earth's crust by weight. Because of their structure, the soil surface chemistry is dominated by the reactions that occur on oxide and hydroxide-rich surfaces. These surfaces are extremely hydrophilic, layers of adsorbed water molecules surrounded by varying volumes of "bulk" water.

Different soils contain different mineral structures and thus exhibit varying properties. Olivines contain elements including iron and magnesium. Other species, such as quartz, contain only silicon and oxygen, whereas feldspar is a mineral containing a fixed 1:4 ratio of aluminum to silicon. Clays, such as kaolinite, are more complex structures that contain layers of minerals arranged in sheets, often with embedded water between the sheets. Soil minerals are important because of their ionic properties; they may be dissolved in water and they may react with other dissolved minerals through exchange reactions.

Exchange reactions are also called metathesis or double-displacement reactions in which ions are interchanged between two partners. The general reaction can be summarized as

$$AB + CD \rightarrow AD + CB$$

where A and C are cations (positively charged species) and B and D are anions (negatively charged species).

An example of an exchange reaction involving two salts soluble in water is the following:

Lead(II) nitrate + potassium chromate → lead(II) chromate + potassium nitrate

The roman numeral II indicates the 2+ charge present on the lead cation, Pb^{2+}. Lead can also exist as lead(I) or Pb^+. Lead(II) nitrate is an inorganic compound, soluble in water, which was historically used in the production of pigments for lead-based paints; such pigments have now been replaced by the less toxic titanium oxide, also found in sunscreen and food coloring. Potassium chromate is a yellow chemical indicator and a carcinogen; when lead nitrate and potassium chromate react, an orange-yellow precipitate of lead(II) chromate is formed. The molecular equation is the following:

$$Pb(NO_3)_2 (aq) + K_2CrO_4 (aq) \rightarrow PbCrO_4 (s) + 2KNO_3 (aq)$$

Because the chromate ions are insoluble in water, lead chromate precipitates and solid particles are produced. A precipitate is an insoluble solid that does not break down into its respective ions in water.

The complete ionic reaction shows the dissociation of all aqueous compounds into their respective ions (since lead chromate is insoluble, it stays as a solid and does not dissociate):

$$Pb^{2+} (aq) + 2NO_3^- (aq) + 2K^+ (aq) + CrO_4^{2-} (aq) \rightarrow PbCrO4 (s) + 2K^+ (aq) + 2NO_3^- (aq)$$

The net ionic reaction is the final reaction showing the removal of all spectator ions (spectator ions are ions present on both sides of the equation):

$$Pb^{2+} (aq) + CrO_4^{2-} (aq) \rightarrow PbCrO_4 (s)$$

Common ions, exchange reactions, and other types of chemical reactions are discussed in detail in Chapter 6.

Organic matter in soil comes from the decomposition of plants and animals. Knowledge of environmental soil chemistry is paramount to predicting the fate, mobility, and potential toxicity of contaminants in the environment. The wide array of organic materials that lead to soil formation produces a largely heterogeneous material containing different particle sizes and organic species with varying solubilities and functional grouping. Because soils come from decayed organic matter, they contain amino (RNH_2 or R_2NH or R_3N), carbonyl (−COH, where C is double bonded to the O), carboxyl (−COOH, where C is double bonded to the first O), and other functional groups that impart added charge and critical properties to the soil structure. The main functional groups and their properties are covered in Chapter 5.

Soil pH is a measure of the soil acidity or soil alkalinity. Because pH can affect the availability of nutrients, it is an important consideration for soil quality. Most crops prefer a neutral or slightly acidic soil. The pH is controlled by the addition of amendments to the soil; the addition of minerals such as lime (calcium hydroxide $Ca(OH)_2$) can increase the soil pH, whereas nitrogen-containing amendments lead to increased acidification. Fertilizer can also be added to the soil to ensure the availability of nutrients, specifically the macronutrients nitrogen, phosphorus, and potassium.

The change in land use has had profound effects on the environment. Intensive urbanization, deforestation, and exhaustive agriculture have impacted biodiversity, air and water quality, and carbon cycles. Soil depletion occurs when the components that contribute to fertility are removed and not replaced, and the conditions that support soil fertility are not maintained. The loss of organic matter means less decomposition of the organic component of the soil, and thus loss of useful land. The combined effects of growing population densities, large-scale industrial logging, and increased agricultural activity have led to almost total nutrient removal in some soils.

The impact of deforestation for use as pasture, urban use, arable land, or wasteland is startling. About 47% of the world's forests, including much of the world's tropical rain forests, have been lost to human use [9]. This affects the world's biodiversity, the water and soil quality, and the ability of plants to promote air purification. Forests provide free ecosystem services in terms of sequestration of carbon, water retention in the soil and groundwater, and atmospheric moisture. In developing countries, the rate of deforestation is driven by population growth and food production (economic incentive as well). In African countries, the rapid human population growth and the need for wood to build houses and for use as a fuel contributes to deforestation, with resulting desertification, water resource degradation, atmospheric pollution, and soil loss. The deforestation rate in the Amazon is setting a new record of 69% in 2008 compared to 2007. Almost 90% of West Africa's rain forest has already been destroyed [11].

Deforestation not only affects the local climate, the water cycle, and the soil but also plays a role in species extinction. Some scientists forecast that up to half of presently existing species may become extinct by 2100 [9], a substantial loss in biodiversity.

Expansion of agricultural land, urbanization, and industrialization are the major causes of land pollution. The intensification of agriculture and mechanization results in loss of habitat and shelter for wildlife. More intensive agriculture also requires more pesticides, herbicides, and insecticides to be used. A pesticide is exactly what it sounds like, a substance designed to prevent, destroy, or repel a pest (generally an insect or other type of organism). Typically, a pesticide is a chemical or biological species designed to act against a specific infestation. After they perform their function

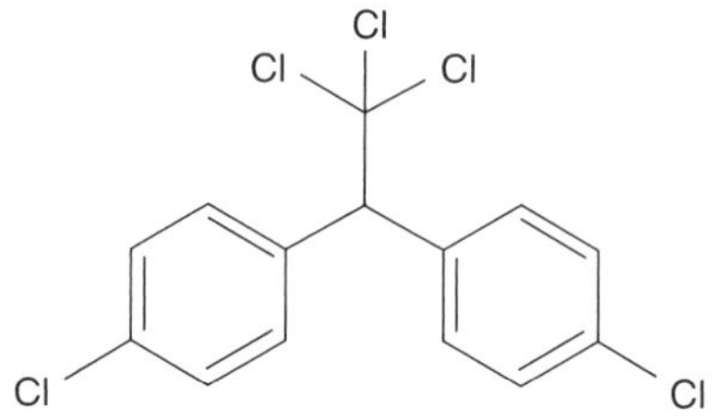

Figure 3.6. Chemical structure of DDT.

against the desired species, they can accumulate in the soil, resulting in long-term decrease of the fertility of the land. Remember, what is toxic to an insect will be toxic to other species in a sufficiently high dose. Pesticides are also problematic, in that only 2% of the applied pesticides actually reach their intended target; the remainder ends up in the air and water. Some chlorinated organic pesticides, containing carbon, hydrogen, and chlorine atoms, are classified as persistent organic pollutants (POPs), materials that resist degradation and remain in the environment for years.

Insecticides have a nonnegligible impact on health and the environment as well. Synthetic insecticides such as organochlorines (particularly the famous DDT (Figure 3.6), dichlorodiphenyltrichloroethane, put into the spotlight by Rachel Carson in her renowned book *Silent Spring*) and organophosphates (such as Parathion) were well known to cause damage to fish, birds, animals, and the nervous systems of humans. DDT, which is very cheap to produce, was banned in the United States and in most developed countries in the 1970s to 1980s, but many developing countries still lack strict pesticide regulations.

Herbicides are designed to kill undesired plants. Herbicides are widely used in agriculture and in landscape turf management. In the United States, they account for about 70% of all agricultural pesticide use. Most herbicides are organic in nature and thus are mostly biodegradable by soil bacteria. Unfortunately, many herbicides also contain chlorinated dioxins such as 2,3,7,8-tetrachlorodibenzo-*p*-dioxin (TCDD) as an impurity. Dioxin is among the most carcinogenic and toxic human-made organic compounds produced. Pollution by dioxin made headlines years ago when the small town of Times Beach in Missouri (mentioned in Chapter 1) had to be completely evacuated in the mid-1980s due to the presence of dioxin in the soil. The town was completely demolished in 1992. Dioxin is also a major impurity in the defoliant Agent Orange that was used extensively to clear the jungles of Vietnam during the U.S. military activity in the early 1970s.

3.1.4 Energy

Our economy runs on energy, very large quantities of energy. The chart in Figure 3.7, prepared from data provided by the Department of Energy's Energy Information Agency in 2010, indicates the total amount of fuel consumption in the United States, using current information and projects the expected consumption for the next 25 years. Today, the United States consumes roughly 100 quadrillion BTUs (also termed

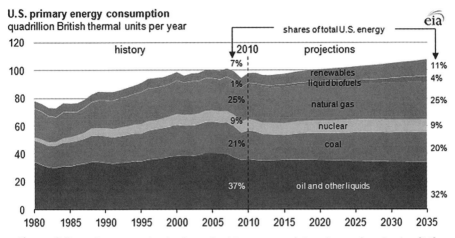

Figure 3.7. Fuel consumption in the United States: actual situation and projection [11].

a quad) of energy (100×10^{12} BTU, or 105×10^{12} MJ). A quad is a unit of energy equal to 10^{15} (a short-scale quadrillion) BTU (British thermal unit) or 1.055×10^{18} joules in SI units. In the United States, energy is produced mainly from coal, natural gas, and liquids (petroleum derived from oil), with nuclear and renewable sources making up the next largest elements. Projections call for an increasing proportion of energy to be supplied from renewable sources, however, even over the next 25 years, 75% of the U.S. energy supply is expected to be derived from fossil resources.

Energy is consumed for nearly every sector of the economy. The greatest consumption is for industrial processes; that is, the manufacture of products, including chemical products. Transportation makes up the next largest sector of the energy consumption, and while industrial consumption is projected to be somewhat level over the next 25 years, the transportation consumption is projected to increase by roughly 15%. Residential use corresponds to the energy used in the home, including lighting, air conditioning, and other appliances, whereas commercial energy consumption corresponds to consumption in stores and offices (Figure 3.8).

Another way of looking at the energy consumption is through the energy flow diagram, which relates the sources of energy production and the areas in which they are consumed. Domestic energy production amounts to approximately 73 quads, compared to domestic consumption of over 99 quads. The roughly 26 quads of energy imports correspond to concerns regarding energy security faced in the United States (Figure 3.9).

The production of energy from fossil resources leads to significant environmental impacts. Figure 3.10 summarizes CO_2 emissions based on source of energy production and sector of usage, both for 2008 and projected for 2035. The current total emissions associated with energy generation are 5.8×10^{12} kg of CO_2, expected to increase to 6.3×10^{12} kg of CO_2 emissions by 2035. The CO_2 production depends greatly on the type of fuel being consumed, and the type of fuel depends heavily on the sector of use. For example, the transportation sector consumes mostly petroleum-based products, whereas coal and some natural gas are used in the generation of

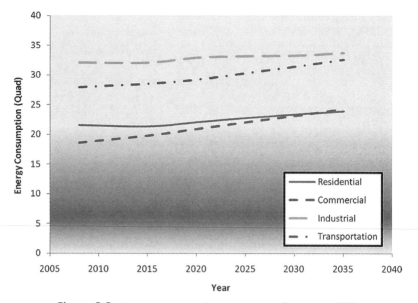

Figure 3.8. Energy consumption per sector of economy [12].

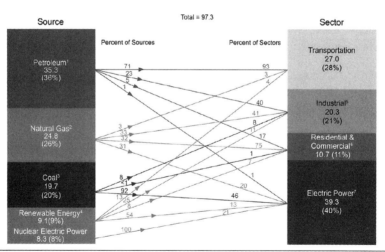

Figure 3.9. Sources of energy production and consumption [13].

electricity. Energy production also produces other emissions, including SO_2 and NO_x. Interestingly, the generation of SO_2 emissions is projected to decrease from roughly 10,000 kg today to approximately 3000 kg in 2035, as greater amounts of natural gas are used to produce the increasing electricity demand.

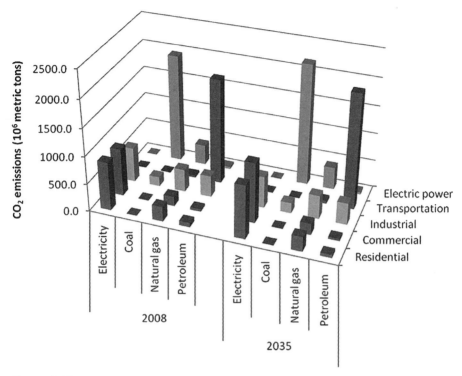

Figure 3.10. Emissions of CO_2 based on source of energy production and sector of usage for 2008 and prediction for 2035 [12].

In addition to the CO_2 impacts, energy generation has significant other environmental concerns. Coal contains substantial quantities of arsenic, mercury, chromium, and cadmium, which are emitted into the atmosphere when coal is burned to produce energy and can be discharged into drinking water through ash and sludge. Waste created by a typical 500-megawatt coal plant includes more than 125,000 tons of ash and 193,000 tons of sludge from the smokestack scrubber each year. In addition, much of the heat produced from burning coal is wasted, since only about 33–35% of the heat of combustion is used to produce electricity. The majority of the heat is released into the atmosphere or absorbed by the cooling water. As much as 2.2 billion gallons of water is used for cooling a typical coal-fired power plant. After use, the heated water is returned to the original water source, and since the return water is hotter (by up to 20–25 °F), the source water becomes hotter, leading to ecosystem malfunction. The production of coal impacts the land as well, as much of the coal is obtained through strip mining.

Combustion of oil produces CO_2, a contributor to global warming, and while oil combustion is more efficient than combustion of coal, it is less efficient than natural gas. Oil is unique as a liquid fuel and thus is used predominantly in the transportation sector, making reduction of emissions more challenging. Oil extraction can be damaging to the environment, and more frequently, offshore sources are being used with increased environmental consequences. Extraction involves drilling and dredging, which disturbs the seabed and damages sensitive coral and other marine life. Spills and leaks from drilling operations and transportation in tankers have produced serious environmental consequences, including oil washing up on beaches and fouling of birds and other wildlife. The disastrous Deepwater Horizon spill in the Gulf of Mexico in April 2010, dumping more than 11 million gallons of oil into the Gulf, has cost over $3 billion in economic claims for cleanup and loss of income, closed 6800 square miles of federal fishing areas, and resulted in countless deaths of sea turtles and other marine wildlife.

Within the context of green chemistry and engineering, the most significant opportunity is in the use of biomass as an energy resource. Biomass is energy converted from sunlight by trees and plants. In general, there are two approaches to generating biomass energy, either growing plants specifically for energy use (energy crops) or using the residues from plants that are used for other needs. Trees and grasses that are native to a region are the most sustainable energy crops since they grow naturally in the environment without the need for fertilizers, pesticides, or other potential contaminants. For example, switchgrass grows quickly in many parts of the country and can be harvested for up to 10 years before replanting. Conversion of biomass energy can be achieved through combustion, usually by co-firing with coal in a modified coal-fired power plant. Other routes for conversion, including thermal or biological conversion, can also be achieved; conversion of biomass to liquid fuels represents one of the few alternatives to the production of petroleum products for use in the transportation sector. The "biorefinery," as depicted in Figure 3.11, describes opportunities for conversion of biomass to useful chemical products.

Because biomass is an inherent part of the carbon cycle, combustion of biomass reduces air pollution. The carbon dioxide that is produced during combustion was recently fixed into the biomass material, and thus from a life cycle perspective, CO_2 emissions are reduced by 90% compared with fossil fuels. Sulfur dioxide and

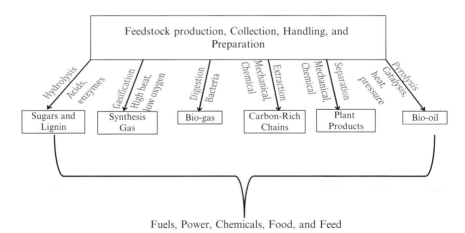

Figure 3.11. Uses for conversion of biomass.

other pollutants are also reduced substantially. Biomass energy also brings other environmental advantages. Water pollution is decreased because less fertilizer is needed. Soil quality is improved and erosion is reduced, and wildlife habitat is restored. Note that these benefits are obtained from the use of native biomass, and the use of energy crops has many of the same challenges as those currently associated with food crops.

3.2 POLLUTION PREVENTION (P2)

As defined earlier and according to Paul Anastas, green chemistry is the "pathway to ensuring economic and environmental prosperity." It is not cleaning the mess with bandages. Pollution prevention provides opportunities to focus on the key concepts of reduce, reuse, and recycle. Reuse has become a key business concept, representing a multimillion-dollar business in the past 15 years. Waste disposal costs money, not the least of which is the cost of cleaning the waste to make sure that it can be disposed of while proper regulations are followed. It also takes time to clean up and it affects productivity. Waste is not an unavoidable cost of doing business. Protecting the environment and making money go hand-in-hand; it is not a trade-off anymore.

According to the EPA, pollution prevention (P2) is about "reducing or eliminating waste at the source by modifying production processes, promoting the use of non-toxic or less-toxic substances, implementing conservation techniques, and re-using materials rather than putting them into the waste stream." The EPA's Office of Pollution Prevention and Toxics (OPPT) investigated approaches on how pollution prevention is applied to the different areas mentioned above such as air and water pollution. Three case studies are presented here [14].

In 2006, Archer Daniels Midland Company in Clinton, Iowa, a major corn products manufacturer, took multiple initiatives to reduce its air emissions. First, a software system was developed to monitor all pollution-related parameters, which was used to optimize operating conditions that would reduce air emissions. Second, after studying the results of the monitoring it was found that about 25–30% of the ash produced by processing boilers could be reused as raw material in cement, concrete blocks, and other products. Archer Daniels Midland Company predicted that approximately 57,500 tons of ash could be reused and potentially $250,000 could be saved annually through this "reuse" strategy. Third, a survey about trash located at the facilities led to a concern about significant quantities of cardboard and steel not being recycled. Recycling cardboard could save an additional $40,000 per year and $104,000 could be gained from metal scrap recovery.

Water conservation represents another area for reuse. Dean Chem in Houston, Texas, an electroplating company, has been collecting rainwater through gutters and uses the collected water in their rinsing processes. Although the total volume of collected water is less than 5000 gallons annually, this company is leading the way toward water conservation. On a smaller scale, rainwater collection can be done through rain barrels connected to the gutters of a house. This allows for watering of shrubs, flowers, and vegetables without the use of well or city water.

Reducing hazardous waste is a priority for MerCruiser in Stillwater, Oklahoma. All new products introduced are evaluated according to a "Green Tag" review. The intended usage, temperature, location, and amount of material to be used as well as Material Safety Data Sheets (MSDS) are assessed. In their painting operations they use powder coatings, which prevent paint from being wasted. They also separate and recycle metals by type, which yields approximately $130,000 estimated savings per year, and have installed equipment to compress metal chips, which are then remelted in furnaces.

These three case studies are only a few among hundreds. Pollution prevention, as one of the pillars of green chemistry, is an essential component of the sustainability pathway.

3.3   ECOTOXICOLOGY

One way to measure the environmental impacts of a process is by analyzing the ecotoxicity of the effluents from the process. The study of ecotoxicity is termed ecotoxicology, which is defined according to *Ecotoxicology Encyclopedia* as "a field of science that studies the effects of toxic substances on ecosystems" [15]. The goals of ecotoxicology are to assess environmental damage from pollution on people, plants, and animals, and to predict the consequences of human actions on different ecosystems. For a population to do well in the environment, they need to survive to be old enough to reproduce, grow to be big enough to reproduce, and eventually reproduce. The goal of ecotoxicity is to understand the concentration of chemicals at which organisms in the environment will be affected.

Assessment tools used in ecotoxicology range from toxicity tests (link the response of damaged biological systems to a particular substance), chemical analyses of substances (proof of the presence of a toxicant), and field surveys (characteristics of the damaged ecosystem). These three tools help in the assessment of risk. Indirect effects of toxicants are more difficult to predict. For example, an organism's response to a toxicant depends on the presence of other organisms in the community and also on the environmental conditions. Most of the relationships between environmental conditions and toxicity have yet to be established.

It is impossible to test all of the millions of species on this planet before a chemical is used. Ecotoxicologists test a few "representative" species in the food chain, starting with the insect eating algae, to the fish eating the insect, through a bigger fish eating the smaller one, and often ending with human consumption. While statistical methods exist to predict the response of organisms, the complexity of ecosystems, the large number of interacting species, and the complex mixture of chemicals released into the environment minimize the effectiveness of simulation models. The U.S. EPA Office of Research and Development (ORD) and the National Health and Environmental Effects Research Laboratory's (NHEERL's) Mid-Continent Ecology Division (MED) created the ECOTOXicology database named ECOTOX. According to the EPA, "ECOTOX integrates three previously independent databases—AQUIRE,

PHYTOTOX, and TERRETOX—into a unique system which includes toxicity data derived predominantly from the peer-reviewed literature, for aquatic life, terrestrial plants, and terrestrial wildlife, respectively" [16]. The database is available on EPA's public webpage and includes more than 620,031 records addressing the adverse effects of 10,101 chemicals to 10,285 species as of June 2012. This database is updated quarterly and contains information about the chemical species impact on plant, terrestrial animal, and aquatic species, as well as test methods.

Two measures are typically used for measurement of ecotoxicity. The first of these measures is the *lethal dose*, usually written as LD_{50}, which is a measure of the short-term poisoning potential or acute toxicity of a substance. The LD_{50} is the amount of a material, given all at once, which causes the death of 50% of a group of test animals. Measurement of LD_{50} provides a description of the terrestrial toxicity; that is, the potential for poisoning associated with chemicals in a land environment. While virtually any type of animal can be used for an LD_{50} test, rats and mice are most common. One can also look at numerous types of exposure, including most commonly tests of dermal (skin) and oral (by mouth) exposure. For smaller animals, the LD_{50} is usually tabulated as the amount of chemical administered (e.g., milligrams) per 100 grams of test animal body weight.

While LD_{50} is a common measure of toxicity, its value is limited in that it only measures acute toxicity and does not provide any information on chronic toxicity (which could lead to health challenges based on exposure to lower doses over a longer period of time). Likewise, it does not take into account toxic effects that are serious but do not result in death. Finally, there can be wide variability between species, with a chemical that is relatively safe for rats perhaps being extremely toxic for humans. Thus, extrapolation of toxicity data may be difficult. It is also important to note that all substances, when administered in high doses, can be toxic. For example, sucrose, otherwise known as table sugar or cane sugar, has an LD_{50} value in rats at 29,700 mg/kg when administered orally [17].

A second measure of toxicity is the LC_{50}, or *lethal concentration* at which 50% of the organisms tested will be killed by exposure to the test chemical. LC_{50} is usually used to measure the toxicity in the environment, such as the concentration in water or air. LC_{50} toxicity of an effluent can be tested in either flow-through or static mode. In flow-through testing, various concentrations of fresh effluent are pumped through test chambers for a predetermined length of time. In a static test, the assay organisms are exposed to the same test samples for the duration of the test. Frequently, a small, developing organism such as *Daphnia*, an organism that is low on the aquatic food chain, is used as a model for testing effluent toxicity. *Daphnia* makes an excellent indicator species because of its short life span and rapid reproductive capabilities. In addition, they are nearly transparent, making their internal organs easy to study in live specimens. The developing organism has high sensitivity to the toxic material because of its rapid growth in the environment, and thus abnormal development, growth or death, is easily observed. The EPA program, ECOSAR, predicts toxicity of chemicals released into water to aquatic life (fish, algae, and invertebrates).

It is noticeable that toxicology is rarely included in the core courses of traditional chemistry curriculum. Even in the eight years (minimum) it requires to get a PhD, students with a PhD are not usually exposed to toxicology. Thus, students wishing to become educated on the environmental aspects of chemistry need to take elective coursework in toxicology, so that they can relate the impacts of a chemical release to its effects in the environment.

3.4 ENVIRONMENTAL ASSESSMENT ANALYSIS

When completing an environmental analysis, it is important to evaluate the full life cycle of a product or process. The life cycle includes all of the elements of a particular product, from the extraction of raw materials, to the manufacture of the product, transportation at all stages, use of the product, and eventually disposal. In the context of a chemical product, one often focuses on the environmental issues associated with manufacture, use, and disposal, and there is an emphasis on the direct environmental impacts of the material. Thus, we generally consider the chemicals that are engaged in the production of the desired chemical product, the harm associated with the use of a specific chemical product, and the disposal of the chemical wastes (including those generated through the manufacture of the product).

The assessment analysis is a long and complex process that follows the procedures established through the International Standards Organization (ISO) through its 14000 series set of standards. The ISO 14000 delineates a series of steps that includes:

- *Classification*—the assignment of environmental impacts into various categories (i.e., air, land, water, greenhouse gas emissions, etc.).
- *Characterization*—a calculation of the specific environmental impact of all contributions in a defined classification.
- *Significance analysis*—a determination of the relative importance of a particular impact.
- *Valuation*—the identification of relative weight of the particular environmental element relative to other elements being conserved.

The overall analysis combines all of these elements into a single measure that would allow one to compare multiple opportunities to determine which of these would have the least environmental harm, as described in Figure 3.12.

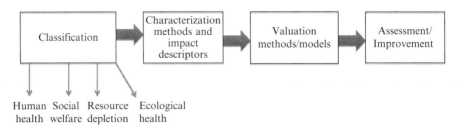

Figure 3.12. Life cycle inventory.

When applied over the entire life cycle of a material, a full analysis of the chemical product can be completed. It is important to note, however, that no analysis can be completely objective, since the valuation phase often comes down to an individual assessment of critical environmental elements. Thus, the valuation phase attempts to compare the impact of loss of biodiversity against emissions of greenhouse gases, a subjective decision.

As seen in the previous sections, chemicals and manufactured products can impact the environment in numerous ways, from greenhouse gas impacts, to toxicity in the water, to land use changes. The minimization of these impacts requires a consistent method of evaluating the impacts of these materials, so that one can conduct an overall assessment of environmental impact analysis. One such example is through the U.S. EPA's Tool for the Reduction and Assessment of Chemical and Other Environmental Impacts (TRACI). The tool "facilitates the characterization of environmental stressors that have potential effects, including ozone depletion, global warming, acidification, eutrophication, tropospheric ozone (smog) formation, ecotoxicity, human health criteria–related effects, human health cancer effects, human health noncancer effects, fossil fuel depletion, and land-use effects" [18], using an overall methodology as seen in Figure 3.13.

Figure 3.13 illustrates how TRACI incorporates the steps of an environmental impact analysis. First, the various stressors are identified. For example, SO_2 may be emitted in a particular process, which would be in the category of chemical emissions. This emission must then be assigned to an impact category or, in general, can be assigned to several impact categories through various effects in the

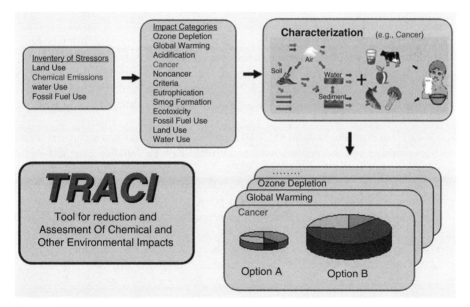

Figure 3.13. TRACI: Tool for the Reduction and Assessment of Chemical and Other Environmental Impacts [19].

environment. Continuing with the SO_2 example, we know that SO_2 can be oxidized by sunlight to produce H_2SO_4, which is then transported through the atmosphere and ultimately returns to earth through rainwater, leading to acid rain and lakes' acidification. The amount of emissions can be quantified based on an analysis of the process, and thus the impacts of the emission leading to the acidification outcome can be quantified.

Any particular process has a range of impacts, and the relative impact of the emission can be related to a variety of characteristics of the emission. Clearly, the amount of the emission is important, although this is not the only criterion of importance. For example, the impact of SO_2 emissions leading to acidification must be related to the impact of SO_2 on climate change issues. Thus, one must not only consider the amount of the emission, but also the severity of the material being emitted. This leads to an analysis of equivalency, which returns us to the concept of LD_{50}, which provides a method of comparing the emissions of one chemical against another. Computerized techniques such as TRACI take all of the inputs from the process and combine these emissions using a consistent set of equivalencies to produce an overall impact analysis for the process, and thus the ability to compare the environmental impacts of one substance against another, in a rigorous and consistent manner. Eventually, one must combine all of the analysis in the evaluation step, during which the various impacts are compared and combined. In this step, one must consider which environmental harm is considered more egregious. Thus, we weight the importance of impacts to human health, ecosystem damage, loss of mineral and fossil resources, and more. The weightings change over time because they are guided by society and vary across regions of the world. The following example provides evidence of the challenge associated with this analysis.

The 2007 Presidential Green Chemistry Challenge Award in Greener Synthetic Pathways for the development of a soy-based wood adhesive to replace formaldehyde-based adhesive used in hardwood panels provides an example that demonstrates significant environmental gain with limited downside [20]. The process, developed by Hercules, Inc. with Professor Kaicheng Li (Oregon State University) and Columbia Forest Products was inspired by natural adhesive found in mussels and tofu loaded with proteins. This is a winning recipe: the environmentally friendly adhesive made from soy flour is stronger and more water resistant than formaldehyde-based glues; it is also cost competitive and the process reduces air emissions by 50–90%. The soy-based adhesives do not contain formaldehyde or use formaldehyde as a raw material. The elimination of formaldehyde improves indoor air quality. The technology also creates a new market for soy flour, which helps to support the soybean industry.

Are there also negative impacts for the technology? Most likely, although these are not quantified in the analysis. If one increases the growth of soybeans for the production of the resin, then one may need to convert forested land into farmland, resulting in a loss of biodiversity. In addition, increased use of fertilizers and pesticides leads to increased eutrophication of lakes and streams. One should also evaluate the ecotoxicity of the pesticides and their impact on wildlife. While a full environmental assessment is needed to properly evaluate this new process, even a preliminary analysis suggests the value of this particular innovation.

However, the analysis is not always as compelling, and quite a large number of "environmental" concepts have later been shown to cause significant environmental concerns. The use of methyl-tertiary-butyl ether (MTBE) as a fuel additive helps to illustrate this point. MTBE was introduced into automotive fuels beginning in 1979 as an antiknocking agent and a replacement for tetraethyl lead. Its use increased dramatically until about 1999, as fuel manufacturers tried to comply with the require-ments of the Clean Air Act amendments that increased the requirements for higher levels of oxygenates in fuel to reduce the emissions of greenhouse gases. However, in 1995, high levels of MTBE were unexpectedly found in drinking water wells throughout California, a result of the combination of MTBE in leaking underground storage tanks and its miscibility with water. Although not labeled as a carcinogen, the presence of MTBE in drinking water lends an unpleasant taste and odor, and thus causes the MTBE-containing water to become nonpotable. As a result, the use of MTBE has been curtailed significantly, and the Energy Policy Act of 2005 reduced the federal requirement for oxygen content in reformulated gasoline.

A similar controversy is associated with the use of ethanol. In Brazil, sugarcane is used as the raw material for production of ethanol. Because of the desire to become self-sufficient, Brazil developed policy expanding the amount of land dedicated to biofuel production. Is this really green? The impacts on the environment, land use practice, ecosystems, and working conditions and the role of international stake-holders have to be taken into account in decision making. Recent data suggests that Brazil risks incurring a 250-year carbon debt based on the deforestation expected by 2020 as it expands production of sugarcane ethanol and soybean biodiesel.

In the United States, corn is the primary source of ethanol. A 2010 *Scientific American* article delineates some of the challenges in evaluating the environmental impacts of this choice [21]. While the federal government has endorsed the use of corn ethanol as a means of reducing greenhouse gas emissions, the State of California has reached a different conclusion. California looked at current data, whereas the federal government used projections for 2022 and assumed higher crop yields, production efficiencies, and other breakthroughs that would mitigate emissions. Since fuel crops displace food crops, there is a short-term pulse of emissions as displaced farmers clear forests and cultivate previously undisturbed land. These emissions are dissipated over time, and when one looks at the overall emissions over the lifespan of the impact, the calculations can vary. Thus, it is clear that the way one calculates the impacts have a significant effect on the decision-making process.

While making an unequivocal environmental choice may not be possible, activities that will contribute to the well-being of both people and the natural environment over the long term should be considered. The following solutions proposed by the EPA suggest opportunities for good environmental stewardship [22, p. 142]:

- "*Air*: sustain clean and healthy air
- *Ecosystems*: protect and restore ecosystem functions, goods, and services
- *Energy*: generate clean energy and use it efficiently
- *Land*: support ecologically sensitive land management and development

- *Materials*: use materials carefully and shift to environmentally preferable materials
- *Water*: sustain water resources of quality and availability for desired uses"

3.5 WHAT CAN YOU DO TO MAKE A DIFFERENCE?

In February 2007 an article published in *C&EN* magazine titled "Avoiding Global Collapse" summarized the status of our society:

> What other societies had faced before us such as the abandoned pyramids of Mayan cities, Easter Island's lonely Moai statues and abandoned Viking settlements in Greenland, we face the same collapse today. These ancient societies failed because of the same problems we are facing today: environmental damage, climate variation, support from trade partners, and a failure of society to respond to problems such as overconsumption of natural resources and overpopulation. We could add human induced climate change and environmental damage from chemical pollution accompanied by a lack in biodiversity. This is a scary perspective. Instead of giving up we should look at solutions to avoid the same mistakes.

Working together toward a more sustainable society includes responsibility, respect of natural resources (no farm and fisheries mining), conservation, preservation, and clever technological development. The three main pillars of sustainability are based on society, economics, and environment. Ray Anderson in his book *Midcourse Correction: Toward a Sustainable Enterprise: The Interface Model* defined three levels in the journey toward sustainability [23].

- The first level is understanding sustainability. Understanding goes beyond knowing the definition of sustainability to identify technologies, attitudes, and practices that support the development of sustainable activities.
- The second level is achieving sustainability. This is the "doing curve" in which engineers and scientists work to overcome "technical challenges." To reach 100% sustainability implies that society really understands sustainable practices. This is not easy, because there is the ultimate need to understand the limits of the natural world.
- The third level is influence. The mission is to inspire others to give back to the earth with credibility and this is limitless.

The resultant curve is an unlimited global benefit curve. Green chemistry and engineering is part of this resultant curve. It should become a standard operating practice, as it has been shown that sustainable business practices boost profits, increase productivity, and reduce waste. Sustainability is a global solution for everybody. It can proliferate from home to school, to our community and work, while shopping and also on the road.

While many of the problems associated with a clean environment are global in nature, the solutions to environmental concerns are the responsibility of every citizen of the earth. "Think globally, act locally" is a phrase that refers to the argument that global environmental problems can only be solved when one considers the ecological, economic, and cultural differences of our local surroundings. While federal and state agencies can enact environmental laws that encourage large organizations to take a more environmental stand, individuals need to come together to protect habitats and the organisms that live within them. As an environmentally conscious citizen, here are some guidelines about what you can do to make a difference.

At home and in the garden protecting the environment starts with:

- Saving energy through the use of Energy Star appliances, which can save families about 30% ($400 a year) of their energy budget.
- Choosing water-efficient products.
- Practicing the three R's: reduce, reuse, recycle.
- Reducing exposure to harmful substances such as pesticides, insecticides, mold, radon, lead, and mercury.
- Practicing sustainable gardening through composting and smart watering.

The EPA's "Schools Chemical Cleanout" campaign looks to reduce the number of hazardous chemicals in local schools and provides tips on responsible chemical management in area schools. And it's not just about chemicals used in chemical laboratories. One also needs to consider the chemicals used in cleaning products in schools (or your home). The chemicals found in some cleaning products can cause health problems, including eye, nose, and throat irritation, as well as headaches. Cleaning products may also contain VOCs that can cause adverse health effects including asthma, upper respiratory irritation, fatigue, nasal congestion, nausea, and dizziness. Green cleaning products contain less of these hazardous materials and reduce the amount of chemicals in the environment.

Green purchasing provides another excellent opportunity to act locally, with global influence. As a consumer, you have the opportunity to purchase green products and Energy Star appliances. Buying energy-efficient products or fuel-efficient vehicles can make a big difference. Understanding the environmental impacts of household products can help inform sustainable choices. The business world has adopted environmentally preferable purchasing practices, which are now required by the federal government. As one of the world's largest consumers, the impact of federal purchasing will propagate throughout the product life cycle. Environmentally preferable means "products or services that have a lesser or reduced effect on human health and the environment when compared with competing products or services that serve the same purpose," according to the Instructions for Implementing Executive Order 13423. As an individual, you can look to make purchases that reduce your environmental footprint. As an example, when considering the purchase of paper, look for paper that has high postconsumer recycled content, is made from certified forest-safe trees, and is processed without chlorine. Use automatic duplexing for your printer, to take advantage of both sides of a sheet of paper, and reduce total paper consumption. Encourage the purchasing agent in your workplace or school to use environmentally preferred products, such as the paper described here, to multiply the impact of your local efforts.

Walmart has become a leading force in the development of environmentally preferable products. According to their website: "Our customers desire products that are more efficient, last longer and perform better. They want to know the product's entire lifecycle. They want to know the materials in the product are safe, that it is made well and is produced in a responsible way" [23]. In order to help quantify the environmental impact of products sold in Walmart stores, they have developed the Sustainability Index, to help promote transparency and support product innovation. In the early stages of development, the index will bring together information from a range of sources to create a single metric that will allow the consumer to identify which product might cause more harm to the environment, and they are using a consortium of universities, suppliers, retailers, nongovernmental organizations, and government officials to develop a global database of information on products' life cycles.

When you travel, you should also consider the opportunities to reduce your environmental impact. The EPA and BlueGreen Meetings provide information on how to make your meeting attendance more environmentally friendly. Many of these tips can be adopted for leisure travel as well. For example, hotels are now using energy-efficient lighting and incorporating linen reuse programs to reduce the amount of water consumption. Hotels use an average of 209 gallons of water each day for each occupied room. A linen reuse program can save up to 30% on water usage (that's roughly 60 gallons of water). Decreasing water consumption reduces operating costs, making the choice of environmental stewardship not only rewarding but also profitable.

In short, there are many opportunities to decrease your personal environmental footprint and adopt more sustainable personal practices. Using the multiplying power associated with large-scale adoption, it is clear that the greatest force leading to a more sustainable society is the power of the consumer. Choose wisely.

REFERENCES

1. http://www.epa.gov/airtrends/images/comparison70.jpg.
2. http://www.epa.gov/airtrends/2007/report/trends_report_full.pdf.
3. http://en.wikipedia.org/wiki/File:Atmospheric_ozone.svg.
4. http://en.wikipedia.org/wiki/File:Atmosphere_composition_diagram.jpg.

5. http://en.wikipedia.org/wiki/File:TOMS_Global_Ozone_65N-65S.png. http://www.cdc.gov/mmwr/preview/mmwrhtml/mm5650a1.htm.

6. http://www.epa.gov/asthma/pdfs/asthma_fact_sheet_en.pdf.

7. http://en.wikipedia.org/wiki/Radon.

8. http://en.wikipedia.org/wiki/Sustainability.

9. http://www.who.int/water_sanitation_health/dwq/wsh0306/en/.

10. http://en.wikipedia.org/wiki/Deforestation.

11. U.S. Energy Information Administration, Annual Energy Outlook 2012 Early Release. Available at http://www.eia.gov/forecasts/aeo/er/.

12. http://www.eia.gov/totalenergy/data/annual/pdf/sec2_3.pdf.

13. http://www.epa.gov/p2/pubs/casestudies/index.htm.

14. http://en.wikipedia.org/wiki/Ecotoxicology.

15. http://www.epa.gov/ecotox.

16. http://chemlabs.uoregon.edu/Safety/toxicity.html.

17. Bare, J.; Norris, G.; Pennington, D.; McKone, T. *J. Indust. Ecol.*, **2003**, 6(3–4), 49–78.

18. http://www.epa.gov/nrmrl/std/sab/traci/.

19. http://www.epa.gov/greenchemistry/pubs/pgcc/winners/gspa07.html.

20. http://www.scientificamerican.com/article.cfm?id=ethanol-corn-climate.

21. "Everyday choices: Opportunities for environmental stewardship" technical report. Available at http://www.epa.gov/environmentalinnovation/pdf/techrpt.pdf.

22. Anderson, R. C. *Mid-course Correction: Toward a Sustainable Enterprise: The Interface Model*, Chelsea Green Publishing Company, White River Junction, VT, 1998.

23. http://www.walmartstores.com/Sustainability/9292.aspx.

4

MATTER: THE HEART OF GREEN CHEMISTRY

Matter surrounds us. It is in everything we taste, touch, smell, and see. It constitutes most of the observable universe. From the time we use a tablespoon to stir sugar or cream in our coffee or tea in the morning, to the time we travel to work, and finally when we sleep, chemistry is occurring and chemical transformations are taking place. Chemistry is the study of matter and its transformations from one form to another. Understanding the properties of matter allows for the design of more environmentally benign materials. In this chapter we discuss matter and its properties, the three states of matter, their applications in green chemistry and green engineering, and how understanding the intrinsic nature of materials can lead to an improved design and a reduction in the environmental impact of the product.

4.1 MATTER: DEFINITION, CLASSIFICATION, AND THE PERIODIC TABLE

Matter is commonly defined as anything that occupies space and has mass. All matter is made up of tiny particles called atoms. An atom is the smallest particle of an element.

Green Chemistry and Engineering: A Pathway to Sustainability,
Anne E. Marteel-Parrish and Martin A. Abraham.
© 2014 American Institute of Chemical Engineers, Inc. Published 2014 by John Wiley & Sons, Inc.

One of the milestones in the study of matter was the theory published by English chemist John Dalton (1766–1844) in 1803, now commonly known as the Dalton's atomic theory. Dalton postulated that all matter could be described according to the following rules:

- All matter is composed of uniform particles called atoms.
- Atoms of a given element have the same chemical properties but have chemical properties different from atoms of other elements (e.g., nonmetallic chlorine and solid sodium have different chemical properties).
- When two or more atoms of different elements combine chemically and in whole-number ratios, they form a compound.
- Chemical changes or reactions involve bonding, dividing, or rearranging atoms but atoms can be neither lost nor created. (At a macroscopic level, this becomes the law of conservation of matter: matter is neither destroyed nor created.)

In 1869, Dmitri Mendeleev (1834–1907) noted the recurring trends in the properties of the elements and used this to classify the elements, leading to the modern periodic table arrangement. Mendeleev arranged the elements in order of increasing atomic weights, putting similar elements in the same column. At that time there were still gaps in the knowledge of the elements, so Mendeleev left space and made predictions for where other elements would place in his chart. The modern periodic table is now based on Moseley's (1887–1915) consideration to arrange the elements in order of increasing atomic numbers. In the periodic table a common representation of an element is shown as below:

Atomic number (whole number)

Symbol

Atomic weight (decimal number; unit in g/mol)

Elements with similar properties such as comparable atomic radius are placed in the same columns called groups or families (1–8 or 1–18). Because of their similar properties, four main regions in the periodic table can be distinguished:

- Main group metals include the first column or group commonly referred to as the alkali metals and the second column commonly referred to as the alkaline earth metals as well as aluminum (Al), gallium (Ga), indium (In), tin (Sn), thallium (Tl), lead (Pb), bismuth (Bi), and polonium (Po).
- Transition metals (columns 3 through 12).
- Nonmetals such as hydrogen (H), helium (He), carbon (C), nitrogen (N), phosphorus (P), oxygen (O), sulfur (S), selenium (Se), the column 17 commonly referred to as the halogens and the column 18 named noble or inert gases.
- Metalloids: there are six of them—boron (B), silicon (Si), germanium (Ge), arsenic (As), antimony (Sb), and tellurium (Te).

TABLE 4.1. Some Good, Bad, and Ugly Common Elements

Good Elements	Bad Elements	Ugly Elements
Nitrogen and explosives	Lead and paint	Arsenic and poisoning
Hydrogen and energy (hydrogen fuel cells)	Mercury and aquatic life	
Fluorine and fluorocarbons	Chlorine and environment	
Calcium and bones	Polonium and radioactivity	
Iron and anemia (iron-poor blood)		
Lithium and batteries		
Oxygen and respiration		
Sodium and food		
Magnesium and enzymes		
Aluminum and household utensils		
Phosphorus and DNA		
Strontium and fireworks		
Antimony and leishmaniasis		
Gold and jewelry		

The transition metals and metalloids, as well as lead, are commonly referred to as heavy metals. The term "heavy metals" is misleading and is not based on any scientific definition. Heavy metals are often used in catalysis and other chemical processes. Many of these metals are also hazardous to the environment and/or human health, leading them to be sometimes known as toxic metals.

The current periodic table contains 118 elements all associated with a unique symbol as of June 2012. All elements from atomic numbers 1 (hydrogen) to 118 (ununoctium) have been isolated and, of these, all up to and including californium exist naturally. Two recently accepted elements were named "flerovium" (114) and "livermorium" (116). The most recent discovery in 2010 of an element was the result of an international collaboration between Oak Ridge National Laboratory (U.S. Department of Energy), the Joint Institute of Nuclear Research (Dubna, Russia), the Research Institute for Advanced Reactors (Dimitrovgrad), Lawrence Livermore National Laboratory, Vanderbilt University, and the University of Nevada Las Vegas. This team claimed to have synthesized six atoms of the temporarily named element of ununseptium (element 117) [1].

Elements in the periodic table can be classified as good, bad, and ugly ones. Categories with listed common elements are presented in Table 4.1. A few of these elements are discussed in more detail here.

4.1.1 Aluminum (Al)

A common element aluminum (with the symbol Al) is present in cooking utensils and is one of the safest metals. Although it was postulated that it was involved in Alzheimer's disease, the correlation has not been confirmed. Aluminum can be recovered from its ore, bauxite, or it can be recycled from recovered aluminum product. Recycling aluminum is more energy efficient than extraction from ore, and aluminum can be recycled many times without a decrease in its quality.

4.1.2 Mercury (Hg)

Like aluminum, mercury (Hg) is also an element naturally found in the environment. But it is among the most toxic elements, through inhalation over long periods of time, through contact with the skin, or if swallowed. It is hard to believe then that mercury has been used for dental fillings for quite a while (Highlight 4.1).

Severe mercury poisoning can lead to Minamata disease, which is a neurological disease exhibited by numbness in the hands and feet, general muscle weakness, and damage to sight, hearing, and speech, eventually leading to death in extreme cases. This disease, first discovered in Minamata City in Japan in 1956, was caused by the release of a derivative of mercury, called methyl mercury, in the wastewater from a local chemical factory. The bioaccumulation of this highly toxic chemical in fish and shellfish in Minamata Bay resulted in mercury poisoning for the local population who ate the fish. It was not until 2001 that the chemical manufacturer, Chisso, initially producing fertilizers, and the local Japanese government finally recognized the extent of the pollution, which caused almost 2000 deaths.

Highlight 4.1   The Dental Amalgam Controversy [2]

Dental amalgam, consisting of mercury alloyed with another element, is still approved for use in most countries, but conflicting views have been expressed over the use of mercury as filling material. Some dentists are still recommending the use of dental amalgam with mercury because of its durability, ease of use, and low cost compared to resin composites (for which concerns have been raised due to the presence of plastic chemicals such as Bisphenol A, well known for its endocrine disruptor effects) and dental porcelain. However, it is believed that leaching of mercury into the mouth and consequential health effects related to mercury toxicity (such as risk of impairment in the central nervous system function, kidney function, immune system, and fetal development) are associated with mercury exposure.

In addition to the health impacts due to the absorption and accumulation of mercury in various organs, the environmental damage caused by mercury disposal is worrisome. A study performed in 2005 by the World Health Organization pointed out that 53% of total mercury emissions come from the disposal of amalgam flushed down the drain and laboratory devices in dental offices.

Highlight 4.2   Lead and Products Recall [3]

The news about recalls of toys or products because of the presence of excessive lead is unfortunately very common. The U.S. Consumer Product Safety Commission in charge of "protecting the public from unreasonable risks of injury or death from thousands of types of consumer products" continues to issue recalls due to the presence of lead in various items. Besides the presence of lead in many painted toys, children's sports, bibs, and furniture products, and jewelry, the newest scare was generated when more than a few dozen handbags from popular retailers were tested for lead levels ranging "from nearly three times to more than 195 times higher than the level agreed to in a 2010 settlement between Center for Environmental Health and dozens of retailers, producers and distributors of the products."

While lead amounts are limited by federal law in children's products, no federal standard exists for lead in adult handbags, purses, and wallets. Health problems linked to lead exposure encompass an increased risk of heart attacks, strokes, and high blood pressure, among other health problems. However, as of June 2012, there is no safe level of lead exposure, especially for pregnant women and young children.

4.1.3   Lead (Pb)

In the same category of elements, lead (Pb) has been the topic of attention since 3800 BCE when the Romans were drinking acidic beverages from lead containers and even adding lead salts to acidic wine to sweeten the taste. As a consequence, lead was present in a higher level in the bones of Romans than those of modern humans. As mentioned in Chapter 1, lead is extremely toxic, especially to children because it accumulates in their growing bodies very fast. In the early 1970s, lead additives were banned from petroleum in the United States. In the United States, lead is banned from paints, but the presence of lead in red and yellow paints used for manufactured wooden toys from China has resulted in numerous product recalls (Highlight 4.2). Lead is naturally present in petroleum and has been one of the major causes of lead pollution.

4.2   ATOMIC STRUCTURE

The reactivity and chemical properties of the elements are highly dependent on the atomic structure. The atom is comprised of two major components:

1. The nucleus, which contains most of the mass of the atom and is made predominantly of two types of subparticles:

- Protons are positively charged particles and define the atomic number of the atom.
- Neutrons, which have no charge.

2. The space around the nucleus containing electrons, which are negatively charged particles with very low mass that orbit in a cloud surrounding the nucleus. Since an atom must be electronically neutral the number of electrons exactly matches the number of protons.

Figure 4.1 shows a representation of atomic structure, illustrating the relationship of the nucleus (containing protons and neutrons) and the space around it that is occupied by electrons.

Table 4.2 summarizes the characteristics of each particle.

The atomic weight of the atom can be calculated by adding together the weights of all of the subparticles that make up the atom. Very often, the atomic weight is given in terms of atomic mass units (amu), equivalent to the mass of one proton. Thus, one also can define the mass of the atom in terms of atomic mass number, which is simply equal to the sum of the number of protons and number of neutrons.

$$\text{Atomic mass number} = \text{number of protons} + \text{number of neutrons}$$
$$(\text{all whole numbers})$$

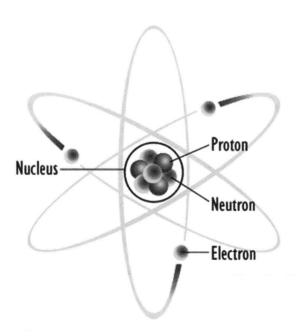

Figure 4.1. Representation of atomic structure [4].

TABLE 4.2. Characteristics of Protons, Electrons, and Neutrons in Terms of Charge, Charge Number, and Mass

Characteristic	Proton	Electron	Neutron
Charge (C)	$+1.602 \times 10^{-19}$	-1.602×10^{-19}	0
Charge number (relative charge)	$+1$	-1	0
Mass (g)	1.673×10^{-24}	9.109×10^{-28}	1.673×10^{-24}

Because we must have whole numbers of subatomic particles, one would expect that the atomic weight of the elements would all be whole numbers. But a quick look at the periodic table indicates that is not the case. How can that be? It turns out that many of the elements may have multiple varieties, or isotopes, that vary only in the number of neutrons the element contains. In total, 80 elements in the periodic table have stable isotopes. In order to calculate the atomic weight (in g/mol) of the element, an average of atomic weight of all isotopes of the same element is calculated, based on the normal abundance of the element as found in nature. As an example, chlorine has two isotopes, one having an atomic mass of 35 and found in 75.8% of the chlorine atoms (this is written as ^{35}Cl), and the other isotope having an atomic mass of 37 (^{37}Cl) and found in 24.2% of the chlorine atoms. When a weighted average of these two isotopes is calculated, the average atomic mass is calculated to be 35.5 (g/mol or amu), which is the number reported in the periodic table.

4.3 THREE STATES OF MATTER

Matter can be classified according to its physical state. The three principal states of matter include the solid, liquid, and gaseous phases, in decreasing levels of structure and atomic order.

Solids have a fixed volume and rigid shape. They cannot be compressed. In a solid, the atoms of an element are highly ordered and are tightly packed. A diamond is an example of a solid containing all atoms of carbon (Figure 4.2). It is the hardest naturally occurring mineral.

Liquids have a definite volume but will take the shape of the container into which they are placed. There is a more random arrangement of particles in liquids. You can pour milk in a glass, in a cup, or in a bowl. Milk is a liquid and it takes the shape of its container. Water, such as that found in a lake or a river, or the water that we drink, is also a liquid.

Gases are the least structured of the phases. A gas will take the shape and volume of the container in which it is placed. Particles in a gas are highly disordered and are relatively far apart. In gases, atoms are moving around randomly and do not occupy any specific position in space.

Figure 4.2. Crystalline structure of diamond (all represented atoms are those of carbon).

These phases define the physical condition of the material, and not the chemical structure. The same material may interchange between the various phases. The same species may exist as either vapor, liquid, or solid, or may coexist as multiple phases, depending on the specific conditions of temperature and pressure. The phase behavior of the species is illustrated in a phase diagram, which illustrates the conditions at which a phase change can occur. The curves in the phase diagram describe the phase boundaries and the point of transition between phases (Figure 4.3). There are two unique points on the phase diagram that bear notice. The first is the triple point, which is the single condition at which a pure species can exist as a solid, liquid, and gas at the same time. The other is the critical point, which marks the end of the liquid–gas phase boundary. Above the critical point, it is no longer possible to induce a phase change and the fluid moves continuously from liquid-like to gas-like properties.

The physical properties of the material depend on the phase of the material. When undergoing a phase change, the substance remains chemically identical, but its appearance (the way it appears to the naked eye) is different. The physical state, size, or shape may have changed but there is no creation of a new substance in a physical change. The identity of the substance is preserved. Important physical properties in engineering and chemistry include density, thermal conductivity (the ability of a substance to transfer heat energy), and specific heat (the ability of the substance to retain energy).

Species (atoms, molecules, and ions) that are not chemically bonded to each other may interact with one another through intermolecular forces. The strength of intermolecular forces dictates the inherent properties of solids, liquids, and gases. Compounds with very strong intermolecular forces are normally solids at room temperature, whereas compounds with intermediate intermolecular forces are liquids, and those with extremely weak intermolecular forces are gases. The strength of these intermolecular forces explains why a solid such as sucrose melts at 185 °C, whereas ice water melts at 0 °C and molecular nitrogen boils at −196 °C.

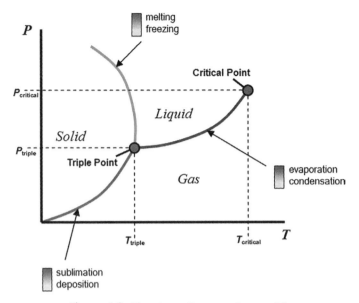

Figure 4.3. The phase diagram of water [5].

The strongest of intermolecular forces within pure substances is hydrogen bonding. Hydrogen bonds are formed between the hydrogen of the H_2O molecule and a highly electronegative atom such as F, N, or O. Hydrogen bonding occurs between molecules of water but also between carboxylic acid molecules, which can often form dimers through hydrogen bonding in the vapor phase.

All metals in the periodic table are solids, except mercury (Hg), which is a liquid. At room temperature, nonmetals can be solids such as phosphorus (P), sulfur (S), and iodine (I_2); liquids such as bromine (Br_2); or gases such as hydrogen (H_2), nitrogen (N_2), and oxygen (O_2). Note that for some elements, we normally write their structure with the subscript 2 attached. That is because these elements (such as iodine, bromine, hydrogen, nitrogen, and oxygen) do not exist in nature as individual atoms, but rather they exist as a diatomic molecule. A diatomic molecule is the simplest molecular form and occurs when two atoms of the same type are connected to one another.

4.4 MOLECULAR AND IONIC COMPOUNDS

When two or more atoms are connected or combine together, they form compounds. The simplest unit of a compound is called a molecule. A molecule can contain identical elements such as in the homonuclear (same nucleus) diatomic molecule of oxygen (O_2) or different elements (heteronuclear) such as in water (H_2O).

Chemical compounds can either consist of individual molecules (formation of molecular compounds) or of positively and negatively charged atoms or groups of atoms called ions (formation of ionic compounds). Ionic bonding is stronger than hydrogen bonding, but it can only occur in heteronuclear molecules; the strength of ionic bonds is the reason that salts (such as sodium chloride, NaCl, ordinary table salt) are normally solids at room temperature.

4.4.1 Molecular Compounds

4.4.1.1 Inorganic Molecules

Recall that the atom is made up of protons and neutrons in the nucleus, and electrons orbiting in an external shell. Because of the way the electrons distribute, there are multiple layers in which the electrons can orbit. The most important electrons are in the outer layer, furthest away from the nucleus, and are denoted as the valence electrons. It is the interaction of these valence electrons that leads to the formation of molecules.

Each grouping of atoms wants to exist in its lowest energy state, which is achieved when it has a complete outer shell of electrons. In order to achieve this completion of the shell, the atom may share electrons with a neighboring atom. When two elements each share one of their valence electrons, they share a pair of electrons and a covalent bond is formed. In some cases, two elements can complete their outer shell if one of the elements donates an electron to a neighboring element. In this case, an ionic bond is formed. Ionic compounds will be described in Section 4.4.2.

The periodic table provides more information about the atomic structure than just the mass number, atomic number, and some of the chemical behaviors of elements. Elements in a column have the same number of valence electrons. The valence electrons are the electrons involved in the formation of covalent bonds. In fact, for the main group elements (group A), the number of valence electrons is equal to the group number (with the exception of helium).

For the first three rows of the periodic table, when molecules are formed (with just a few exceptions), the complete outer shell of each atom involved in the formation of a covalent bond contains eight electrons. For example, the diatomic molecule of fluorine (F_2) is made from two atoms of fluorine (F) (each with seven valence electrons). When these two atoms combine, they each share a pair of electrons to form a covalent bond and six electrons remain on each fluorine atom. The outer shell of each fluorine atom contains eight electrons, the six original electrons remaining with each atom and the two that are now shared, thus forming a stable molecule.

One way to represent the sharing of electrons is through the Lewis dot representation. Here, the nucleus of the atom is represented by the letter representation from the periodic table, then the electrons in the outer shell are indicated as dots surrounding the nucleus. The two atoms that share an electron are shown with the two electrons between the nuclei, providing a visual clarification that a stable material is produced. When more than two types of atoms are involved in a molecule, it is often necessary to define a central atom, with the other atoms in the molecule being strategically placed to illustrate how the electron sharing occurs (Figure 4.4).

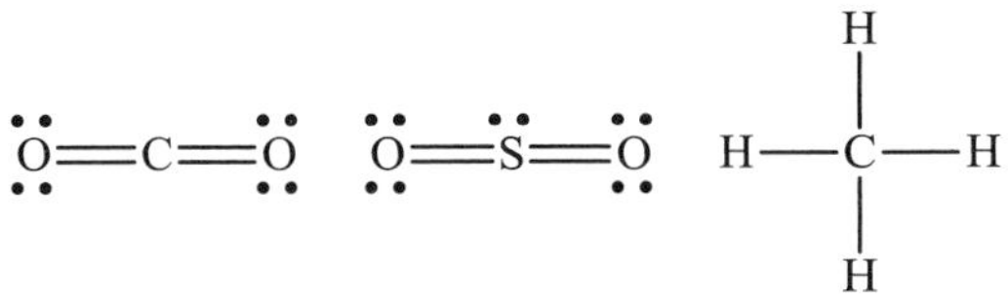

Figure 4.4. Examples of Lewis structures of CO_2, SO_2, and CH_4.

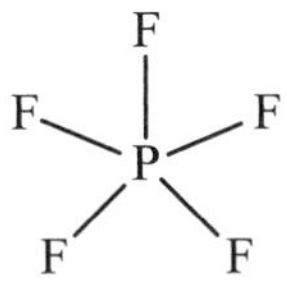

Figure 4.5. Example of Lewis structure for which the central atom has more than eight electrons.

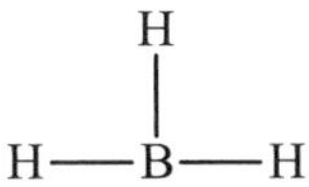

Figure 4.6. Lewis structure for the borane molecule.

There are numerous exceptions to the octet rule, particularly as you move down the periodic table to the higher molecular weight elements. Elements in period three and beyond have more than eight valence electrons. In this case, they form molecules or ions with an expanded octet. One such example is the element phosphorus. Phosphorus is in group 5A and therefore has a total of five valence electrons. It can form five covalent bonds such as in the molecule of phosphorus pentafluoride (PF_5), a highly reactive colorless gas at room temperature and pressure. Fluorine, which is in group 7, has seven valence electrons. So phosphorus may share one of its valence electrons with each of the F atoms, creating a shell with eight electrons for each fluorine atom, and thus a stable molecular species. The phosphorus atom is an exception to the octet rule in this case, because it has 10 shared electrons, as seen in the Lewis structure of Figure 4.5.

On the other hand some central atoms have fewer than eight valence electrons and still form stable molecules. Boranes, whose parent member is BH_3, are extremely reactive and are involved in the formation of cage-like materials used in catalysis. Boron is in group 3A and only has three valence electrons to share. When B forms three bonds with three hydrogen atoms, it has a total of six electrons around it, as shown in Figure 4.6. Therefore, boron cannot achieve the octet rule since it will never have a total of eight valence electrons around it. Similar elements to B are found in group 3A, such as aluminum (Al) and gallium (Ga).

For transition elements also known as transition metals (periods 4 through 7), the outer valence shell may contain as many as 18 electrons, distributed into sub-orbitals. Transition metals are involved in the formation of coordination complexes,

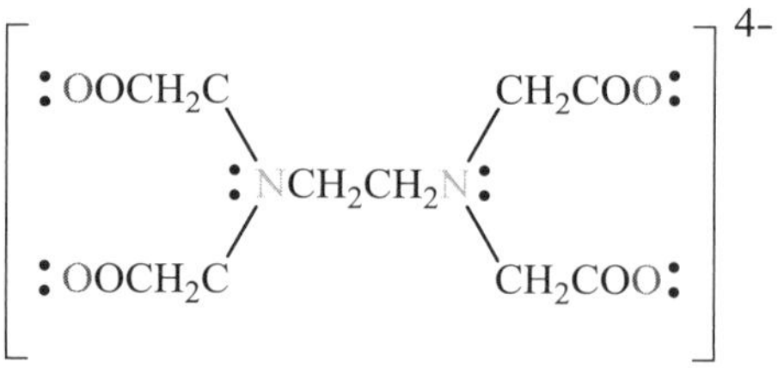

$$\begin{bmatrix} :OOCH_2C & & CH_2COO: \\ & :NCH_2CH_2N: & \\ :OOCH_2C & & CH_2COO: \end{bmatrix}^{4-}$$

Figure 4.7. The EDTA ligand.

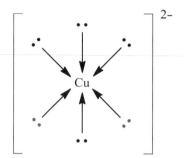

Figure 4.8. Binding of EDTA ligand to copper.

in which the transition metal is the central atom and the ions or molecules sur-rounding the central atom are called ligands. The type of bond connecting the central atom to the ligands is known as a coordinate covalent bond, where donation of electrons occurs from the ligands to an empty metal orbital. Coordination complexes are used extensively as catalysts in a variety of organic reactions. Examples of coordination complexes include the anticancer drug cisplatin $[Pt(NH_3)_2Cl_2]$, where Pt^{2+} is the central metal cation surrounded by NH_3 and Cl^- ligands. This complex is neutral since the two negative charges from Cl^- balance the two positive charges from Pt^{2+}. Another well-known example of a ligand is $EDTA^{4-}$, drawn as shown in Figure 4.7.

Because $EDTA^{4-}$ binds strongly with many metal species, it is often added to commercial salad dressing to remove traces of metal ions from solution (the presence of metal ions will accelerate the oxidation of oils in salad dressings). The other useful need for $EDTA^{4-}$ is in the treatment of lead and mercury poisoning since $EDTA^{4-}$ anions can form coordination complexes with the metal cations of Cu^{2+}, Pb^{2+} and Hg^{2+} (Figure 4.8). When bound to a metal atom, the six available lone pairs of elec-trons can wrap around the central metal ion, creating a stable coordinate covalent molecule, as shown here for Cu(EDTA).

Inorganic compounds typically do not have any carbon covalently bonded to oxygen, hydrogen, or nitrogen. Now that we know a little about how inorganic compounds are formed, it is also important to know how to name them. For binary inorganic compounds some rules apply.

- An element farther left in the periodic table is written first.
- If both elements are in the same group, the lower element is written first.
- The second element gets the "-ide" suffix.
- Greek prefixes indicate numbers of atoms, but do not use "mono-" for the first element.

The most useful Greek prefixes are the following:

2—di
3—tri
4—tetra
5—penta

Typical inorganic compounds include titanium tetrachloride ($TiCl_4$), dinitrogen pentoxide (N_2O_5), phosphorus pentachloride (PCl_5), and aluminum trichloride ($AlCl_3$).

4.4.1.2 Organic Molecules

Molecules that contain only the elements C, H, O, N, S, and P may be organic compounds. Not all molecules containing these elements are organic, but those derived from living species and based on the structure of carbon as a building block are generally included. Nearly all materials derived from fossil fuels are organic compounds. Through derivatization (replacement of a hydrogen with another element), halogens (Cl, Br, etc.) may also be incorporated into an organic compound.

Organic compounds can be classified based on the types of bonds and atoms that they contain. Hydrocarbons are a large class of organic compounds that contain only the elements hydrogen and carbon, and may be subclassified into groups of alkanes such as propane (C_3H_8), a gas used in household heating, and alkenes such as ethene or ethylene, the basic building block of polyethylene, a major group of plastics. Alcohols contain an oxygen atom bound to a hydrogen as an end group, such as ethanol (C_2H_5OH) commonly found in alcoholic beverages such as wine and beer (as a side note, ethanol was also added to gasoline for many years). Carboxylic acids contain an oxygen double-bonded to carbon plus an OH group; an example is ethanoic or acetic acid (CH_3COOH) commonly found in salad dressings. There are many other classes of organic compounds.

There are three different ways to represent a molecular compound, as described for propane, an alkane containing three carbon atoms linked together with hydrogen atoms forming the remaining bonds.

- One can use the molecular formula, which represents the type and number of atoms in a molecule. The molecular formula for propane is C_3H_8.
- On the other hand, the structural formula shows exactly how the atoms are connected to one another.

$$
\begin{array}{ccc}
\text{H} & \text{H} & \text{H} \\
| & | & | \\
\text{H}-\text{C}-\text{C}-\text{C}-\text{H} \\
| & | & | \\
\text{H} & \text{H} & \text{H}
\end{array}
$$

- Finally, the condensed formula shows the grouping of atoms or groups of atoms connected through each C atom. The condensed formula for propane is $CH_3CH_2CH_3$.

The simplest hydrocarbon structure is that of the alkane, formed from simple bonds of carbon and hydrogen atoms. Alkanes are made up of a chain of carbon atoms bound to each other, with only hydrogen atoms surrounding the carbon atoms. When two carbon atoms bond together in a single manner, and each of the carbon atoms is then bound to three additional hydrogen atoms, this is the two-carbon alkane known as ethane. Similarly, three carbon atoms in a chain make up propane, and similarly named substances exist for every possible alkane. The name, molecular formula, and structural formula of the first ten alkanes are summarized in Table 4.3; note that the structure of an alkane always has a formula that can be represented as C_nH_{2n+2}.

It is also possible to make an alkane that is nonlinear, otherwise known as a branched alkane. Such species have a central carbon atom that is bound to three carbon atoms, and thus three branches are created. The simplest of the branched alkanes is methylpropane, also called isobutane (Figure 4.9).

When naming a branched alkane, you start with the longest single chain to identify the root structure, and then add the side chain with the proper term as if you were naming the sidechain as a separate species. Some examples of branched alkanes are shown, with the proper name for the species also listed (Figure 4.10).

TABLE 4.3. The First Ten Alkanes

Name of Alkane	Molecular Formula	Structural Formula
Methane	CH_4	$\begin{array}{c}\text{H} \\ | \\ \text{H}-\text{C}-\text{H} \\ | \\ \text{H}\end{array}$
Ethane	C_2H_6	$\begin{array}{cc}\text{H} & \text{H} \\ | & | \\ \text{H}-\text{C}-\text{C}-\text{H} \\ | & | \\ \text{H} & \text{H}\end{array}$

Propane C_3H_8

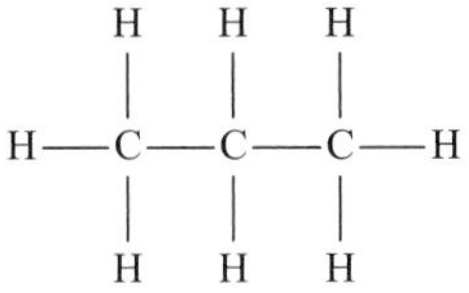

```
       H   H   H
       |   |   |
   H — C — C — C — H
       |   |   |
       H   H   H
```

Butane C_4H_{10}

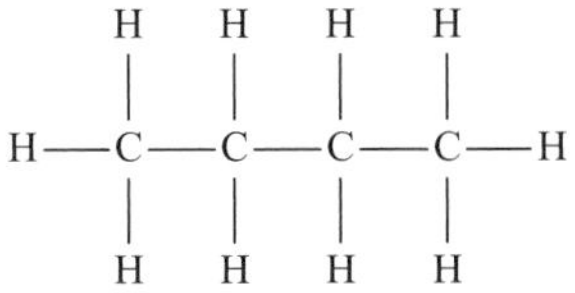

```
       H   H   H   H
       |   |   |   |
   H — C — C — C — C — H
       |   |   |   |
       H   H   H   H
```

Pentane C_5H_{12}

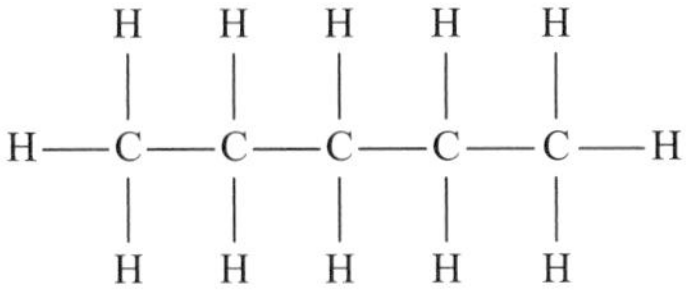

```
       H   H   H   H   H
       |   |   |   |   |
   H — C — C — C — C — C — H
       |   |   |   |   |
       H   H   H   H   H
```

Hexane C_6H_{14}

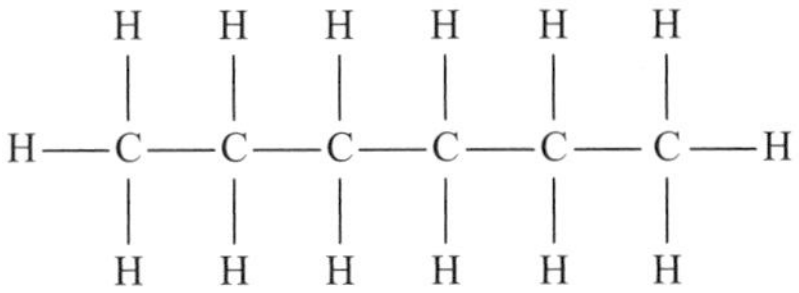

```
       H   H   H   H   H   H
       |   |   |   |   |   |
   H — C — C — C — C — C — C — H
       |   |   |   |   |   |
       H   H   H   H   H   H
```

Heptane C_7H_{16}

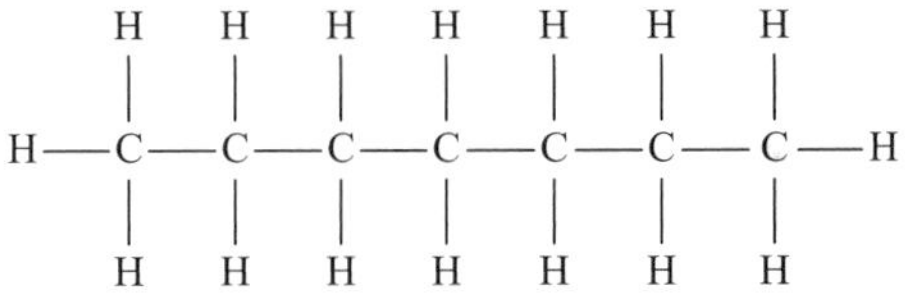

```
       H   H   H   H   H   H   H
       |   |   |   |   |   |   |
   H — C — C — C — C — C — C — C — H
       |   |   |   |   |   |   |
       H   H   H   H   H   H   H
```

Octane C_8H_{18}

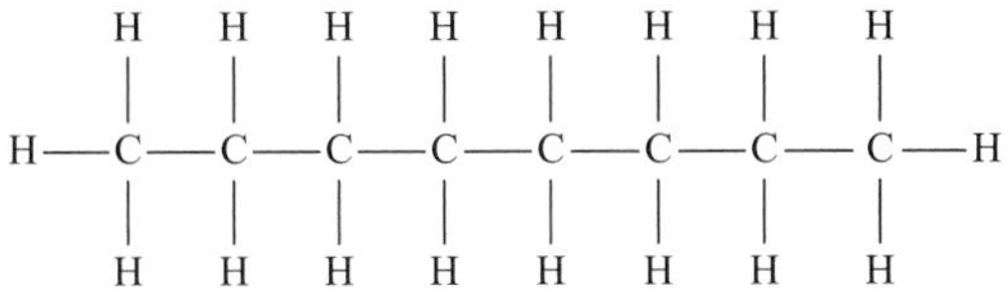

```
       H   H   H   H   H   H   H   H
       |   |   |   |   |   |   |   |
   H — C — C — C — C — C — C — C — C — H
       |   |   |   |   |   |   |   |
       H   H   H   H   H   H   H   H
```

Nonane C_9H_{20}

```
       H   H   H   H   H   H   H   H   H
       |   |   |   |   |   |   |   |   |
   H — C — C — C — C — C — C — C — C — C — H
       |   |   |   |   |   |   |   |   |
       H   H   H   H   H   H   H   H   H
```

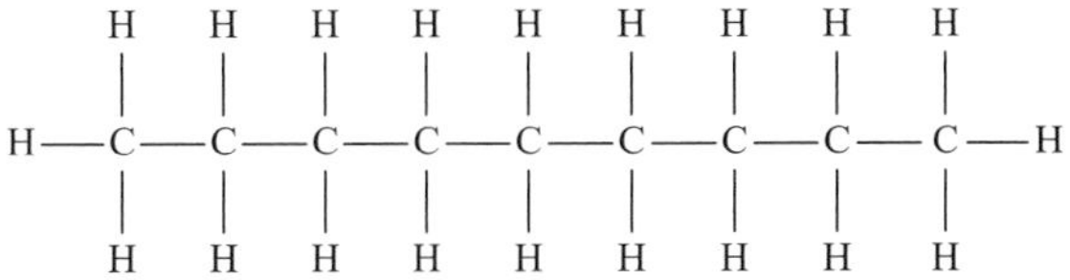

Decane $C_{10}H_{22}$

```
       H   H   H   H   H   H   H   H   H   H
       |   |   |   |   |   |   |   |   |   |
   H — C — C — C — C — C — C — C — C — C — C — H
       |   |   |   |   |   |   |   |   |   |
       H   H   H   H   H   H   H   H   H   H
```

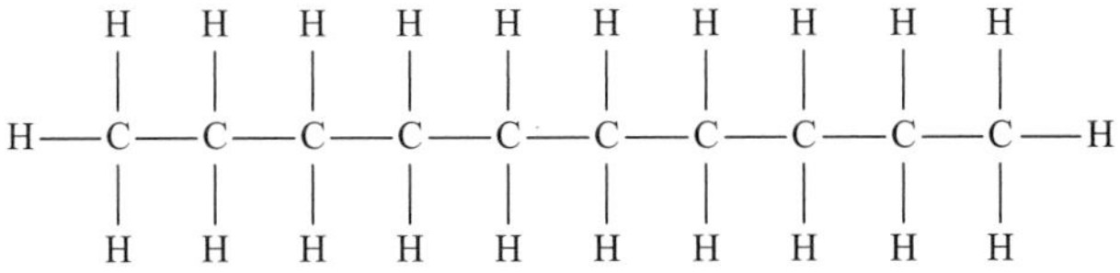

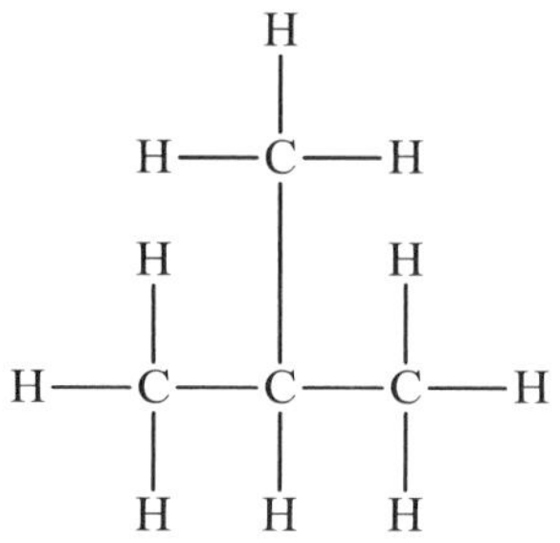

Figure 4.9. Structural formula of methylpropane.

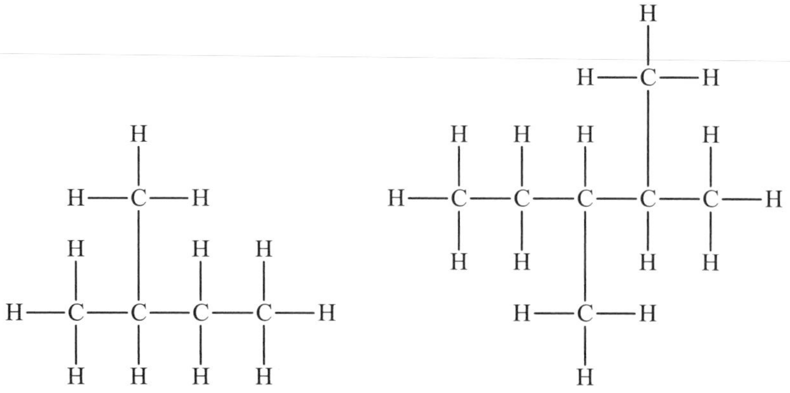

2-Methylbutane (or isopentane)   2,3-Dimethylpentane

Figure 4.10. Examples of branched alkanes and structural formulas.

When you look at the carbon element present in group 14 or 4A, the number of valence electrons is four. In other words, carbon can form a maximum of four bonds. As you probably noticed all C–C bonds in alkanes are single bonds. But it is also possible for carbon atoms to share multiple electrons, and form double or triple bonds. When two carbon atoms share two electrons, they form a double bond to produce a molecule known as an alkene. The simplest of the alkenes is ethene, or ethylene, which contains two C atoms and four H atoms. All alkenes can be represented by the general structure C_nH_{2n}, and the molecules are named in a manner analogous to alkanes. Thus, a three-carbon molecule containing a double bond would be propene (or propylene, an older form of the name).

A common colorless and flammable gas at room temperature and pressure, propene is generally produced from coal or petroleum and is a nonrenewable resource. However, it can be used in combination with hydrogen peroxide (H_2O_2) and employed as an inexpensive rocket fuel propellant. Alkenes can also form branched compounds, just as the alkanes (Figure 4.11). However, in the case of the alkene, the main part of the name is always the portion containing the double bond. And, it is important to identify the carbon atom from which the branch comes. Some examples of the names

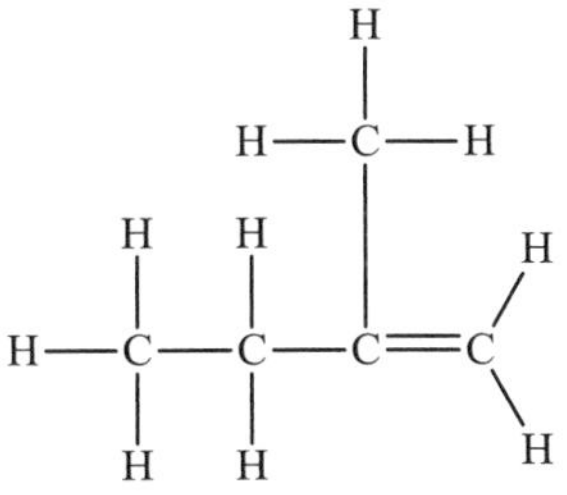

(a) 2-Methylbut-1-ene
or 2-methylbutene

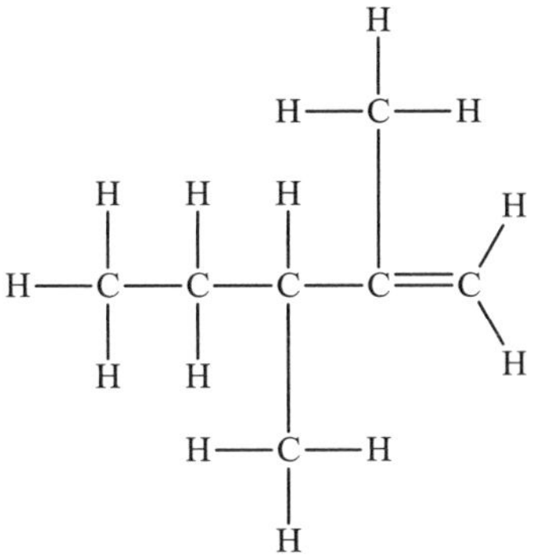

(b) 2,3-Dimethylpent-1-ene or
2,3-dimethylpentene

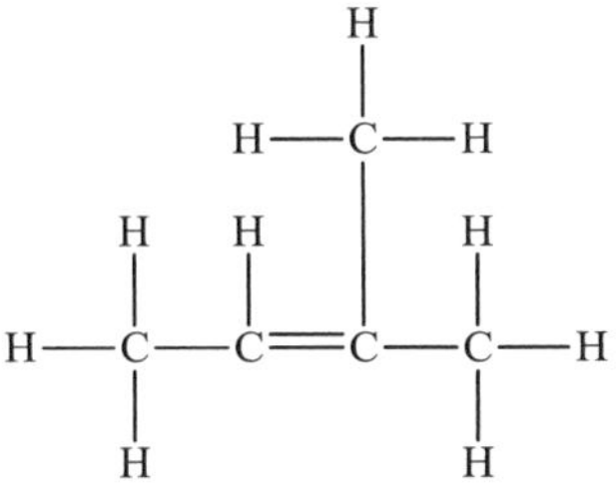

(c) 2-Methyl-but-2-ene

Figure 4.11. Examples of branched alkenes.

of alkenes are provided here. Note the convention on naming, but also know that the naming conventions become very complicated as the molecules become larger and more complex.

Note that in the examples in Figure 4.11a,b, since all double bonds start on the first carbon (counting from the right), it is not essential to include the 1 in the name of the alkene. However, since the double bond starts on carbon number two (counting from all possible starting carbons) in the example in Figure 4.11c, it is necessary to mention the position of the double bond in the name.

Since carbon can only form a total of four bonds, it is possible for a carbon atom to contain one triple bond and one single bond; hydrocarbons containing a triple bond are termed alkynes. Alkynes are named in parallel to the alkanes and alkenes, as described previously, and can also be branched and linear. Propyne, also known as methylacetylene, has the condensed formula $CH_3C \equiv CH$ and is a component of MAPP gas, a tradename of the Petromont, previously of the Dow Chemical Company. MAPP stands for *methylacetylene-propadiene propane*. Propyne is usually used in combination with oxygen for welding and is about two to four times more expensive than propane.

That takes care of the molecules made up exclusively of carbon and hydrogen, but there are also many important organic compounds that contain oxygen (and other

TABLE 4.4. Useful Functional Groups in Organic Chemistry

Class of Organic Compound	Functional Group
Alcohols	R–OH
Aldehydes	$\underset{R}{\overset{\displaystyle O}{\underset{\displaystyle}{\|}}}\underset{H}{C}$
Carboxylic acids	$\underset{R}{\overset{\displaystyle O}{\underset{\displaystyle}{\|}}}\underset{OH}{C}$
Ketones	$\underset{R}{\overset{\displaystyle O}{\underset{\displaystyle}{\|}}}\underset{R}{C}$
Ethers	R–O–R′
Esters	$\underset{R}{\overset{\displaystyle O}{\underset{\displaystyle}{\|}}}\underset{O\text{-}R}{C}$
Amines	$NR_1R_2R_3$

atoms, but we'll focus on the oxygenated molecules for now). In order to distinguish one compound from others and to be able to name it, we look at their specific functional groups. Useful functional groups covered in this chapter are summarized in Table 4.4.

R and R′ are the hydrocarbon frameworks with one hydrogen removed for each functional group added. R_1, R_2, and R_3 in amines are alkyl groups, typically $-CH_3$ (methyl) or $-CH_2CH_3$ (ethyl).

The general structures and names are important to know, since chemists use the names to properly describe molecules of importance. The names of alcohols, aldehydes, and ketones are derived directly from the names of their respective alkanes. Alcohols end in "ol," aldehydes end in "al," and ketones end in "one." In naming alcohols, the longest chain of carbon atoms is numbered so that the C atom attached to the –OH functional group has the lowest number.

For example, the simplest alcohol, methanol, which is a flammable and poisonous liquid, but which is also used in the production of biodiesel, is derived from methane and has the structural formula shown in Figure 4.12.

Butanol or butan-1-ol, which is based on butane (having four C atoms in its hydrocarbon chain), would be represented as in Figure 4.13.

Methanal, the methane-derived aldehyde also called formaldehyde, would be drawn as in Figure 4.14.

For ketones, a number is used to designate the position of the –C=O (also called carbonyl) group, keeping in mind that the numbering of the longest chain of carbons should give the carbonyl group the smallest number.

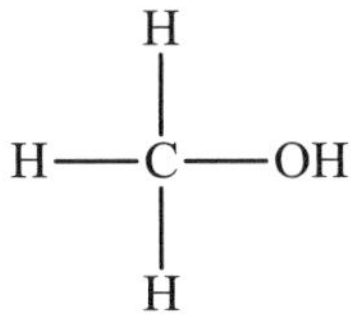

```
        H
        |
 H —— C —— OH
        |
        H
```

Figure 4.12. Structural formula of methanol.

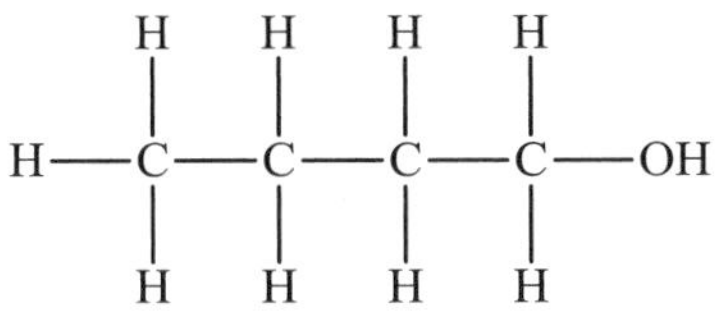

```
     H    H    H    H
     |    |    |    |
 H — C  — C  — C  — C — OH
     |    |    |    |
     H    H    H    H
```

Figure 4.13. Structural formula of butanol.

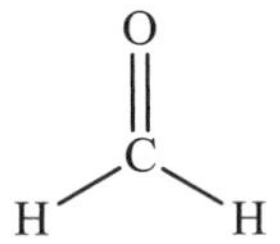

Figure 4.14. Structural formula of methanal.

Butan-2-one would have the structural formula shown in Figure 4.15.

Carboxylic acids are named after dropping the final "e" of the corresponding alkane and adding "oic acid." Butanoic acid would be represented as in Figure 4.16.

Ethers are named by identifying the two alkyl sections on either side of the oxygen and then adding the name ether. Thus, methylethylether (also called methoxyethane) is written as $CH_3OCH_2CH_3$ and has the structural formula shown in Figure 4.17.

An ester is a compound that is similar to a carboxylic acid, except the hydrogen atom bound to the oxygen is replaced with another alkyl group. The naming of an ester derives from the name of the two alkyl groups in the molecule, with the second group identified through the suffix "oate" at the end (Figure 4.18). It is important to keep in mind that for aldehydes and carboxylic acids, the carbon atom part of the functional group is always numbered one.

Functional groups provide the relationship between the chemical structure and the compound activity. Often, the physical properties of similar types of molecules may be related, and thus it is possible to substitute an undesired molecule for a less hazardous one. These structure–activity relationships give the green chemist the opportunity to "tune" the molecule to achieve a more environmentally responsible process. For example, the use of a solvent from the same functional group but of a higher molecular weight may prove advantageous if the volatility of the solvent is a concern. Specifically, diethanolamine is seen to be a better solvent for CO_2 capture than is monoethanolamine because of its reduced volatility.

Classes of raw materials and reactions can help to eliminate exposure to toxic functional groups such as isocyanate (whose functional group is R–N=C=O), thereby

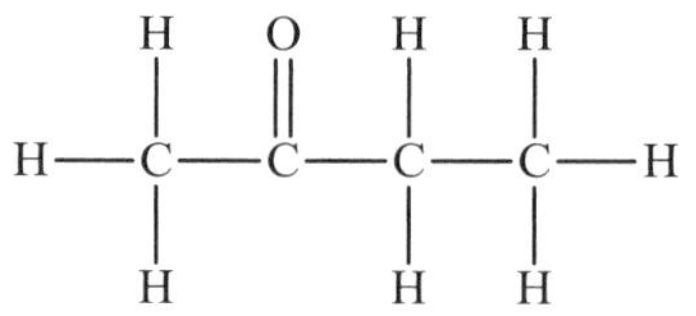

Figure 4.15. Structural formula of butan-2-one.

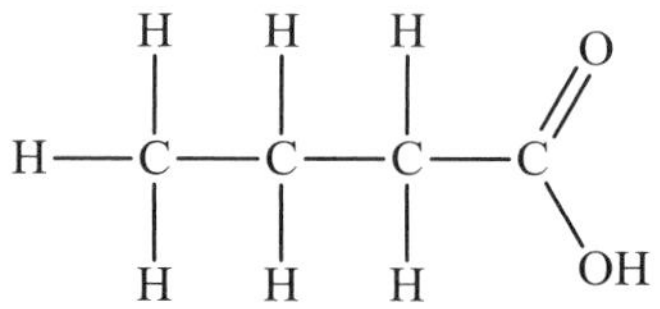

Figure 4.16. Structural formula of butanoic acid.

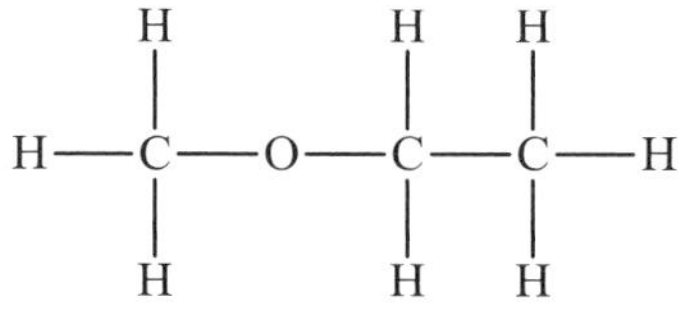

Figure 4.17. Structural formula of methylethylether.

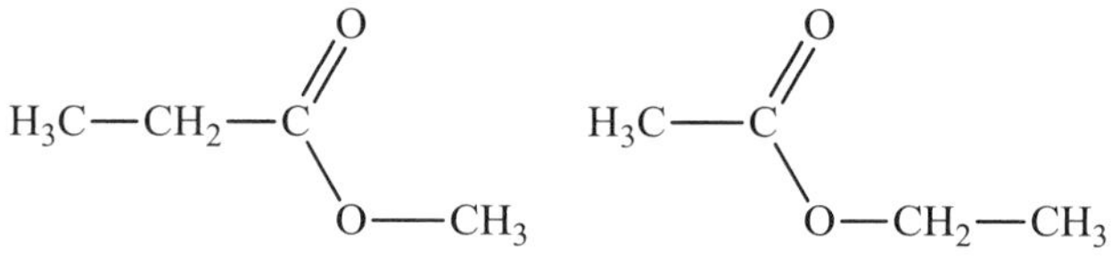

Figure 4.18. Examples of esters (methylpropanoate and ethylethanoate).

preventing harm to humans or the environment. Alternative elements and structures are required that provide the functionality of the desired group but not the toxicity. Researchers at NanoTech Industries were able to produce hybrid non-isocyanate polyurethane (HNIPU) through reaction of cyclocarbonate with an amine, avoiding the formation of the undesirable isocyanate species. HNIPU is currently marketed in the United States and Europe as an alternative to polyurethane foam used in paints, coatings, sealers, and adhesives as well as in insulation and packaging.

Organic molecules can be much more complicated than the linear species already considered, forming ring-like structures and multiple rings. When carbon atoms in a molecule are connected in a ring, the term cyclo- is added to the front of the structure. So cyclohexane, containing six carbon atoms all connected in a ring (with two hydrogen atoms attached to each carbon), is drawn as in Figure 4.19.

Cyclic molecules can also contain double bonds or may be functionalized. If two of the carbon atoms in a six-carbon ring are connected with a double bond, then we would have cyclohexene. And we could also have functionality in a cyclic ring, creating cyclohexanol or cyclohexanone, as shown in Figure 4.20.

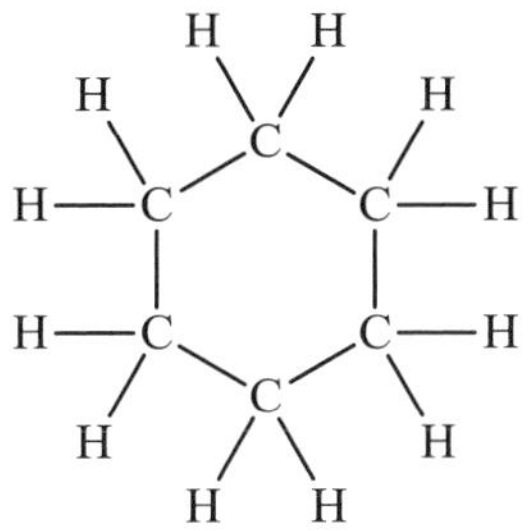

Figure 4.19. Structural formula of cyclohexane.

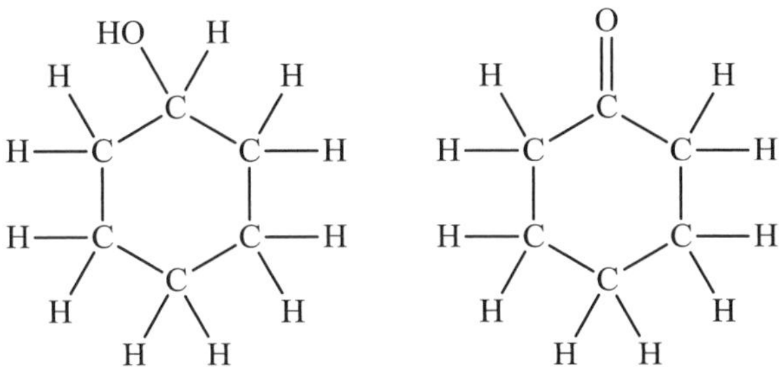

(a)   Cyclohexanol   (b)   Cyclohexanone

Figure 4.20. Structural formula of (a) cyclohexanol and (b) cyclohexanone.

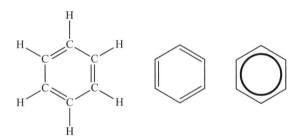

Figure 4.21. Structural formula of benzene.

As we begin to form multiple double bonds in a ring structure, the nature of the double bond changes, forming a stable ring with shared single and double bonds. Benzene is the most common of these materials, with each carbon atom containing one single bond and one double bond (plus one bond to a hydrogen atom). Because these molecules share the double bonds, its properties are unlike those of other cyclic compounds they look like. It is these types of complicated molecules that make up the structure of oil and coal, as well as biomass. The molecule of benzene is drawn in Figure 4.21. Note that all these different representations are equivalent.

Benzene is a colorless, highly flammable liquid and is a known carcinogen. It is a natural constituent of crude oil and also an important industrial solvent and precursor in the production of drugs, plastics, synthetic rubber, and dyes. In March 2006, the official Food Standards Agency in Britain found that four brands of soft drinks among 150 studied contained benzene levels above World Health Organization limits. One way that pure benzene can cause harm is when it oxidizes in the body to produce benzene oxide, which is not excreted readily and can cause DNA strand breaking as well as chromosomal damage.

The Lewis representation of a molecule, and all other methods of describing the molecules mentioned previously, does not give any clue about the three-dimensional nature of the molecule. The atoms in the molecule form complex shapes based on the repulsive and attractive forces that exist between pairs of bonding and lone pairs of electrons. The specific positions of each of the atoms in the molecule are critical determinants of the properties and behavior of molecules in living organisms, in the environment, and in drugs. Because of the importance of the shape of molecules, a model is necessary to predict the shapes of molecules. This model is called VSEPR for valence-shell electron-pair repulsion. The geometries predicted according to the VSPER model are summarized in Table 4.5.

Before leaving the subject of covalent bonding structure, we should look at one more type of molecule. In some cases, a semistable molecule may be formed, which has an overall odd number of valence electrons. These molecules, containing an unpaired electron, are called free radicals. These species appear during chemical reactions, including reactions used by nature. For example, nitric oxide or nitrogen monoxide (NO) is an important intermediate molecule involved in physiological and pathological processes in mammalian bodies including humans. Nitrogen oxide NO has a total of 11 valence electrons. Nitrogen is in group 5A and therefore has five valence electrons. N uses two of its five valence electrons to make a double bond with the O and three valence electrons remain on the N. Free radicals are extremely reactive and can react with the oxygen in the air and lead to the formation of tropospheric ozone, O_3, an air pollutant well-known to affect the respiratory system (Chapter 3).

4.4.2   Ionic Compounds

In a molecular compound, covalent bonding happens when two nonmetals share electrons. Covalent bonds are defined not only by the number of electrons shared but also by their length and strength. Covalent bonds have a specific bond length and bond energy. In a homonuclear diatomic molecule where both atoms are identical, the pair or pairs of electrons is/are shared equally between the two atoms. However, in a heteronuclear diatomic molecule, one shared pair(s) will be more attracted to one atom than the other.

The tendency of an atom to attract shared electrons is called electronegativity. The more electronegative, the higher the electron density, and the higher the probability of finding the bonding electrons near this atom. Again, the periodic table arrangement describes this trend with the more electronegative atoms on the right side (except for the noble gases at the far right). Electronegativity decreases along a group and increases across periods. Fluorine is the most electronegative element whereas francium (Fr) is the least electronegative.

TABLE 4.5. Geometries Predicted by the VSEPR Model

Type (X = atoms bonded to central atom A; E = lone electron pairs on central atom)	Number of Bonding Electron Pairs	Number of Lone Pairs on Central Atom	Electron-Pair Geometry	Molecular Geometry	Examples
AX_2E_0	Two	None	Linear	Linear	CO_2 $O{=}C{=}O$ $BeCl_2$ $Cl{-}Be{-}Cl$
AX_2E_1	Two	One	Triangular planar	Angular	$SnCl_2$ 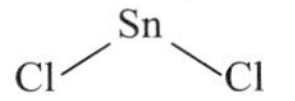 Sn Cl Cl
AX_2E_2	Two	Two	Tetrahedral	Angular	H_2O 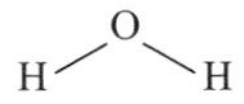 O H H OCl_2 O Cl Cl 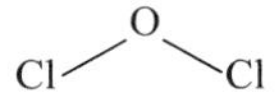
AX_2E_3	Two	Three	Triangular bipyramidal	Linear	XeF_2 $F{-}Xe{-}F$
AX_3E_0	Three	None	Triangular planar	Triangular planar	BCl_3 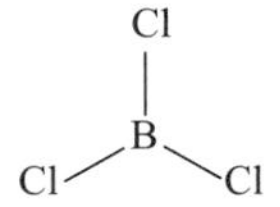 Cl B Cl Cl CO_3^{2-} 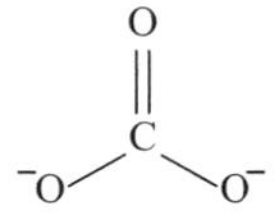 O=C O$^-$ O$^-$

(Continued)

95

TABLE 4.5. (*Continued*)

Type (X=atoms bonded to central atom A; E=lone electron pairs on central atom)	Number of Bonding Electron Pairs	Number of Lone Pairs on Central Atom	Electron-Pair Geometry	Molecular Geometry	Examples
AX_3E_1	Three	One	Tetrahedral	Triangular pyramidal	NCl_3 
AX_3E_2	Three	Two	Triangular bipyramidal	T-shaped	ClF_3 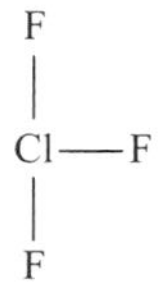
AX_4E_0	Four	None	Tetrahedral	Tetrahedral	CH_4 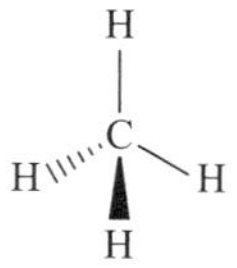 $SiCl_4$ 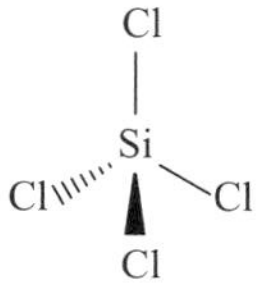

AX_4E_1	Four	Triangular bipyramidal	One	Seesaw	SF_4 
AX_4E_2	Four	Octahedral	Two	Square planar	XeF_4 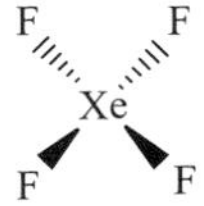
AX_5E_0	Five	Triangular bipyramidal	None	Triangular bipyramidal	PF_5 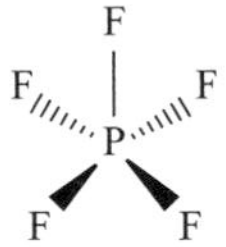
AX_5E_1	Five	Octahedral	One	Square pyramidal	BrF_5 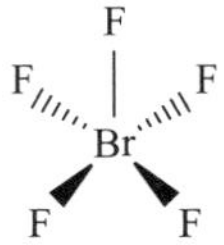
AX_6E_0	Six	Octahedral	None	Octahedral	SF_6 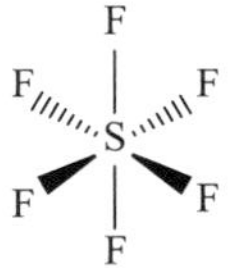

When the difference in electronegativity between two elements is less than about 0.5, the bond is considered nonpolar. When the difference is between 0.5 and 2, the bond is defined as polar. In each case, the type of bond is still a covalent bond; in a homonuclear molecule, the covalent bond is nonpolar; in a heteronuclear molecule, the covalent bond is polar if the electronegativity difference is sufficiently large. In a polar molecule, one region of the molecule has a partial positive charge, whereas the opposite region has an equivalent partial negative charge, and thus the molecule has a net zero charge.

When the difference in electronegativity is greater than 2.0, the electrons are so greatly associated with one of the atoms that a transfer of electrons to the more electronegative species is now involved. Under these conditions, the molecule is no longer considered polar, but rather an ionic compound is formed. One of the most common ionic species is NaCl, common table salt. In this species, the sodium atom (Na) "donates" an electron to the chlorine, leaving it with an empty outer shell and the chlorine with an outer shell containing eight electrons. Because of the extra electron, the chlorine atom contains a strong negative charge, whereas the sodium, which is missing an electron, has an equally strong positive charge.

Metals that donate electrons to an ionic pair form cations with a positive charge (fewer electrons than protons) whereas nonmetals tend to gain electrons to form anions with a negative charge (more electrons than protons). In NaCl, the metal sodium is the cation Na^+ and the chlorine atom becomes the chloride anion, Cl^-. For metals, the (positive) charge is equal to the group number. For nonmetals, the (negative) charge is equal to: 8 − group number. Thus, in the example above, the sodium atom has a +1 charge and is written as Na^{+1} whereas the chlorine has a −1 charge and is written as Cl^{-1}; together these ions form the stable species NaCl. Ionic compounds consist of arrays of cations and anions.

There are two types of ions: monoatomic (single atom) and polyatomic (two or more atoms possessing an overall charge). The naming protocols for these species are precise, but difficult to learn. The following rules should help you.

For monoatomic cations:

- Metals with only one kind of cations: use metal name + "ion."
  Example: K^+, "potassium ion"
- Metals with more than one kind of cations: need roman numerals.
  Examples: Fe^{2+}, "iron(II) ion"
  Fe^{3+}, "iron(III) ion"
- Older (commonly used) naming method: use Latin root and
  "-ous" suffix for lower charged ion (Fe(II) or ferrous)
  "-ic" suffix for higher charged ion (Fe (III) or ferric)

For monoatomic anions: replace ending from element name with "-ide" suffix.

  Example: F (fluorine) becomes F^- or fluoride

For our example of table salt (NaCl), the proper chemical name is sodium chloride.
For polyatomic anions, some guidelines are useful.

Oxoanions: polyatomic ions that contain oxygen.

• If only two oxoanions of a given nonmetal exist: "-ate" suffix for the oxoanion with greater number of O or "-ite" suffix for the oxoanion with fewer number of O.
  Examples: SO_4^{2-}, sulfate; SO_3^{2-}, sulfite
  $\qquad\quad$ NO_3^-, nitrate; NO_2^-, nitrite

• If four oxoanions: prefix "per" and "hypo."
  Examples: ClO_4^-, perchlorate
  $\qquad\quad$ ClO_3^-, chlorate
  $\qquad\quad$ ClO_2^-, chlorite
  $\qquad\quad$ ClO^-, hypochlorite

• Oxoanions with hydrogen: add "hydrogen" before the name of the oxoanion.
  Examples: HSO_4^-, hydrogen sulfate ion
  $\qquad\quad$ $H_2PO_4^-$, dihydrogen phosphate
  $\qquad\quad$ HPO_4^{2-}, monohydrogen phosphate

To name ionic compounds, always name the cation first, followed by the proper form of the anion.

Some alternative solvents in green chemistry are based on complex ionic species. Ionic liquids are ionic salts whose melting point is relatively low (typically below 100 °C) and thus form a liquid near room temperature. The cation is usually a large nitrogen-containing organic-type species, whereas the anion is a smaller inorganic ion. A typical cation would be 1-butyl-3-methyl imidazolium (BMIM) whereas traditional anions would include a halogen such as in $[AlCl_4]^-$ and $[PF_6]^-$ (Figure 4.22).

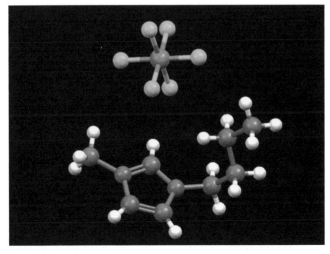

Figure 4.22. Example of ionic liquid containing BMIM cation and $[PF_6]^-$ anion.

Because of the mismatch in size, structure, and shape, these species are less rigid than a simple ionic species, leading to their low melting points. Ionic liquids are important green solvents, because they do not vaporize upon heating (and thus produce no toxic air emissions, in contrast to organic solvents). However, not all ionic liquids are environmentally friendly, as expected from the anion structures indicated previously. New anions are now being developed based on organic species which do not contain any halide and therefore prevent the halide waste formation.

4.5   CHEMICAL REACTIONS

Matter is not inert. It undergoes chemical and/or physical changes. Chemical changes called reactions involve transformations of one or more substances (the reactants) into one or more different substances (the products). It is through these chemical reactions that all chemicals are derived from their raw materials, or one chemical is converted into another.

A chemical reaction is represented as

$$Reactants \rightarrow products$$

with the arrow meaning "forming" or "transforming into" or "leads to."

One fairly common reaction is the oxidation reaction, in which an oxygen atom is added to the chemical structure of the molecule. The most common of the oxidation reactions is combustion, in which a molecule is heated to a high temperature in the presence of air and decomposed to produce CO_2 and water. This reaction forms the basis by which energy is generated from fossil resources, with the production of CO_2 as a by-product. Combustion of fossil resources has led to an increase in the atmospheric content of CO_2. However, under conditions of restricted oxygen, the oxidation of hydrocarbons C_xH_y to carbon dioxide is not complete, and the toxic gas carbon monoxide may also be formed (1 atom of C and 1 atom of O forms CO) (Chapter 3).

Renewable raw materials such as wheat straw, corn, and sugar are highly complex materials containing structures similar to those present in the simpler molecules that have been described above. Biomass may be burned through combustion to produce energy. Biomass combustion also produces CO_2 as a by-product, but a life cycle analysis reveals that the carbon was originally derived from atmospheric CO_2 and thus the net CO_2 generation is negligible. Because additional energy is required for the production of the biomass, and fertilizer is consumed to grow the crop, these activities do add to the overall atmospheric CO_2 content, but this is certainly a reduced quantity compared with fossil resources.

In order for renewable materials to serve as a feedstock for chemical processes, it is important to break them apart. Thermal routes to chemical products involve the partial combustion of the renewable material into CO and H_2, a reaction that is partially controlled by the amount of oxygen present in the reactor. This mixture of gases is commonly referred to as synthesis gas (or syngas) and can be used as the raw feed to a number of other chemical processes that produce compounds that would traditionally have been produced from nonrenewable fossil resources. It may also be possible to break down the biomass directly into functional molecules using enzymatic methods, simplifying and greening the production of desired molecules.

Oxidation of individual compounds can also be completed under controlled conditions to produce molecules of commercial interest. You may have noticed that in alcohols there is a single C–O bond, in aldehydes there is a double C=O bond, and in carboxylic acids there are three C–O bonds (two from the C=O double bond and one from the C attached to the OH group). We can conclude that the oxidation of primary alcohols (where the –OH group is at the end of the hydrocarbon chain producing a straight chain and not a branch) gives aldehydes, which are then oxidized into carboxylic acids.

Primary alcohol → aldehyde → carboxylic acid in the presence of oxidizing agents such as potassium permanganate ($KMnO_4$) (Figure 4.23).

In the case of a secondary alcohol such as 2-butanol or butan-2-ol, the product of oxidation would be a ketone (Figure 4.24). In this case, the oxidation of 2-butanol produces 2-butanone or butan-2-one.

Chemists have the power to use and control chemical reactions. As described previously in the 12 principles of green chemistry, chemists can modify chemical reactions to take advantage of renewable resources, use more environmentally benign solvents in which to carry out reactions, and employ catalysts to promote the formation of desired products.

In Chapter 2, we discussed one example in which chemists and engineers are looking to use renewable feedstocks to produce chemicals of interest. Other examples include the production of polymers commonly used in consumer products like automobiles and food packaging from renewable feedstocks such as soy and corn.

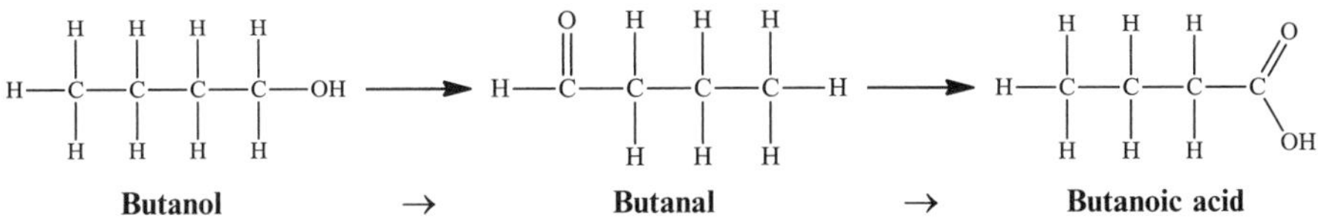

Figure 4.23. Conversion of butanol to butanal to butanoic acid.

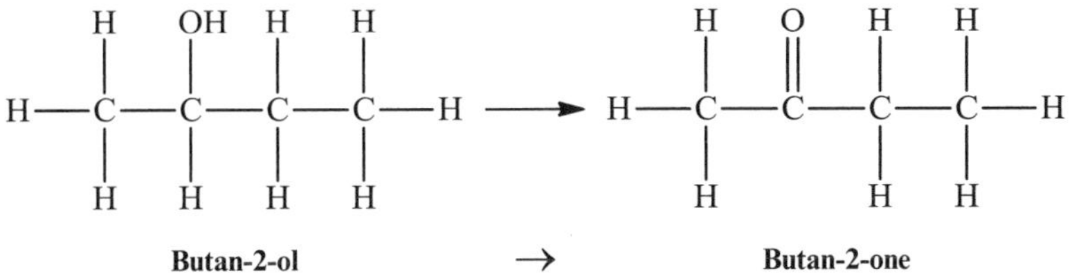

Figure 4.24. Conversion of the secondary alcohol of butan-2-ol to the ketone butan-2-one.

Biodegradable polymers can be made from the microbial fermentation of glucose, a common compound with the formula $C_6H_{12}O_6$.

A lot of research is being devoted to the replacement of volatile organic compounds (VOCs) or hazardous air pollutants (HAPs), which are flammable, toxic, or carcinogenic and frequently used as solvents for chemical reactions. For example, carbon dioxide in its supercritical state (i.e., at a temperature above its critical temperature of 31.1 °C and pressure above its critical pressure of 74 bar) has several advantages as a benign solvent. It is nontoxic, nonflammable, and inexpensive. Supercritical CO_2 has properties midway between a gas and a liquid and therefore has found many applications in chemical extractions such as the decaffeination of coffee and in the dry cleaning industry where it replaced perchloroethylene as a solvent.

A catalyst can sometimes be used to provide an alternative reaction step that reduces the energy barrier to reaction, or directs the reaction to produce more of the desired product without the formation of undesired by-products. Enzymes are natural catalysts that are often highly selective and are thus excellent green chemistry tools. Catalysis will be covered in Chapter 6. Types of chemical reactions and processes will be discussed in detail in Chapter 5.

4.6   MIXTURES, ACIDS, AND BASES

Matter can exist in the form of either a pure substance or a mixture. Pure substances can be either elements that contain only one type of atoms or compounds that are comprised of at least two elements. Compounds in a mixture are not chemically bonded to one another. Although chemicals are often identified as pure, pure substances are actually quite rare. More often, even a pure substance contains small amounts of impurities. Most of the substances surrounding us are mixtures rather than pure substances.

Mixtures can be classified as homogeneous or heterogeneous. In a homogeneous mixture, the composition is uniform throughout the entire sample; in a heterogeneous mixture, the composition is nonuniform and each individual component can be distinguished from others. Depending on particle sizes, homogeneous mixtures can be further classified as solutions, colloids, and suspensions (Figure 4.25).

Drinking water is a common substance that is a homogeneous mixture. Analysis of drinking water may reveal the presence of dissolved calcium cations, Ca^{2+}, or chloride anions, Cl^-, added for disinfection in water treatment plants. An analysis of any volume of the water would reveal an identical composition. Because these components are evenly distributed, they cannot be separated through mechanical means.

On the other hand, a heterogeneous mixture is one in which different compounds and/or substances are mechanically blended together and create a nonuniform distribution of substances. This is the case for concrete. Concrete is usually made by mixing cement, limestone or granite, and water, just to name a few concrete components. Limestone is made up of calcium carbonate ($CaCO_3$), which is decomposed to calcium oxide, releasing CO_2 in the process. Recently, the use of recycled materials to make concrete has gained popularity. Fly-ash, a by-product of coal-fired electric

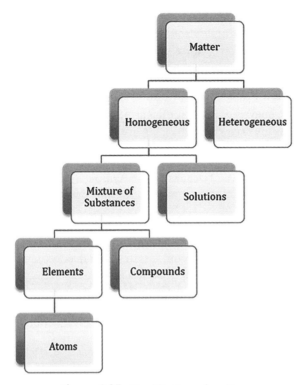

Figure 4.25. Classification of matter.

generating power plants, is now used to supplement the commonly used Portland cement. The impact of cement-replacement technology such as this will play an important role in cutting further the carbon dioxide emissions as cement production generates massive amounts of carbon dioxide.

Solutions are generally defined as clear mixtures, meaning the components are dissolved. Our naked eye cannot observe a particle settling out over time. Components in a solution are either the solute (minor component) or the solvent (major component). For example, when sugar is put in a hot cup of water, the sugar dissolves in the hot water. The hot water is termed the solvent, the sugar is the solute, and the overall mixture is an aqueous solution (aqua for water).

Colloids have larger particles and often appear cloudy. Our naked eye cannot distinguish suspended particles, which are too small to be filtered. A typical example of a colloid is paint, which is an emulsion that encompasses tiny droplets containing the pigment randomly dispersed in a liquid solvent. The use of water-based paints has led to substantial decreases in the emission of hazardous air pollutants.

Suspensions contain particles with different densities. When particles are larger than about 1000 nm in diameter, they can separate from solution upon standing.

Because a mixture is made from multiple species, there are other intermolecular forces present, depending on the types of molecules present in the mixture. For example, polar molecules may interact through hydrogen bonding, whereas ionic species may dissociate in a highly polar solvent such as water.

A general rule of thumb is "like dissolves like." An organic solvent is generally needed for an organic molecule, whereas ionic species will dissolve in a highly polar solvent such as water. On the other hand, nonpolar solutes usually do not dissolve in polar solvents such as water. Water and oil do not form a solution, but with proper additional ingredients and sufficient mixing, you can get them to form a homogeneous mixture. The example of sugar dissolving in a cup of coffee leads us to assume that sugar is a polar solute. As a matter of fact, sugar has several hydroxyl groups, –OH, and is therefore a polar molecule. The molecules of sugar form a solution with the molecules of water through hydrogen bonding. The challenge of dissolving organic molecules into nonhazardous solvents is one of the central concerns of green chemistry.

Acids and bases are an important class of compounds dissolved in aqueous solution and often found in common household products as well as in many drugs. When you eat a grapefruit or a lemon, it tastes sour due to the presence of acids. Coffee and tonic water, on the other hand, taste bitter, due to the presence of bases. The definition of an acid is based on the Brønsted–Lowry theory, which specifies that acids produce protons (H^+) in aqueous solution, that is, they are proton donors.

The dissociation of acids in water can be represented according to the following chemical equation:

$$HA\,(aq) \rightarrow A^-\,(aq) + H^+\,(aq)$$

or also written as

$$HA\,(aq) + H_2O \rightarrow A^-\,(aq) + H_3O^+\,(aq)$$

On the other end, Brønsted–Lowry bases produce hydroxide ions (OH^-), that is, they are proton acceptors.

The dissociation of bases in water can be represented as follows:

$$BOH\,(aq) \rightarrow B^+\,(aq) + OH^-\,(aq)$$

Acids and bases can be classified as strong and/or weak acids/bases depending on their tendency to ionize. Strong acids completely ionize into H^+, whereas weak acids only partially ionize or dissociate in water.

The acid present in vinegar is acetic acid, which is a weak acid. Hydrochloric acid, found in our stomach, is a strong acid, meaning that it completely dissociates in water.

$$HCl\,(aq) \rightarrow H^+\,(aq) + Cl^-\,(aq)$$

Complete dissociation is written as a chemical reaction involving a single arrow. When the amounts of products and reactants remain constant (which means the system has reached equilibrium) no HCl will be left.

Acetic acid or ethanoic acid only partially dissociates in water, the partial dissociation being symbolized by the double arrow. At equilibrium, there would be molecules of acetic acid as well as acetate CH_3COO^- and hydronium cations H_3O^+. Acetic acid is a weak acid.

$$CH_3COOH \text{ (aq)} \Leftrightarrow CH_3COO^- \text{(aq)} + H^+ \text{(aq)}$$

or also written as

$$CH_3COOH \text{ (aq)} + H_2O \text{ (l)} \Leftrightarrow CH_3COO^- \text{(aq)} + H_3O^+ \text{(aq)}$$

Strong bases completely ionize into OH^-, whereas weak bases only partially ionize in water. Ammonia, found mostly in window cleaners, is a weak base whereas sodium hydroxide, found in oven cleaners, is a strong base.

The most common strong base is sodium hydroxide NaOH. Sodium hydroxide is used mainly in the manufacture of pulp and paper, soaps, and detergents, but also in the production of biodiesel, where sodium hydroxide is used as a catalyst.

Sodium hydroxide is a very strong base and completely decomposes into sodium cations and hydroxide anions. At equilibrium no NaOH remains.

$$NaOH \text{ (aq)} \rightarrow Na^+ \text{(aq)} + OH^- \text{(aq)}$$

Ammonia is a weak base and only partially dissociates into NH_4^+ and OH^-. At equilibrium molecules of NH_3, NH_4^+, and OH^- will coexist.

$$NH_3 \text{(aq)} + H_2O \text{ (l)} \Leftrightarrow NH_4^+ \text{(aq)} + OH^- \text{(aq)}$$

Aspirin and ibuprofen are acids; caffeine and nicotine are bases. More examples of acid–base reactions will be seen in Chapter 5.

In order to quantitatively measure the strength of acids and bases, a scale called the pH scale was developed. pH values are numbers between 0.0 and 14.0. An acidic solution is defined when $pH < 7.0$, a basic solution when $pH > 7.0$, and a neutral solution when $pH = 7.0$.

The pH value is directly related to the concentration of protons, H^+, as follows:

$$pH = -\log[H^+]$$

the concentration or molarity of any substance being expressed in moles of solute/liters of solution symbolized as M.

Therefore, to express the concentration of protons in base 10: $[H^+] = 10^{-pH}$.

For aqueous solutions containing acids and bases, a general expression links the concentrations of protons and hydroxides as follows:

$$\text{At } 25\,^\circ C, [H^+] \times [OH^-] = 1 \times 10^{-14} \text{ M}$$

Common chemicals have a specific pH as shown in Table 4.6.

TABLE 4.6. Common Chemicals and Their Respective pH [6]

Name	pH
Hydrochloric acid (HCl)	0
Battery acid (H_2SO_4 sulfuric acid)	1.0
Lemon juice	2.0
Vinegar	2.2
Apples	3.0
Wine and beer	4.0
Tomatoes	4.5
Milk	6.6
Pure water	7.0
Human blood	7.4
Baking soda ($NaHCO_3$ sodium bicarbonate)	8.3
Milk of magnesia	10.5
Ammonia	11.0
Lime ($Ca(OH)_2$ calcium hydroxide)	12.4
Lye	13.0
Caustic soda (NaOH sodium hydroxide)	14.0

Brønsted–Lowry acids and bases are well-known and are generally naturally occurring substances. In 1923, Gilbert Lewis extended the definition of acids and bases when he published the following definition of an acid: "An acid substance is one which can employ an electron lone pair from another molecule in completing the stable group of one of its own atoms" [7]. Similarly, a Lewis base is one which can donate a pair of electrons when reacting with a Lewis acid and therefore forming a Lewis complex or adduct.

Lewis acids are used as homogeneous catalysts in fine and specialty manufacturing processes. Aluminum chloride ($AlCl_3$), where the central atom of aluminum is covalently bonded to three chlorine atoms, is the most widely used Lewis acid. It is very stable, reactive, soluble in many organic solvents, and inexpensive. However, because of its strength, it often promotes unwanted side reactions and the release of hazardous substances in the environment. In addition, it is difficult to separate aluminum chloride from the reaction products and large volumes of hazardous waste can be generated during product recovery. The performance of Lewis acids can be improved by immobilizing them on supports such as silica (SiO_2) or zirconia (ZrO_2), which results in the formation of reusable heterogeneous catalysts discussed in detail in Chapter 7. The outcomes of using this type of catalysts are consistent with the principles of green chemistry: the selectivity of reactions is improved, the volume of hazardous waste is significantly reduced, and moderate reaction conditions can be used.

REFERENCES

1. Oganessian, Y. T. et al. Synthesis of a new element with atomic number Z=117, *Phys. Rev. Lett.*, **2010**, *104*, 142502.

2. http://en.wikipedia.org/wiki/Dental_amalgam_controversy.

3. http://www.ceh.org/making-news/press-releases/29-eliminating-toxics/568-progress-made-but-some-handbags-still-pose-lead-threats.

4. http://en.wikipedia.org/wiki/File:Atom_diagram.png.

5. http://www.marspedia.org/index.php?title=File:Phase_diagram_water.png.

6. http://chemistry.about.com/od/acidsbases/a/phtable.htm.

7. Lewis, G. N. *Valence and the Structure of Atoms and Molecules*, The Chemical Catalog Company, 1923, p. 142.

5

CHEMICAL REACTIONS

5.1 DEFINITION OF CHEMICAL REACTIONS AND BALANCING OF CHEMICAL EQUATIONS

The concept of molecules and compounds was introduced in Chapter 4. Chemists and chemical engineers manipulate compounds to form new materials that achieve a specific goal or purpose. It is of crucial importance that compounds be mixed in the proper ratios so that the desired reaction can be achieved. The relationship between the compound and its atoms is displayed through the molecular formula, and the chemist uses this information to develop a balanced chemical equation that can properly describe the breaking and forming of bonds in chemical reactions.

As mentioned in Chapter 4, a chemical reaction is "a process in which substances (reactants) change into other substances (products) by rearrangement, combination, or separation of atoms." A chemical reaction is represented by a chemical equation with two sides: one for the reactants and one for the products.

Green Chemistry and Engineering: A Pathway to Sustainability,
Anne E. Marteel-Parrish and Martin A. Abraham.
© 2014 American Institute of Chemical Engineers, Inc. Published 2014 by John Wiley & Sons, Inc.

$$A \rightarrow B$$

$$\text{Reactant(s)} \rightarrow \text{Product(s)}$$

The arrow means "forms," "yields," "changes to," or "is/are converted into."

The law of conservation of matter, "matter is neither destroyed nor created," applies to all atoms in a chemical reaction. As a result, a balanced chemical reaction requires that there must be an equivalent number of atoms of each type on both sides of the equation. Consider the reaction between hydrogen and oxygen to form water, as shown in Equation 5.1.

$$n\,H_2\,(g) + m\,O_2\,(g) \rightarrow p\,H_2O\,(g) \tag{5.1}$$

In Equation 5.1, we have placed a letter in front of each molecule to represent the stoichiometric coefficient, a multiplying number assigned to the species in a chemical equation in order to balance the equation. Now the challenge is to determine what number each letter in Equation 5.1 represents.

Let's suppose we were producing one molecule of water. In this case, there would be one oxygen atom on the right-hand side of the equation, and to have a balanced chemical equation would require that we have one oxygen atom on the left side. But diatomic oxygen is shown on the left side, and it is not possible to have fractional numbers of molecules. So the smallest number of oxygen molecules on the left side would be one. Based on this analysis, let's tentatively state that $m=1$.

Now, let's go back to the right-hand side. We have two oxygen atoms on the left side (since there is one molecule of O_2), so we need two oxygen atoms on the right side. That means we need two water molecules on the right side, or $p=2$. Now the equation is balanced in terms of oxygen.

What about hydrogen? There are two water molecules on the right side, and each one has two hydrogen atoms, which means there are four hydrogen atoms total. When a stoichiometric coefficient is placed in front of a molecular formula, the stoichiometric coefficient acts as a multiplying number and applies to ALL subscripts present in the molecular formula. Hydrogen is present as diatomic hydrogen on the left side, so each molecule of H_2 contains two hydrogen atoms. That means we need two hydrogen molecules (H_2), or $n=2$. Therefore, the equation is balanced in terms of hydrogen atoms.

$$2H_2\,(g) + O_2\,(g) \rightarrow 2\,H_2O\,(g) \tag{5.2}$$

One must keep in mind that the molecular formula for each molecule MUST NOT be changed, because that would change the nature of the molecule. In other words, a water molecule can never be changed to H_2O_2 to balance the equation. It would not be the molecule of water anymore but a completely different species, hydrogen peroxide, used as a powerful bleaching agent. The only route to balance Equation 5.1 is to use a stoichiometric coefficient in front of each molecule to make sure that the number of atoms of each type is equal on both sides of the equation (Highlight 5.1).

Highlight 5.1   Balancing Chemical Equations

Balance the following typical equations:

(a) The combustion of pentane (liquid) in air.
(b) The synthesis of titanium dioxide from titanium tetrachloride and water.
(c) The reaction of ammonia gas with oxygen to produce gaseous nitrogen monoxide and water vapor.

Solutions:

(a) Pentane is an important component of some fuels and is mostly used as a laboratory solvent. Its molecular formula is C_5H_{12}. Pentane is a volatile hydrocarbon that is liquid at room temperature. It burns in air by reacting with oxygen and forms carbon dioxide and water, the two products of combustion reactions.

$$C_5H_{12}\,(l) + 8\,O_2\,(g) \rightarrow 5\,CO_2\,(g) + 6\,H_2O\,(l)$$

(b) Titanium dioxide, TiO_2, is produced from the purification of titanium tetrachloride, $TiCl_4$, and is accompanied by the release of hydrochloric acid:

$$TiCl_4\,(l) + 2\,H_2O\,(l) \rightarrow TiO_2\,(s) + 4\,HCl\,(aq)$$

TiO_2 is one of the most prominent photocatalysts and has been used for the oxidation of numerous hazardous materials in either air or water. New technology places a very thin layer of TiO_2 on the outside of glass, which causes soot that would otherwise build up on the surface to be oxidized in sunlight, keeping the window clean.

(c) Ammonia gas reacts with oxygen to produce gaseous nitrogen monoxide and water vapor:

$$NH_4\,(g) + \tfrac{3}{2}O_2\,(g) \rightarrow NO\,(g) + 2\,H_2O\,(g)$$

While this is properly balanced, it is preferable to have whole numbers instead of fractions for stoichiometric coefficients, because it is not possible to have fractional molecules. Let's multiply all coefficients by 2:

$$2\,NH_4\,(g) + 3\,O_2\,(g) \rightarrow 2\,NO\,(g) + 4\,H_2O\,(g)$$

When all atoms are accounted for in a chemical equation, the masses represented by the equation are also balanced (remember the law of conservation of mass). The second section illustrates this concept.

5.2 CHEMICAL REACTIONS AND QUANTITIES OF REACTANTS AND PRODUCTS

Equation 5.2 shows that two diatomic molecules of hydrogen and one diatomic molecule of oxygen react to produce two molecules of water. But chemists and engineers don't work with individual molecules; rather, we work with large quantities of substances. In order to relate the number of molecules to mass quantities, we first start by defining the quantity of matter as a mole, which is identically equal to 6.022×10^{23} molecules (or atoms). Then, to determine the quantities of reactants and products, we know that each mole of atoms has an atomic weight expressed in g/mol. By adding the atomic weight of each atom in a molecule, we can deduce the molecular weight or molar mass of the molecule.

For example, the molar mass of the diatomic molecule of hydrogen would be

$$2 \times \text{atomic weight of an atom of hydrogen} = 2 \times (1.0079 \, \text{g/mol}) = 2.0158 \, \text{g/mol}$$

Remember, a hydrogen molecule contains two hydrogen atoms, as indicated by the subscript 2, and thus the molar mass is two times the atomic mass of a hydrogen atom (as written in the periodic table).

We can do the same with the diatomic molecule of oxygen:

$$2 \times \text{atomic weight of an atom of oxygen} = 2 \times (15.9994 \, \text{g/mol}) = 31.9988 \, \text{g/mol}$$

As well as for the molecule of water:

$$2 \times \text{atomic weight of an atom of hydrogen} + 1 \times \text{atomic weight of an atom of oxygen}$$
$$= 2 \times (1.0079 \, \text{g/mol}) + 1 \times (15.9994 \, \text{g/mol}) = 18.0152 \, \text{g/mol}$$

The molar masses show that 1 mol of H_2 is equivalent to 2.0152 g of H_2, that 1 mol of O_2 is equivalent to 31.9988 g of O_2, and 1 mol of water is equivalent to 18.0152 g of H_2O.

However, in the *balanced* equation, 2 diatomic molecules of hydrogen react with 1 diatomic molecule of oxygen to produce 2 molecules of water. We can also say 2 mol of H_2 react with 1 mol of O_2 to produce 2 mol of H_2O.

$$\textit{Total mass of reactants in g} = (2 \, \text{mol } H_2 \times \text{molar mass of } H_2)$$
$$+ (1 \, \text{mol } O_2 \times \text{molar mass of } O_2)$$
$$= (2 \, \text{mol} \times 2.0158 \, \text{g/mol}) + (1 \, \text{mol} \times 31.9988 \, \text{g/mol})$$
$$= \textit{36.0304 g}$$

Let's calculate the total mass of products and confirm that the conservation of mass is respected:

$$Total\ mass\ of\ products\ in\ g = 2\ mol\ H_2O \times molar\ mass\ of\ H_2O$$
$$= 2\ mol \times 18.0152\ g/mol$$
$$= \boldsymbol{36.0304\ g}$$

Note that when using metric units, atomic weight, molecular weight, and molar mass are expressed in grams per mole (g/mol) and then the mass of the species comes out in grams (g). When looking at the units, it makes sense:

$$mol \times g/mol = g$$

We will need these conversions between molar quantities and mass quantities to measure the efficiency of chemical reactions in Section 5.4.

For Highlights 5.2 and 5.3, you will need a periodic table.

Highlight 5.2 Calculations of Molecular Weights

Consider the following compounds and calculate their molecular weights.

(a) PCl_5
(b) NaH_2PO_4
(c) $C_6H_{12}O_6$
(d) $Mg_3(PO_4)_2$
(e) $Mg(NO_3)_2$

Solutions:

(a) PCl_5

$$M.W. = (1\,P \times 30.9738\,g/mol) + (5\,Cl \times 35.453\,g/mol) = 208.2388\,g/mol$$

The answer should be expressed with five significant figures (since there are five significant figures in the measure of the molecular weight of the chlorine atom): 208.24 g/mol.

(b) NaH_2PO_4

$$M.W. = (1\,Na \times 22.9898\,g/mol) + (2\,H \times 1.0079\,g/mol)$$
$$+ (1\,P \times 30.9738\,g/mol) + (4\,O \times 15.9994\,g/mol)$$
$$= 119.977\,g/mol$$

The answer should be expressed with five significant figures: 119.98 g/mol.

(c) $C_6H_{12}O_6$

$$M.W. = (6\,C\times12.0107\,g/mol)+(12\,H\times1.0079\,g/mol)$$
$$+(6\,O\times15.9994\,g/mol)$$
$$=180.1554\,g/mol$$

The answer should be expressed with five significant figures: 180.16 g/mol.

(d) $Mg(NO_3)_2$

$$M.W. = (1\,Mg\times24.3050\,g/mol)+(2\,N\times14.0067\,g/mol)$$
$$+(6\,O\times15.9994\,g/mol)$$
$$=148.3148g/mol$$

The answer should be expressed with six significant figures: 148.315 g/mol.

Attention: $Mg(NO_3)_2$ is a compound containing 2 anions of nitrate NO_3^- to balance the 2+ charge from Mg^{2+}. Thus, there are 2 N and (3×2) O total.

(e) $Mg_3(PO_4)_2$

$$M.W. = (3\,Mg\times24.3050\,g/mol)+(2\,P\times30.9738\,g/mol)$$
$$+(8\,O\times15.9994\,g/mol)$$
$$=262.8578\,g/mol$$

The answer should be expressed with six significant figures: 262.858 g/mol.

Highlight 5.3   Calculations of Masses

Calculate the mass (in g) for the following:

(a) 0.5 mol of aspirin®, acetylsalicylic acid with the molecular formula $C_9H_8O_4$
(b) 1.1×10^{-3} mol of vitamin C also known as L-ascorbate with the molecular formula $C_6H_8O_6$

Solutions:

(a) In this example we know the number of moles and we have the molecular formula of aspirin. The goal is to find the mass in grams of aspirin. We know that by multiplying the number of mol by the molecular weight, mol × g/mol, we get a mass in g. The molecular weight of aspirin, $C_9H_8O_4$, is calculated as we did in the previous examples above:

$$M.W. = (9\,C\times12.0107\,g/mol)+(8\,H\times1.0079\,g/mol)$$
$$+(4\,O\times15.9994\,g/mol)$$
$$=180.1571\,g/mol$$

$$\text{Mass of aspirin} = 180.1571 \text{ g/mol of } C_9H_8O_4 \times 0.5 \text{ mol}$$
$$= 90.07855 \text{ g of } C_9H_8O_4$$

With only one significant figure (as present in 0.5 mol) the answer is expressed correctly as 90 g.

(b) We will proceed the same way we did for example (a).

$$\text{M.W. of vitamin } C = (6C \times 12.0107 \text{ g/mol}) + (8H \times 1.0079 \text{ g/mol})$$
$$+ (6O \times 15.9994 \text{ g/mol})$$
$$= 176.1238 \text{ g/mol of vitamin C}$$

Knowing the molecular weight of vitamin C and the number of moles, we can determine its mass.

$$\text{Mass of vitamin } C = (1.1 \times 10^{-3} \text{mol}) \times 176.1238 \text{ g/mol}$$
$$= 0.193736 \text{ g of vitamin C}$$

With only two significant figures as present in 1.1×10^{-3}, the answer is expressed correctly as 0.19 g.

5.3 PATTERNS OF CHEMICAL REACTIONS

The most common patterns of chemical reactions are summarized in this section. Examples of typical reactions involving inorganic and organic compounds are discussed and the concept of atom economy (introduced in Chapter 2) is used to describe the efficiency of the reaction.

5.3.1 Combination, Synthesis, or Addition Reactions

$$A + B \rightarrow AB \tag{5.3}$$

Combination or addition reactions involve the addition of a compound B to a compound A (or the addition of a reagent to a substrate) and results in the complete consumption of both compounds (or both reagent and substrate). Therefore, this type of reaction is 100% atom economical and efficient with no generation of waste or by-product.

5.3.1.1 Examples with Inorganic Compounds

These three examples illustrate the different types of combination reactions.

1. Two elements can combine to form a compound:

$$C(s) + O_2(g) \rightarrow CO_2(g) \tag{5.4}$$

2. Two compounds can combine to form a different one:

$$CaO \ (s) + H_2O \ (l) \rightarrow Ca(OH)_2 \ (s) \tag{5.5}$$

3. One element and one compound can combine to form a different compound:

$$O_2 (g) + 2SO_2 (g) \rightarrow 2SO_3 (g) \tag{5.6}$$

Recall from Chapter 2 the definition of atom economy, the ratio of the mass of the species in the desired product relative to the mass of all of the reactants. In all of these reactions the atoms present in the reactants are also present in the products. In this case it is implied that these reactions are 100% atom economical. Remember from Chapter 2 that atom economy is a mass utilization number only; it does not take into account the energy usage, the toxicity of the products, or whether the feedstocks are derived from renewable materials.

5.3.1.2 Examples with Organic Compounds

There are two main types of addition reactions, involving either an electrophile or a nucleophile as one of the starting compounds. All addition reactions of organic compounds involve the breaking of a double bond (π bond) and the formation of two new single covalent bonds (σ bonds).

In electrophilic addition, one of the starting compounds must have a double or a triple bond; in other words, it must possess an electron-rich unsaturated C=C or C≡C bond. An electrophile seeks out an unshared electron pair. The reaction in Scheme 5.1 shows an electrophilic addition to ethylene.

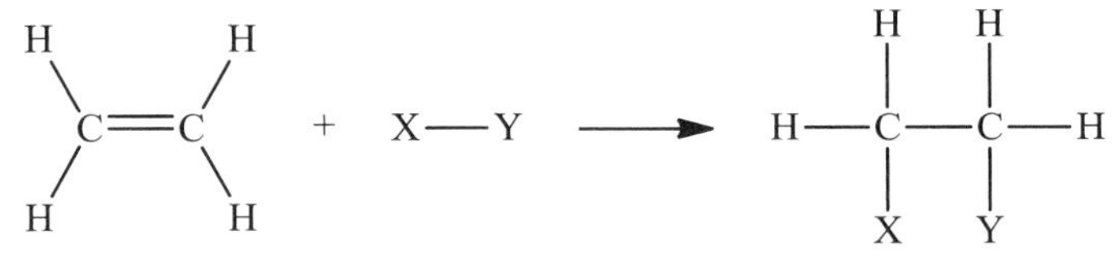

Scheme 5.1. Electrophilic addition to ethylene.

Nucleophilic addition involves the addition of a nucleophile across a double bond. The nucleophile can be any negative ion or any neutral molecule that has at least one unshared electron pair. The reaction in Scheme 5.2 shows nucleophilic addition to a ketone.

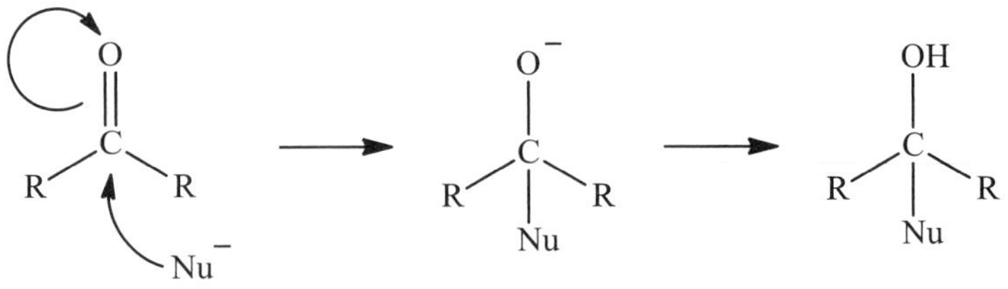

Scheme 5.2. Nucleophilic addition to a ketone.

Lewis bases are nucleophiles whereas Lewis acids are electrophiles.

5.3.2 Decomposition Reactions

Decomposition reactions should be 100% economical but the environmental impact of the decomposition products should be examined closely.

$$AB \rightarrow A + B \tag{5.7}$$

A typical example of a decomposition reaction is the electrolysis of water, which is the decomposition of water into hydrogen and oxygen gases using an electric current passing through the water or using very high temperatures.

$$2\,H_2O\,(l) \xrightarrow{\text{Direct current or high } T} 2\,H_2\,(g) + O_2\,(g) \tag{5.8}$$

This reaction has significance as a clean way to produce hydrogen for use in fuel cells. However, a broader analysis of the reaction requires one to consider the source of energy used to produce the current. Many researchers are experimenting with photovoltaics, which convert sunlight into electricity, as a means of creating the current needed to drive the reaction. Some researchers at the University of Berkeley in California are looking at the use of hydrogenase, an enzyme that can split water into oxygen and hydrogen.

5.3.3 Elimination Reactions

Elimination reactions are the opposite of addition reactions. This type of reaction generates unsaturation of the starting material and two substituents are usually removed from the starting molecule. A typical example is a dehydration reaction where there is a loss of a water molecule from the starting reagent. Alcohols can be dehydrated to alkenes accompanied by the production of a molecule of water, as shown in Scheme 5.3.

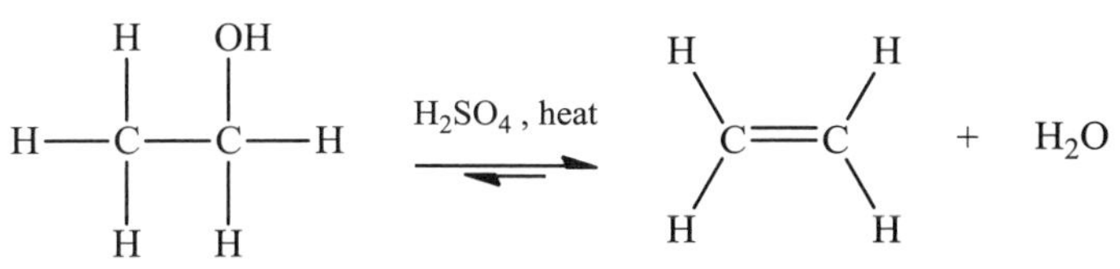

Scheme 5.3. Dehydration of alcohols to alkenes.

In this example, if the alcohol is propanol then the alkene is propene. The atom economy can be obtained as the mass of a propene molecule (42 g/mol) divided by the original mass of the propanol molecule (60 g/mol) for an atom economy of 70%. However, in this case the undesired product is water, which is innocuous, so except for the effort required to recover the alkene product, the atom economy less than 100% is not as significant from a green standpoint.

5.3.4 Displacement Reactions

This very common type of reaction is one in which a portion of one molecule is replaced with another species, as shown here:

$$A + BC \rightarrow AC + B \tag{5.9}$$

Typical displacement reactions encompass oxidation–reduction (redox) reactions:

$$Zn\,(s) + CuSO_4\,(aq) \rightarrow Cu\,(s) + ZnSO_4\,(aq) \tag{5.10}$$

Note that all soluble ionic compounds such as $CuSO_4$ and $ZnSO_4$ dissociate into their respective ions in aqueous solution. When dissolved in water, $CuSO_4$ (aq) ionizes to produce Cu^{2+} (aq) and SO_4^{2-} (aq). Likewise, $ZnSO_4$ (aq) dissociates to form Zn^{2+} (aq) and SO_4^{2-} (aq) when immersed in water.

If the displacement reaction displayed in Equation 5.10 is carried out in water (as it must be, since two solids sitting next to each other would not react), then recognizing that each of the species dissociates provides

$$Zn\,(s) + Cu^{2+}\,(aq) + SO_4^{2-}\,(aq) \rightarrow Cu\,(s) + Zn^{2+}\,(aq) + SO_4^{2-}\,(aq) \tag{5.11}$$

Note that the sulfate anion appears on both sides of the reaction. In this case, sulfate is called a spectator ion and is not involved in the overall process. This allows us to rewrite Equation 5.10 in terms of only the Zn and the Cu:

$$Zn\,(s) + Cu^{2+}\,(aq) \rightarrow Cu\,(s) + Zn^{2+}\,(aq) \tag{5.12}$$

The net effect of an oxidation–reduction reaction is a transfer of electrons from one reactant to another. The substance that accepts electrons incurs a reduction of the electric charge on an atom of the substance and is said to be reduced. The other substance loses electrons, has an increase of the electric charge on an atom of the substance, and is oxidized. In all oxidation–reduction reactions, one reactant is oxidized and another is reduced.

Because there is a transfer of electrons between two species, oxidation–reduction reactions can be divided into two half-reactions, one for each of the species undergoing a transfer of electrons. In these cases, it is important that we keep track not only of the number of atoms but also of the number of electrons, since the conservation of mass law also requires that charge is conserved. We designate an electron as e^-, and thus we can write an oxidation reaction as

$$M \rightarrow M^{n+} + ne^- \tag{5.13}$$

Likewise, we can express the reduction reaction by recognizing that the electrons are on the reactants side:

$$M^{n+} + ne^- \rightarrow M \tag{5.14}$$

Now we can go back to the specific redox reaction between $CuSO_4$ and $ZnSO_4$ that we were considering previously.

The oxidation reaction occurs for the zinc atom, as it goes from solid Zn with a charge of zero to the aqueous Zn ion with a charge of +2. In this case, the half-reaction is written as

$$Zn\ (s) \rightarrow Zn^{2+}(aq) + 2e^- \qquad (5.15)$$

The reduction reaction occurs for the copper, which is reduced from Cu ion with a charge of +2 to solid copper with no charge. The reduction reaction can be written as

$$Cu^{2+}(aq) + 2e^- \rightarrow Cu\ (s) \qquad (5.16)$$

If we add these two half-reactions together, we recombine to obtain the entire redox reaction,

$$Zn\ (s) \rightarrow Zn^{2+}(aq) + 2e^-$$
$$+\ Cu^{2+}(aq) + 2e^- \rightarrow Cu\ (s)$$
$$\overline{Zn\ (s) + Cu^{2+}(aq) \rightarrow Zn^{2+}(aq) + Cu\ (s)} \qquad (5.17)$$

The $2e^-$ appear on both sides of the equation and cancel out when adding the two half-reactions together to recover the full redox reaction. In some cases, the half-reactions don't balance precisely, and it is important to consider that one half-reaction occurs multiple times for every time the other half-reaction occurs. That will ensure that the electrons always balance, and charge is conserved. Highlight 5.4 explains what to do in case of unbalanced numbers of electrons.

Highlight 5.4   Balancing Redox Reactions

Consider the following equation:

$$Fe\ (s) + NaCl\ (aq) \rightarrow FeCl_2\ (aq) + Na\ (s)$$

(a) Balance this equation in terms of atoms and calculate the atom efficiency.
(b) Determine which element is reduced and which one is oxidized.
(c) Write the overall redox reaction.

Solutions:

(a) In this equation, the number of Fe atoms and the number of Na atoms are correctly balanced. However, there are 2 Cl atoms in the iron(II) chloride molecule, $FeCl_2$, whereas there is only 1 Cl in sodium chloride, NaCl. When adding a stoichiometric coefficient of 2 in front of the NaCl molecule to obtain 2 Cl atoms, we are also changing the number of Na atoms to 2. Remember when balancing a chemical equation, we cannot change the intrinsic formula of the compound and when adding a stoichiometric coefficient in front of a formula, it applies to all atoms in the formula. So we now have 2 Na atoms on the reactants side. We need to add a stoichiometric coefficient of 2 in front of Na (s) in the products.

$$Fe \text{ (s)} + 2NaCl \text{ (aq)} \rightarrow FeCl_2 \text{ (aq)} + 2Na \text{ (s)}$$

This equation is now balanced with respect to all atoms.

Assuming that the $FeCl_2$ is the desired product, we can calculate its molecular weight as 126.751 g/mol. The reactants are Fe and 2 moles of NaCl, which has a total mass of 172.73 g/mol. Taking the ratio of the desired product to the total mass of reactants provides an atom efficiency of 73.4%.

(b) When looking at the above equation, it looks like a displacement reaction. Let's see if it is a redox reaction. Remember in a redox reaction, there must be a transfer of electrons.

On the reactants side, Fe has a charge of 0. On the product side $FeCl_2$ (aq) is an ionic compound consisting of Fe^{2+} (aq) and 2 Cl^- (aq). So Fe switches from a charge of 0 to a charge of +2 on the products side, which means that Fe loses 2 electrons and is oxidized. We can write that as an oxidation half-reaction, as follows:

$$\text{Oxidation: } Fe \text{ (s)} \rightarrow Fe^{2+}(aq) + 2e^-$$

On the reactants side, NaCl (aq) is an ionic compound consisting of Na^+ (aq) and Cl^- (aq). Na^+ has a charge of +1 on the reactants side; on the product side the charge of Na in Na (s) is 0. So Na switches from a charge of +1 to a charge of 0, which means that Na gains 1 e^- and is reduced to Na (s). We can write this as a reduction half-reaction, as follows:

$$\text{Reduction: } Na^+(aq) + 1e^- \rightarrow Na \text{ (s)}$$

(c) In a redox reaction, the number of electrons lost should be equal to the number of electrons gained. If we add the two half-reactions above, we get

$$\begin{array}{l} Fe \text{ (s)} \rightarrow Fe^{2+}(aq) + 2e^- \\ + Na^+(aq) + 1e^- \rightarrow Na \text{ (s)} \\ \hline Fe \text{ (s)} + Na^+(aq) + 1e^- \rightarrow Fe^{2+}(aq) + Na \text{ (s)} + 2e^- \end{array}$$

which could be rewritten as

$$Fe \text{ (s)} + Na^+(aq) \rightarrow Fe^{2+}(aq) + Na \text{ (s)} + 1e^-$$

This looks different from what we've seen before, because we still have an electron remaining on the right-hand side. If we do one more reduction reaction, we can eliminate the remaining electron:

$$\begin{array}{l} Fe~(s) + Na^+~(aq) \rightarrow Fe^{2+}~(aq) + Na~(s) + 1e^- \\ \underline{+ Na^+(aq) + 1e^- \rightarrow Na~(s)~(\times 2)} \\ Fe~(s) + 2\,Na^+(aq) \rightarrow Fe^{2+}(aq) + 2\,Na\,(s) \end{array}$$

Now the equation is properly balanced, and the electrons do not appear in the final equation. We can get to the same end result, without the intermediate equation, simply by multiplying all of the stoichiometric coefficients in the reduction reaction by 2, giving the following single set of redox reactions:

$$\begin{array}{l} Fe~(s) \rightarrow Fe^{2+}(aq) + 2e^- \\ \underline{+ 2Na^+(aq) + 2e^- \rightarrow 2Na~(s)} \\ Fe~(s) + 2Na^+(aq) \rightarrow Fe^{2+}(aq) + 2Na~(s) \end{array}$$

This allows for the number of electrons to cancel out so that both atoms and charges are balanced correctly.

Many students have a difficult time remembering which reaction is oxidation and which is reduction. If you recall that reduction reduces the charge of the ion, that will be helpful, as long as you can remember that zero is less than +2, and −2 is less than zero. A useful mnemonic to remember the definitions of oxidation and reduction is the following:

OIL RIG (Oxidation is Loss; Reduction is Gain)

One can also remember that oxidation starts with an O and there is an "o" in loss.

Another problem that is frequently encountered is to know what the appropriate charge can be on the ions under consideration. For *monoatomic* cations and anions, the common charges are called oxidation numbers, and they can often be identified on a periodic table (see Chapter 4). Polyatomic ions are more challenging and often must be derived from the oxidation numbers of the underlying monatomic components. Knowing some rules to determine the oxidation numbers of elements will allow you to calculate the charge for polyatomic ions.

Rule 1: The oxidation number of an atom of a pure element is zero.

Example: Oxygen in O_2, sulfur in S_8, iron in metallic Fe, or chlorine in Cl_2 all have an oxidation number of zero.

Rule 2: Because atoms want to have a complete outside electron shell, they will normally donate electrons if they have only a few extra in the outside shell, or acquire some to complete the shell. As a result, the oxidation number of many elements can be determined from their placement in the periodic table.

Example: Alkali metals in group 1(A) have an oxidation state of +1, and metals in group 2A have an oxidation state of +2. Specifically, Na^+ has an oxidation number of +1; Ca^{2+} has an oxidation number of +2.

Likewise, the oxidation number of elements in groups 6A and 7A are generally −2 and −1, respectively.

Example: Cl^- has an oxidation number of −1; Se^{2-} has an oxidation number of −2.

One needs to use Rule 2 with caution, since elements of higher periods may have multiple oxidation states. In addition, the transition elements usually have multiple oxidation states and can accept or donate electrons in multiple arrangements. In such cases, the oxidation state of the unknown element can be determined based on the oxidation state of a known species.

Rule 3: The sum of the oxidation numbers in a neutral compound is zero.

For example, to calculate the oxidation number of P in PCl_5, one starts from knowing that the oxidation state for Cl is −1. PCl_5 is a neutral molecule. There are 5 Cl atoms, which contributes a total charge of −5. Thus, the oxidation state of P in this molecule must be +5, in order for the charges to balance. See Highlight 5.5.

Highlight 5.5 Balancing Redox Reactions

Chlorine (Cl_2) as an oxidizing agent used in water and sewage treatment plants. Consider the following equation, which demonstrates how chlorine can be used to eliminate hydrogen sulfide coming from the decay of organic matter:

$$Cl_2 (g) + H_2S (aq) \rightarrow S_8 (s) + HCl (aq)$$

(a) Balance this equation in terms of atoms.
(b) Determine which element is reduced and which one is oxidized.
(c) Write the overall redox reaction.

Solutions:

(a) Since elemental sulfur takes the form of S_8, the balanced equation requires 8 molecules of H_2S, which means 16 molecules of HCl on the right-hand side and finally 8 chlorine molecules on the reactant side. In summary, the balanced equation is written as

$$8Cl_2 (g) + 8H_2S (aq) \rightarrow S_8 (s) + 16HCl (aq)$$

(b) In a redox reaction, one element is oxidized (it loses electrons) and one is reduced (it gains electrons). In the balanced equation, we first look at the reactants. Both Cl_2 and H_2S are neutral molecules. The H_2S can dissociate in water, and since the oxidation state of hydrogen is +1, then the oxidation state of sulfur must be −2. Likewise, looking at the products, we know the oxidation states of hydrogen and chlorine are +1 and −1, respectively. As a result, we can write the complete ionic equation:

$$8Cl_2(g) + 16H^+(aq) + 8S^{2-}(aq) \rightarrow S_8(s) + 16H^+(aq) + 16Cl^-(aq)$$

After canceling the spectator ions, the net ionic equation is the following:

$$8Cl_2(g) + 8S^{2-}(aq) \rightarrow S_8(s) + 16Cl^-(aq)$$

Chlorine goes from zero as a reactant to −1 as a product, so it is reduced. Similarly, sulfur goes from −2 as a reactant to zero as a product, so it must be oxidized.

(c) To write the overall redox equation, let's write the two half-equations first.

$$\text{Reduction: } 8Cl_2(g) + 16e^- \rightarrow 16Cl^-(aq)$$

(Because there are 16 Cl^- on the products side and there are 8 Cl_2 on the reactants side, the number of electrons gained also needs to be multiplied by 8.)

$$\text{Oxidation: } 8S^{2-}(aq) \rightarrow S_8(s) + 16e^-$$

(Because there are 8 S^{2-}, the number of electrons also needs to be multiplied by 8, which gives a total of 16.)

In the two half-reactions the number of electrons lost is equal to the number of electrons gained. The 16 e^- cancel each other out. We can add the two half-reactions to obtain the overall redox equation:

$$8Cl_2(g) + 16e^- \rightarrow 16Cl^-(aq)$$

$$8S^{2-}(aq) \rightarrow S_8(s) + 16e^-$$

$$8Cl_2(g) + 8S^{2-}(aq) \rightarrow S_8(s) + 16Cl^-(aq)$$

This equation is balanced with respect to the atoms and the charges.

All the examples above involved inorganic compounds; oxidation and reduction reactions can also take place with organic compounds. For example, methane can be oxidized to carbon dioxide (CO_2) (effectively the oxidation number of C changes from −4 to +4). As mentioned in Chapter 4 primary alcohols can be oxidized to aldehydes (the oxidation number of C changes from −2 to 0). Typical reductions include the conversion of an alkene to an alkane. In organic reactions, an oxidation involves the removal of hydrogen atoms whereas the reverse reduction adds hydrogen atoms to an organic molecule.

5.3.5   Exchange or Substitution Reactions

Substitution reactions can involve inorganic or organic compounds and follow the typical chemical path:

$$AD + BC \rightarrow AC + BD \qquad\qquad (5.18)$$

5.3.5.1 Examples with Inorganic Compounds

Typical exchange reactions result in three possible products:

1. The formation of a precipitate or insoluble ionic compound (case for precipitation reactions).
2. The formation of a molecular compound remaining in aqueous solution (case for the production of H_2O (l) in acid–base neutralization reactions).
3. The formation of a gaseous molecular compound.

PRECIPITATION REACTIONS. Exchange reactions involve ionic compounds. Some ionic compounds are soluble in water and others are not. When an ionic compound dissolves in water, it dissociates into ions; when an ionic compound is completely converted to ions and forms an aqueous solution, it is referred to as a strong electrolyte. An electrolyte is a substance whose aqueous solution contains ions and conducts electricity.

The solubility of a compound in water depends on the relative force of attraction between the ions that make up the compound and the counteracting force between the water molecules and the ions. Since it is very difficult to estimate these relationships, one commonly refers to a set of rules that describe which compounds are soluble, slightly soluble, and insoluble. Table 5.1 provides these rules [1, p. 165].

When a compound contains at least one ion that is not soluble (insoluble), the compound does not dissociate and remains in its solid state. When the chemical reaction is written with the phase designation(s) included, this indicates the formation of a precipitate or that the product is a solid; it does *not* signify that it is soluble.

For example, let's look at the reaction between ammonium sulfide, $(NH_4)_2S$, and copper nitrate $Cu(NO_3)_2$. Will a precipitate form?

The first step in working through this reaction is to determine what products will be formed, independent of whether they are soluble. Since this is an exchange reaction, the cation of the first compound reacts with the anion of the second compound and vice versa. Two compounds are formed: ammonium nitrate NH_4NO_3 and copper

TABLE 5.1. Solubility Rules

Usually Soluble Ions	Exceptions
Group 1A alkali metals Li^+, Na^+, K^+, Rb^+, Cs^+, and ammonium NH_4^+	None
Nitrates, NO_3^-	None
Chlorates, ClO_3^-	None
Perchlorates, ClO_4^-	None
Acetates, CH_3COO^-	None
Sulfates, SO_4^{2-}	$CaSO_4$, $SrSO_4$, $BaSO_4$, and $PbSO_4$ are *insoluble*
Chlorides, Cl^-, bromides, Br^-, and iodides, I^-	$AgCl$, Hg_2Cl_2, $PbCl_2$, $AgBr$, Hg_2Br_2, $PbBr_2$, AgI, Hg_2I_2, and PbI_2 are *insoluble*

Usually Insoluble Ions	Exceptions
Phosphates, PO_4^{3-}	Phosphates of group 1A alkali metals and of ammonium are *soluble*
Carbonates, CO_3^{2-}	Carbonates of group 1A alkali metals and of ammonium are *soluble*
Hydroxides, OH^-	Hydroxides of group 1A alkali metals and of ammonium are *soluble*; $Sr(OH)_2$, $Ba(OH)_2$, and $Ca(OH)_2$ are *slightly soluble*
Oxalates, $C_2O_4^{2-}$	Oxalates of group 1A alkali metals and of ammonium are *soluble*
Sulfides, S^{2-}	Sulfides of group 1A alkali metals and of ammonium are *soluble*; sulfides of group 2A, MgS, CaS, and BaS, are *sparingly soluble*

sulfide CuS. Without regard to the phase state of the compounds, the overall balanced chemical reaction looks like this:

$$(NH_4)_2S + Cu(NO_3)_2 \rightarrow 2NH_4NO_3 + CuS \qquad (5.19)$$

Note that the stoichiometric coefficient for the ammonium nitrate product must be two, since there are two ammonium ions and two nitrate ions present in the reacting species.

Next, we can determine which of these products is soluble. According to the solubility rules, all nitrates are soluble, so the ammonium nitrate product will dissolve in water and we designate its phase as aqueous (aq). On the other hand, sulfides are insoluble, and copper sulfide is not an exception. Thus, copper sulfide will precipitate and the phase designation is listed as solid (s). Now, combining the phase designations with the balanced overall equation provides

$$(NH_4)_2S\,(aq) + Cu(NO_3)_2\,(aq) \rightarrow 2NH_4NO_3\,(aq) + CuS\,(s) \qquad (5.20)$$

This equation is commonly referred to as the molecular equation.

As we demonstrated in the case of displacement reactions, it is possible to write the equation involving compounds that dissociate. It consists in dissociating all aqueous compounds into their respective ions and leaving solids as they are. This constitutes the complete ionic equation. Common ions that appear in both reactants and products will cancel each other out; they are spectator ions. The resulting equation is the net ionic equation.

The complete ionic equation is

$$2NH_4^+(aq) + S^{2-}(aq) + Cu^{2+}(aq) + 2NO_3^-(aq) \rightarrow$$
$$2NH_4^+(aq) + 2NO_3^-(aq) + CuS(s)$$

Remember that the number 2 in front of NH_4NO_3 is a stoichiometric coefficient, and upon dissociation, it generates two ammonium ions and two nitrate ions. The ammonium and nitrate ions are spectator ions, which appear on both sides of the reaction equation. They cancel each other out, leaving the following net ionic equation:

$$S^{2-}(aq) + Cu^{2+}(aq) \rightarrow CuS(s) \tag{5.21}$$

While some of this discussion brings back memories of the redox analysis described in the previous section, you should be careful to note that the current case is not redox chemistry. The sulfur ion in the reactant has a charge of −2, as described in the equation. The sulfur in the product CuS also has a charge of −2, but it is not shown that way since it is an element in a neutral molecule. Thus, there is no change in the oxidation of the sulfur, and this is not a redox reaction.

Another example looks at the reaction between calcium acetate, $Ca(CH_3COO)_2$, and iron sulfate, $FeSO_4$. The exchange reaction between calcium acetate and iron(II) sulfate results in the formation of calcium sulfate, $CaSO_4$, and iron(II) acetate, $Fe(CH_3COO)_2$. According to the solubility rules, calcium sulfate is an exception and is insoluble. Iron(II) acetate is soluble. So the molecular equation can be written as

$$Ca(CH_3COO)_2(aq) + FeSO_4(aq) \rightarrow CaSO_4(s) + Fe(CH_3COO)_2(aq) \tag{5.22}$$

Dissociation of soluble ionic compounds into their respective ions gives the complete ionic equation:

$$Ca^{2+}(aq) + 2CH_3COO^-(aq) + Fe^{2+}(aq) + SO_4^{2-}(aq) \rightarrow$$
$$CaSO_4(s) + Fe^{2+}(aq) + 2CH_3COO^-(aq)$$

Fe^{2+} and CH_3COO^- are spectator ions; the net ionic equation can be written as follows:

$$Ca^{2+}(aq) + SO_4^{2-}(aq) \rightarrow CaSO_4(s) \tag{5.23}$$

See Highlight 5.6.

Highlight 5.6   Precipitation Reactions

In each of the following cases predict if a precipitate would form and justify your answer. In the case of precipitation, write the molecular, complete ionic, and net ionic equations. Only the names of the compounds will be given. Make sure the ionic formula is correct before you start balancing the molecular equation.

   (a)   Reaction between lead nitrate and potassium iodide.
   (b)   Reaction between calcium nitrate and potassium chloride.
   (c)   Reaction between nickel chlorate and barium phosphate.

Solutions:

   (a) Lead nitrate has the following ionic formula: $Pb(NO_3)_2$. Two nitrate ions are necessary to balance the +2 charge from Pb^{2+}. Potassium iodide is KI.
   If an exchange reaction occurs, lead would react with iodide to form lead iodide, PbI_2, and potassium would react with nitrate to form potassium nitrate, KNO_3.
   The balanced molecular equation, without regard to the phase of the compounds, is the following:

$$Pb(NO_3)_2 + 2KI \rightarrow PbI_2 + 2KNO_3$$

According to the solubility rules, all nitrates are soluble and all iodides are soluble but PbI_2 is an exception and will therefore form a precipitate:

$$Pb(NO_3)_2\,(aq) + 2KI\,(aq) \rightarrow PbI_2\,(s) + 2KNO_3\,(aq)$$

The complete ionic equation shows the dissociation of all soluble species into their respective ions:

$$Pb^{2+}(aq) + 2\cancel{NO_3^-}(aq) + 2\cancel{K^+}(aq) + 2I^-(aq) \rightarrow PbI_2(s) + 2\cancel{K^+}(aq) + 2\cancel{NO_3^-}(aq)$$

The net ionic equation eliminates all the spectator ions:

$$Pb^{2+}(aq) + 2I^-\,(aq) \rightarrow PbI_2(s)$$

   (b) Calcium nitrate has the following ionic formula: $Ca(NO_3)_2$. Two nitrate ions are necessary to balance the +2 charge from Ca^{2+}. Potassium chloride is KCl.

   If an exchange reaction occurs, calcium would react with iodide to form calcium iodide, CaI_2, and potassium would react with nitrate to form potassium nitrate, KNO_3. Thus, the balanced reaction, without regard to phase, can be written as

$$Ca(NO_3)_2 + 2KCl \rightarrow CaI_2 + 2KNO_3$$

According to the solubility rules, all nitrates are soluble and all iodides are soluble. CaI_2 is *not* an exception and therefore *no precipitation* will occur. Thus, the final reaction expression is written as

$$Ca(NO_3)_2(aq) + 2KCl(aq) \rightarrow CaI_2(aq) + 2KNO_3(aq)$$

(c) Nickel chlorate has the following ionic formula: $Ni(ClO_3)_2$. Two chlorate ions are necessary to balance the +2 charge from Ni^{2+}. Barium phosphate has the following formula: $Ba_3(PO_4)_2$. Barium cation has a +2 charge. The polyatomic anion, phosphate, is PO_4^{3-}. To balance the positive and negative charges, we need three barium cations (total charge of +6) and two phosphate anions (total charge of −6).

If an exchange reaction occurs, nickel would react with phosphate to form nickel phosphate, $Ni_3(PO_4)_2$ (the reasoning is the same as for barium phosphate). Barium would react with chlorate to form barium chlorate, $Ba(ClO_3)_2$. Two chlorate anions are necessary to balance the +2 charge from the barium cation.

The balanced molecular equation, without regard to phase, is the following:

$$3Ni(ClO_3)_2 + Ba_3(PO_4)_2 \rightarrow Ni_3(PO_4)_2 + 3Ba(ClO_3)_2$$

According to the solubility rules, all phosphates are insoluble (nickel is not a group 1A element) and all chlorates are soluble. Thus, the solubility definition provides the following result:

$$3Ni(ClO_3)_2(aq) + Ba_3(PO_4)_2(aq) \rightarrow Ni_3(PO_4)_2(s) + 3Ba(ClO_3)_2(aq)$$

The complete ionic equation looks like this:

$$3Ni^{2+}(aq) + 6ClO_3^-(aq) + 3Ba^{2+}(aq) + 2PO_4^{3-}(aq) \rightarrow$$
$$Ni_3(PO_4)_2(s) + 3Ba^{2+}(aq) + 6ClO_3^-(aq)$$

Remember:

1. When there is a stoichiometric coefficient in front of the formula for a compound, the stoichiometric coefficient is multiplied by the subscript to get the total number of ions.
2. Polyatomic anions do not dissociate further. Thus, phosphates stay as PO_4^{3-} and chlorates remain as ClO_3^-.

The net ionic equation is the following:

$$3Ni^{2+}(aq) + 2PO_4^{3-}(aq) \rightarrow Ni_3(PO_4)_2(s)$$

ACID–BASE NEUTRALIZATION REACTIONS. As defined in Chapter 4, acids and bases can either be weak or strong. In the case of the reaction between a strong acid and a strong base, an exchange reaction occurs. For example, hydrochloric acid and potassium hydroxide react in an exchange manner to form hydrogen hydroxide (WATER!!!!) and potassium chloride. Potassium chloride is not an exception and is therefore soluble. The molecular equation can be written as

$$HCl(aq) + KOH(aq) \rightarrow HOH(l) + KCl(aq)$$

or

$$HCl(aq) + KOH(aq) \rightarrow H_2O(l) + KCl(aq) \qquad (5.24)$$

Since both hydrochloric acid and potassium hydroxide are strong electrolytes, they dissociate completely in solution, allowing one to write the following complete ionic equation:

$$H^+(aq) + \cancel{Cl^-(aq)} + \cancel{K^+(aq)} + OH^-(aq) \rightarrow H_2O(l) + \cancel{K^+(aq)} + \cancel{Cl^-(aq)}$$

Recognizing that the chloride and the potassium ions are spectator ions, the net ionic equation can be written as follows:

$$H^+(aq) + OH^-(aq) \rightarrow H_2O(l) \qquad (5.25)$$

This net ionic equation represents the formation of water from its respective ions, hydrogen and hydroxide. The formation of water results from the neutralization of the strong acid with the strong base. Only in the case of the reaction between a strong acid and a strong base will this neutralization occur. The ability to reduce the net ionic reaction into the formation reaction for water is an indication that the reaction is a neutralization reaction.

The reaction between a weak acid, HCN, and a strong base, NaOH, also results in the formation of water and a soluble salt, in this case, sodium cyanide, NaCN (aq):

$$HCN(aq) + NaOH(aq) \rightarrow NaCN(aq) + H_2O(l) \qquad (5.26)$$

The strong base of NaOH completely dissociates into Na^+ and OH^-. HCN is a weak acid, and recalling that the definition of a weak acid is one which does not completely dissociate, it is written in the reaction equation in its molecular form. Thus, the complete ionic equation is written as

$$HCN(aq) + \cancel{Na^+(aq)} + OH^-(aq) \rightarrow \cancel{Na^+(aq)} + CN^-(aq) + H_2O(l)$$

Now, canceling the Na^+ spectator ions gives the net ionic equation:

$$HCN(aq) + OH^-(aq) \rightarrow CN^-(aq) + H_2O(l) \qquad (5.27)$$

See Highlight 5.7.

Highlight 5.7   Acids and Bases and Neutralization Reactions

In the following examples, identify the strength of acids and bases and derive the molecular, complete ionic, and net ionic equations.

   (a) Reaction between nitric acid and calcium hydroxide. (*Note*: Even if calcium hydroxide is slightly soluble, it is considered as completely ionized.)
   (b) Reaction between acetic acid and barium hydroxide. (*Note*: Even if barium hydroxide is slightly soluble, it is considered as completely ionized.)

Solutions:

   (a) Nitric acid, HNO_3, is a strong acid and calcium hydroxide, $Ca(OH)_2$, is a strong base. This reaction is therefore a neutralization reaction, which produces a salt, calcium nitrate, $Ca(NO_3)_2$, and water.
   The balanced molecular equation is

$$2HNO_3 (aq) + Ca(OH)_2 (aq) \rightarrow Ca(NO_3)_2 (aq) + 2H_2O (l)$$

The complete ionic equation is

$$2H^+ (aq) + 2\cancel{NO_3^-}(aq) + \cancel{Ca^{2+}}(aq) + 2OH^- (aq) \rightarrow$$
$$\cancel{Ca^{2+}}(aq) + 2\cancel{NO_3^-}(aq) + 2H_2O (l)$$

The net ionic equation is the formation of water from its respective ions:

$$2H^+ (aq) + 2OH^- (aq) \rightarrow 2H_2O (l)$$

Simplifying the equation by dividing all stoichiometric coefficients by 2 gives

$$H^+ (aq) + OH^- (aq) \rightarrow H_2O (l)$$

   (b) Acetic acid, CH_3COOH, is a weak acid whereas barium hydroxide, $Ba(OH)_2$, is a strong base. The products are a salt, barium acetate, $Ba(CH_3COO)_2$, and water.
   The balanced molecular equation is

$$2CH_3COOH (aq) + Ba(OH)_2 (aq) \rightarrow Ba(CH_3COO)_2 (aq) + 2H_2O (l)$$

The complete ionic equation shows the dissociation of the strong base, barium hydroxide, whereas the weak acid does not dissociate:

$$2CH_3COOH(aq) + \cancel{Ba^{2+}}(aq) + 2OH^- (aq) \rightarrow$$
$$\cancel{Ba^{2+}}(aq) + 2CH_3COO^- (aq) + 2H_2O (l)$$

The net ionic equation is

$$2CH_3COOH\,(aq) + 2OH^-\,(aq) \rightarrow 2CH_3COO^-\,(aq) + 2H_2O\,(l)$$

Dividing all stoichiometric coefficients by 2 gives

$$CH_3COOH\,(aq) + OH^-\,(aq) \rightarrow CH_3COO^-\,(aq) + H_2O\,(l)$$

This equation is balanced with respect to the atoms and charges.

GAS-FORMING EXCHANGE REACTIONS. The last type of exchange reaction involves the formation of a gas. Common gas-forming reactions involve the reaction between an acid and a metal carbonate. The common products are a salt and carbonic acid, H_2CO_3, which decomposes into CO_2 (g) and water H_2O (l). Gas-forming reactions can also occur with sulfates, in which case SO_2 is produced.

As an example, let's consider the reaction that occurs when you ingest a tablet of Alka-Seltzer made of sodium carbonate, $NaHCO_3$, to relieve excess hydrochloric acid in your stomach.

$$NaHCO_3\,(aq) + HCl\,(aq) \rightarrow NaCl\,(aq) + \cancel{H_2CO_3\,(aq)}$$
$$\cancel{H_2CO_3\,(aq)} \rightarrow H_2O\,(l) + CO_2\,(g)$$
$$\overline{NaHCO_3\,(aq) + HCl\,(aq) \rightarrow NaCl\,(aq) + H_2O\,(l) + CO_2\,(g)} \qquad (5.28)$$

The complete ionic equation for this reaction is

$$\cancel{Na^+(aq)} + HCO_3^-\,(aq) + H^+\,(aq) + \cancel{Cl^-(aq)} \rightarrow \cancel{Na^+(aq)} + \cancel{Cl^-(aq)} + H_2O\,(l) + CO_2\,(g)$$

Canceling the sodium and chloride spectator ions gives the following net ionic equation:

$$HCO_3^-\,(aq) + H^+\,(aq) \rightarrow H_2O\,(l) + CO_2\,(g) \qquad (5.29)$$

See Highlights 5.8 and 5.9.

5.3.5.2 Examples with Organic Compounds

Substitution reactions in organic chemistry involve the replacement of a functional group from one reagent with another functional group from the second reagent. In this case a leaving group will be produced. A typical example is the conversion of benzyl chloride to benzyl cyanide as follows:

$$C_6H_5CH_2Cl \quad + \quad KCN \rightarrow C_6H_5CH_2CN \quad + \quad KCl \qquad (5.30)$$

Benzyl chloride　　　　　　　　　　Benzyl cyanide

In this case the K^+ from the second reagent is exchanged with the Cl^- from benzyl chloride.

Highlight 5.8 Prediction of Products for Exchange Reactions

For the following reactions, predict the products and write the molecular, complete ionic, and net ionic equations:

(a) Iron(II) sulfide reacts with hydrochloric acid.
(b) Lithium carbonate reacts with sulfuric acid.

Solutions:

(a) Iron(II) sulfide, FeS, reacts with hydrochloric acid, HCl. This is an exchange reaction, which leads to the production of the naturally occurring toxic gas H_2S, which usually has a very unpleasant "rotten egg" odor, and the salt $FeCl_2$.
   The balanced molecular equation can be written as

$$FeS\,(s) + 2HCl\,(aq) \rightarrow H_2S\,(g) + FeCl_2\,(aq)$$

The strong ionic species HCl and $FeCl_2$ completely dissociate, so the complete ionic equation is

$$FeS(s) + 2H^+(aq) + 2\,Cl^-(aq) \rightarrow H_2S\,(g) + Fe^{2+}(aq) + 2\,Cl^-(aq)$$

Canceling the spectator ion Cl^- provides the net ionic equation:

$$FeS\,(s) + 2H^+(aq) \rightarrow H_2S\,(g) + Fe^{2+}(aq)$$

(b) Lithium carbonate, Li_2CO_3, reacts with sulfuric acid, H_2SO_4. When a metal carbonate reacts with an acid, the products of this exchange reaction are a salt (in this case, Li_2SO_4) and carbonic acid, H_2CO_3.
   The molecular equation is

$$Li_2CO_3(aq) + H_2SO_4(aq) \rightarrow H_2CO_3(aq) + Li_2SO_4(aq)$$
$$H_2CO_3(aq) \rightarrow H_2O\,(l) + CO_2\,(g)$$
$$Li_2CO_3(aq) + H_2SO_4(aq) \rightarrow H_2O\,(l) + CO_2\,(g) + Li_2SO_4(aq)$$

The complete ionic equation is

$$2Li^+(aq) + CO_3^{2-}(aq) + 2H^+(aq) + SO_4^{2-}(aq) \rightarrow$$
$$H_2O\,(l) + CO_2\,(g) + 2Li^+(aq) + SO_4^{2-}(aq)$$

The net ionic equation is then

$$CO_3^{2-}(aq) + 2H^+(aq) \rightarrow H_2O\,(l) + CO_2\,(g)$$

This equation is balanced with respect to the charges and atoms.

Highlight 5.9 Identification of Chemical Reactions and Prediction of Products

Identify each of these exchange reactions as a precipitation reaction, an acid–base reaction, or a gas-forming reaction. Predict the products of each reaction and write the molecular, complete ionic, and net ionic equations.

(a) Magnesium hydroxide reacts with nitric acid.
(b) Potassium phosphate reacts with magnesium nitrate.
(c) Nitric acid reacts with strontium carbonate.

Solutions:

(a) Magnesium hydroxide, $Mg(OH)_2$, is a strong base whereas nitric acid, HNO_3, is a strong acid. This exchange reaction is an acid–base neutralization reaction. The products are a salt, $Mg(NO_3)_2$, and water. The balanced molecular equation is

$$Mg(OH)_2\,(aq) + 2HNO_3\,(aq) \rightarrow Mg(NO_3)_2\,(aq) + 2H_2O\,(l)$$

The complete ionic equation is

$$\cancel{Mg^{2+}(aq)} + 2OH^-(aq) + 2H^+(aq) + 2\cancel{NO_3^-(aq)} \rightarrow$$
$$\cancel{Mg^{2+}(aq)} + 2\cancel{NO_3^-(aq)} + 2H_2O\,(l)$$

Eliminating the spectator ions leaves us with the net ionic equation:

$$2OH^-(aq) + 2H^+(aq) \rightarrow 2H_2O\,(l)$$

Dividing all coefficients by 2 simplifies the equation:

$$OH^-(aq) + H^+(aq) \rightarrow H_2O\,(l)$$

(b) Potassium phosphate, K_3PO_4, reacts with magnesium nitrate, $Mg(NO_3)_2$, in an exchange manner to give two products: potassium nitrate, KNO_3, and magnesium phosphate, $Mg_3(PO_4)_2$. *Remember*: It is crucial to have the correct ionic formulas before writing the molecular equation. All nitrates are soluble so potassium nitrate is in aqueous solution and all phosphates are insoluble (Mg is not an exception), so this is a precipitation reaction.

The balanced molecular equation is

$$2K_3PO_4\,(aq) + 3Mg(NO_3)_2\,(aq) \rightarrow 6KNO_3\,(aq) + Mg_3(PO_4)_2\,(s)$$

The complete ionic equation shows the dissociation of all aqueous species into their respective ions:

$$6\cancel{K^+(aq)} + 2PO_4^{3-}\,(aq) + 3Mg^{2+}(aq) + 6\cancel{NO_3^-(aq)} \rightarrow$$
$$6\cancel{K^+(aq)} + 6\cancel{NO_3^-(aq)} + Mg_3(PO_4)_2\,(s)$$

Eliminating the spectator ions gives the net ionic equation:

$$2PO_4^{3-}(aq) + 3Mg^{2+}(aq) \rightarrow Mg_3(PO_4)_2(s)$$

This equation is balanced with respect to the charges and the atoms.

(c) Nitric acid, HNO_3, reacts with strontium carbonate, $SrCO_3$. When an acid reacts with a metal carbonate, an exchange reaction forming a gas occurs. In this case, the two products are a salt, strontium nitrate, $Sr(NO_3)_2$, and carbonic acid, which decomposes readily into water and carbon dioxide. Strontium carbonate is insoluble (Sr is not an exception).

The balanced molecular equation is

$$2HNO_3(aq) + SrCO_3(s) \rightarrow Sr(NO_3)_2(aq) + CO_2(g) + H_2O(l)$$

The complete ionic equation gives

$$2H^+(aq) + 2\cancel{NO_3^-(aq)} + SrCO_3(s) \rightarrow Sr^{2+}(aq) + 2\cancel{NO_3^-(aq)} + CO_2(g) + H_2O(l)$$

Eliminating the spectator ions leads to

$$2H^+(aq) + SrCO_3(s) \rightarrow Sr^{2+}(aq) + CO_2(g) + H_2O(l)$$

This equation is balanced with respect to the charges and atoms.

The efficiency and usefulness of substitution reactions in organic reactions really depends on the nature of the leaving group generated.

When evaluating the previously detailed chemical transformations in terms of intrinsic atom economy, we can conclude that:

- Combination, synthesis, or (also called) addition reactions incorporate all reactant atoms into the product(s); therefore, they are 100% atom economical.
- Decomposition reactions are atom economical in theory but the amount of heat or current needed for the reaction to occur prevents the reaction from being "green."
- Displacement reactions (in particular, redox reactions) are usually not atom economical. There is always the presence of an unwanted by-product. Furthermore, many of the strong oxidizing agents are highly environmentally harmful. However, Terry Collins and his research group at Carnegie Mellon University have developed a new catalyst known as TAML (short for tetra-amido macrocyclic ligand) that catalyzes the decomposition of hydrogen peroxide, creating a green oxidizing agent.

- Exchange or substitution reactions are not atom economical because the substituting group displaces a leaving group, which becomes a wasted by-product.

5.4 EFFECTIVENESS AND EFFICIENCY OF CHEMICAL REACTIONS: YIELD VERSUS ATOM ECONOMY

There are many ways to define the efficiency of a chemical reaction: the most common one being percent yield and selectivity. However, these metrics do not include how much waste is produced in a process (i.e, do not include formation of side products, unwanted isomers, etc.)

From an environmental and economic point of view, it would be useful to know how many atoms present in the reactants end up in the desired products and how many end up as waste. This concept was developed by Barry Trost from Stanford University and published in *Science* in 1991. The concept of atom economy or atom utilization allowed Barry Trost to begin "to change the way in which chemists measure the efficiency of the reactions they design" (award ceremony by Paul Anderson 1997 ACS President). Professor Trost received the Presidential Green Chemistry Challenge Award in 1998.

Atom economy answers the following: "How much of what you put into your pot ends up in your product?" Atom economy can be defined as:

- The mass of desired product divided by the total mass of all reagents, times 100.

Percent atom economy = (mass of desired product / total mass of all reagents)×100

Or

- The mass of desired product divided by the total mass of all products and by-products produced, times 100.

$$\text{Percent atom economy} = \frac{\text{mass of desired product}}{\substack{\text{total mass of all products and} \\ \text{by-products produced}}} \times 100$$

Or

- A measure of the efficiency of a reaction.

The use of atom economy requires conversion of a molecular equation into the mass quantities of all species. An example could be steam reforming of methane to produce hydrogen, an important step in the processing of natural gas and other fossil resources. The reaction is carried out at elevated temperature in the gas phase, and the reaction equation is written as

$$CH_4 + 2H_2O \rightarrow CO_2 + 4H_2 \tag{5.31}$$

The mass of the starting materials can be obtained as

$$\text{Mass of } CH_4 = \text{mass of } C + 4 \times \text{mass of } H = 12.01 + 4(1.008) = 16.04 \text{ g}$$

$$\text{Mass of water} = 2 \times \text{mass of } H + \text{mass of } O = 2(1.008) + 15.9994 = 18.02 \text{ g}$$

Thus, the total mass of reactants can be obtained as

$$\text{Mass of } CH_4 + 2 \times \text{mass of water} = 16.04 \text{ g} + 2(18.02 \text{ g}) = 52.08 \text{ g}$$

Hydrogen is the desired product, so we need to calculate the mass of hydrogen:

$$\text{Mass of } H_2 = 2 \times \text{mass of } H = 2.016 \text{ g}$$

And then the total mass of the desired product is

$$4 \times \text{mass of } H_2 = 8.06 \text{ g}$$

Finally, the atom efficiency is the mass of the desired product relative to the total mass of the reactants, which is simply

$$\text{Atom efficiency} = (8.06 \text{ g} / 52.08 \text{ g}) \times 100\% = 15.5\%$$

This is a very low atom efficiency, reflecting the large mass of undesired product that is formed, and the strong desire for carbon capture as an opportunity to minimize air pollution. In addition, the atom economy concept does not take into account the use of energy, auxiliaries, or catalysts and the toxicity of the waste, and in this case, the reaction is impacted by the high temperature required.

While atom economy measures efficiency of the reaction, percent yield answers: "What is the maximum possible quantity of product formed by this reaction?" If the answer is 100%, then the theoretical yield (calculated) and the actual yield (determined through experiment) are identical.

$$\text{Percent yield} = (\text{actual yield} / \text{theoretical yield}) \times 100$$

The actual yield is defined as the quantity of product actually obtained from a synthesis in a laboratory or industrial chemical plant; the theoretical yield is the maximum possible amount of product that can be formed when the limiting reactant is completely used.

Highlight 5.10   Calculations with Reactants Being Completely Consumed

How many grams of oxygen are required to react *completely* with 0.12 mol of propane?

   (a) Write a balanced chemical equation: $C_3H_8(g) + 5O_2(g) \rightarrow 3CO_2(g) + 4H_2O(g)$.
   (b) Using the stoichiometry of the balanced chemical equation determine the number of moles of O_2 needed to react completely with propane and using the molar mass of O_2 as a conversion factor, find the mass of O_2.

Solutions:

$$0.12 \text{ mol } C_3H_8 \times 5 \text{ mol } O_2/1 \text{ mol } C_3H_8 \times 32.0\text{g } O_2/1 \text{ mol } O_2 = 19.2 \text{ g } O_2$$

Thus, 19 g (with 2 significant figures) of O_2 is needed to react completely with propane.

When calculating percent yield, two types of reactions can be discussed.

REACTIONS WITH REACTANTS COMPLETELY CONSUMED. See Highlight 5.10.

REACTIONS WITH ONE REACTANT IN LIMITED SUPPLY. In the previous chemical reactions we assumed that the reactants were completely consumed. No reactant was left when the reaction was over. However, that is usually not the case when chemists carry out an actual synthesis in a laboratory or in industry.

A limiting reactant is the reactant that is completely converted to products during a reaction. Once the limiting reactant has been used up, no more products can form and the reaction will stop. The *limiting* reactant *limits* the amount of product(s) that can be formed.

The moles of product formed are always determined by the starting number of moles of the limiting reactant.

To give you a comparison, when you make sandwiches with bread, lettuce, tomatoes, ham, and cheese, if you run out of bread you cannot make any more sandwiches. In this case the bread is your limiting reactant and the number of sandwiches made is based on how much bread you have. In some exercises the data will state that one reactant was used in excess. Therefore, no calculation is necessary to determine which reactant is the limiting one (the opposite of the reagent in excess). However, in reality we often have to determine which reactant is the limiting one by calculating how much product will be formed (Highlight 5.11).

Highlight 5.11   Calculations with a Reactant in Limited Supply

The synthesis of aspirin by the reaction of salicylic acid, $C_7H_6O_3$, with acetic anhydride, $C_4H_6O_3$, is represented by the following reaction:

$$2\,C_7H_6O_3\,(s) \quad + \quad C_4H_6O_3\,(l) \quad \rightarrow \quad 2\,C_9H_8O_4\,(s) \quad + \quad H_2O\,(l)$$

| Salicylic acid (SA) | Acetic anhydride (AA) | 180.2 g/mol |
| 138.1 g/mol | 102.1 g/mol | |

If we use 10.0 g of salicylic acid and 5.4 g of acetic anhydride, what is the limiting reactant and what is the maximum amount of aspirin produced?

Solutions:

There are many methods to get to the answer. The following method is called the mass method.

Calculate the number of moles of each reactant available and compare their mole ratio with the mole ratio from the stoichiometric coefficients.

$$10.0\,g\ SA \times 1\,mol\ SA / 138.1\,g\ SA = 0.0724\ mol\ SA$$

$$5.4\,g\ AA \times 1\,mol\ AA / 102.1\,g\ AA = 0.0529\ mol\ AA$$

The limiting reagent is the one that produces the least amount of aspirin, which is not necessarily the reactant with the smallest mass.

The number of moles of aspirin produced from 0.0724 mol SA is

$$0.0724\ mol\ SA \times 2\,mol\ aspirin / 2\,mol\ SA = 0.0724\ mol\ aspirin$$

The number of moles of aspirin produced from 0.0529 mol AA is

$$0.0529\ mol\ AA \times 2\,mol\ aspirin / 1\,mol\ AA = 0.1058\ mol\ aspirin$$

$$\Rightarrow AA\ is\ in\ excess;\ SA\ is\ the\ limiting\ reagent!$$

The mass of aspirin produced from 0.0724 mol of SA is

$$0.0724\ mol\ aspirin \times 180.2\,g\ aspirin / 1\,mol\ aspirin = 13.0\,g\ aspirin$$

REFERENCE

1. Moore, J; Staniski, C.; Jurs, P. *Chemistry: The Molecular Science*, 4th edition, Brooks Cole Cengage Learning, Belmont, CA, 2011.

KINETICS, CATALYSIS, AND REACTION ENGINEERING

6.1 BASIC CONCEPT OF RATE

6.1.1 Definition of Reaction Rate

Molecules are in constant molecular motion. The molecules may collide at either a high or low speed depending on the concentration of species present and the temperature of the medium. The electrons from one molecule interact with those of another molecule, leading to the formation of a new chemical species. The amount of time it takes for that change to occur is called the *rate of the reaction*, and the study of reaction rate is termed *kinetics*. In other words, kinetics is just a fancy way of describing how fast reactions happen.

To quantify the speed at which a reaction occurs, it is necessary to determine the *reaction rate* defined as the change in the concentration of reactants or products over a period of time.

For example, the following non-phosgene isocyanate synthesis leading ultimately to the production of polyurethanes used in solvent-based adhesive systems can be represented as follows:

Green Chemistry and Engineering: A Pathway to Sustainability,
Anne E. Marteel-Parrish and Martin A. Abraham.
© 2014 American Institute of Chemical Engineers, Inc. Published 2014 by John Wiley & Sons, Inc.

$$RNH_2 + CO_2 \rightarrow RNCO + H_2O \tag{6.1}$$

Amine + carbon $\rightarrow$ isocyanate + water
dioxide

This synthesis was developed by a group at Monsanto Company.

The rate can be measured as the change in the concentration of the amine per unit of time:

$$\text{Rate} = -\Delta[RNH_2]/\Delta t = -\{[RNH_2]_f - [RNH_2]_i\}/\{t_f - t_i\} \tag{6.2}$$

with i and f standing for initial and final, respectively.

The rate can also be expressed as the change in the concentration of the isocyanate per unit of time:

$$\text{Rate} = +\Delta[RNCO]/\Delta t = +\{[RNCO]_f - [RNCO]_i\}/\{t_f - t_i\} \tag{6.3}$$

In this case, the change in the concentration of a product is positive whereas the change in the concentration of a reactant is negative. Since products are formed during a chemical reaction, their concentration will increase over time. Reactants are consumed during a chemical reaction; therefore, the concentrations of the reactants decrease over time. Since rates of reactions are expressed as positive quantities, it is necessary to add a negative sign in front of the change of concentrations of reactants.

In many cases, it will be easier to refer to the *conversion* of a particular reactant, rather than looking at molar amounts or flow rates. We define the conversion of a reactant as the fraction of material that is used up in the reaction. This provides

$$X = \frac{N_{j,0} - N_j}{N_{j,0}} = 1 - \frac{N_j}{N_{j,0}} \tag{6.4}$$

where X is the conversion and N is the number of moles of the species. Here, the subscript j is used to indicate any species in the reaction, with the subscript 0 (zero) indicating the initial amount. If one were to divide the number of moles by the volume of the system, then Equation 6.4 would represent the change in concentration of the species, similar to that described by Equation 6.2.

Equation 6.4 is always correct but most easily applied to a static system. In some cases, reactions are carried out in flowing systems, in which case we substitute the molar flow rate in place of the amount, to provide

$$X = \frac{F_{j,0} - F_j}{F_{j,0}} = 1 - \frac{F_j}{F_{j,0}} \tag{6.5}$$

The only difference between Equation 6.5 and Equation 6.4 is the use of the molar flow rate (F) in place of the number of moles of the species.

We also like to normalize the composition of the species that are produced in the reaction, and define the *yield* of a material as

$$Y_i = \frac{N_i}{N_{j,0}} = \frac{F_i}{F_{j,0}} \tag{6.6}$$

where component i is the species of interest and component j is the *limiting* reactant. The limiting reactant is the one that will be used up first if the reaction is allowed to proceed for a long enough time.

When a chemical reaction involves stoichiometric coefficients that are different from one (1), it is necessary to take the stoichiometry into account in the expression of the reaction rate.

For example, ozone decomposes naturally to form oxygen as follows:

$$2O_3(g) \rightarrow 3O_2(g) \tag{6.7}$$

In this case, the reaction rate can be expressed as

$$\text{Rate} = -\tfrac{1}{2}\Delta[O_3]/\Delta t \quad \text{(with respect to the reactant)} \tag{6.8}$$

or

$$\text{Rate} = +\tfrac{1}{3}\Delta[O_2]/\Delta t \quad \text{(with respect to the product)} \tag{6.9}$$

The rate will be the same regardless of whether the product concentrations or the reactant concentrations are used in the calculations.

For a general reaction involving A and B as reactants and C and D as products,

$$aA + bB \rightarrow cC + dD \tag{6.10}$$

the rate is defined as

$$\begin{aligned}
\text{Rate} &= -(1/a)\Delta[A]/\Delta t = -(1/b)\Delta[B]/\Delta t \\
&= +(1/c)\Delta[C]/\Delta t = (1/d)\Delta[D]/\Delta t
\end{aligned} \tag{6.11}$$

Engineers use an algebraic method to keep track of the interdependence between the amounts of the various materials participating in the chemical reaction. Since the number of moles of the product is related to the conversion of the reactant, a single stoichiometric variable is used. The bookkeeping system is called the *stoichiometric table*. We will use the conversion as a basis for calculation.

Let's start by defining the stoichiometric coefficient, v_i, which is the number of moles of species i that reacts (or is formed) according to the chemical equation. By convention, we define v_i as being *positive* for material produced by the reaction. We start by defining the extent of the reaction,

$$\xi = \frac{N_i - N_{i,0}}{v_i} = \frac{F_i - F_{i,0}}{v_i}$$

(6.12)

In developing this definition, we have used the stoichiometric coefficient, v_i, which is the number of moles of species i that reacts or is formed according to the chemical equation. For a constant volume system, we can easily convert this expression from moles to concentration simply by dividing the molar terms in the numerator by the volume,

$$\xi = \frac{N_i/V - N_{i,0}/V}{v_i} = \frac{C_i - C_{i,0}}{v_i}$$

(6.13)

The extent of reaction is related to the conversion according to

$$\xi = \frac{-N_{i,0}}{v_i} X$$

(6.14)

You can see that this method provides the same relationship provided in Equation 6.12.

This method is a very powerful tool, enabling one to keep track of many components at the same time. It can be extended easily to account for inert species, multiple reactions, and other more complicated systems. See Highlight 6.1.

6.1.2   Parallel Reactions

In the real world, several reactions usually occur at the same time. In this case, we must write the various reactions as separate, independent reactions. So, for example, in the case of combustion, carbon monoxide can be produced through incomplete combustion of a fuel. In order to describe the production of both CO and CO_2, we write each reaction separately:

$$C + \tfrac{1}{2}O_2 \rightarrow CO$$

(6.15)

$$C + O_2 \rightarrow CO_2$$

(6.16)

By controlling the rate of formation of the desired product relative to the rate of formation of the undesired product, we control the selectivity of the reaction and the amount of waste produced.

Highlight 6.1 Expressing and Determining the Reaction Rate

Consider the following reaction between nitrogen dioxide and fluorine:

$$2NO_2(g) + F_2(g) \rightarrow 2NO_2F(g)$$

In the first 10.0 seconds of the reaction, the concentration of fluorine dropped from 0.10 M to 0.082 M.

(a) Calculate the average rate of this reaction in the first 10.0 seconds.

(b) Calculate the amount of nitrogen dioxide consumed in this time interval.

Solution:

(a) Using Equation 6.11 the average rate of this reaction is expressed as

$$\text{Rate} = -\Delta[F_2] / \Delta t = -(0.082\,M - 0.10\,M)/(10.0\,s) = 0.0018\,M/s$$
$$= 1.8 \times 10^{-3}\,M/s$$

(b) The extent of reaction can be calculated from the consumption of F_2, According to Equation 6.13

$$\xi = \frac{C_i - C_{i,0}}{v_i} = \frac{0.082\,M - 0.10\,M}{-1} = 0.018\,M$$

The amount of NO_2 can now be obtained by a second application of this equation, except now where the extent of reaction is known and the change in NO_2 concentration is the unknown:

$$\xi = \frac{C_i - C_{i,0}}{v_i}$$

$$0.018\,M = \frac{\Delta C}{-2}$$

$$\Delta C = -0.036\,M$$

This algebraic relationship allows easy conversion between the concentrations of each species in the reaction that is either consumed or produced.

In the following discussion, we generalize this situation to that of reactant A, which is converted to two products, the desired product D and the undesired product U. We can write these reactions as

$$A \rightarrow D \qquad\qquad (6.17)$$

$$A \rightarrow U \qquad\qquad (6.18)$$

TABLE 6.1. Example Stoichiometric Table for Parallel Reactions

Species	Inlet	Stoichiometric Coefficient Reaction 1	Stoichiometric Coefficient Reaction 2	Outlet
A	A_i mol	-1	-1	$A_i - \xi_1 - \xi_2$
D	0 mol	$+1$		ξ_1
U	0		$+1$	ξ_2

In order to solve problems of this type, we need to define two different extents of reaction—the extent of reaction 1 describes the amount of A that is converted to product D while the extent of reaction 2 describes the amount of A that is converted to product U. If we want to know the total amount of A converted, it can be obtained as the extent of reactions 1 and 2, combined. In terms of the stoichiometric table (Table 6.1) we include these elements within our table as separate columns, and then proceed with our calculations as we would for a single reaction.

If we knew two pieces of information (e.g., the conversion and the amount of D produced), then we could solve this problem to find, for example, the amount of U produced, or the composition of the product.

In practical applications, the important question to ask is not only how much D can be produced, but also how much U is formed (and how much we will need to throw away). We can characterize this parameter as the *selectivity* (*S*), or the ratio of the yield (as defined in Equation 6.6) of one product relative to another. In a general sense, this can be written as

$$S = \frac{Y_{\text{desired product}}}{Y_{\text{undesired product}}} \tag{6.19}$$

An alternate definition provides selectivity as the ratio of the amount of the desired product relative to the total conversion (*X*, defined in Equation 6.5)

$$S = \frac{Y_{\text{desired product}}}{X} \tag{6.20}$$

These two definitions provide alternative measures of the same concept and can be used interchangeably. Note that the latter method requires that selectivity be between 0 and 1, whereas in the former method the selectivity may approach infinity. See Highlight 6.2.

Highlight 6.2   Parallel Reactions

Steam reforming of natural gas (90% methane and the remainder CO_2) is being considered for the production of hydrogen. The conversion of methane can produce either CO or CO_2, depending on the amount of steam included in the process. In one particular case, a steam to carbon ratio of 5 is used, resulting in a selectivity to CO_2 (relative to CO) of 3. What is the percentage of hydrogen in the product gas at complete methane conversion?

Solution:

Let's begin by writing the two parallel reactions:

$$CH_4 + H_2O \rightarrow CO + 3H_2$$

$$CH_4 + 2H_2O \rightarrow CO_2 + 4H_2$$

and then put the information into the stoichiometric table (Table 6.2).

TABLE 6.2. Stoichiometric Table for Parallel Reactions in Steam Reforming

Species	Inlet	Stoichiometric Coefficient Reaction 1	Stoichiometric Coefficient Reaction 2	Outlet
CH_4	9 mol	−1	−1	$9 - \xi_1 - \xi_2$
H_2O	45 mol	− 1	− 2	$45 - \xi_1 - 2\,\xi_2$
CO	0	+1		ξ_1
CO_2	1 mol	0	+1	$1 + \xi_2$
H_2	0	+3	+4	$3\,\xi_1 - 4\,\xi_2$

We have two unknowns, and two pieces of information that we can use to solve the problem. First, complete conversion of methane provides

$$9 - \xi_1 - \xi_2 = 0$$

The selectivity describes the amount of CO_2 produced relative to the amount of CO and is given as 3. This can be expressed mathematically as

$$3 = \frac{\xi_2}{\xi_1}$$

and then substitution into the conversion equation provides the unknown extents of reaction.

$$0 = 9 - \xi_1 - 3\zeta_1$$
$$\xi_1 = 2.25$$

From which we can directly evaluate $\xi_2 = 6.75$. Finally, substituting numbers into Table 6.2 provides the number of moles for each product species:

$$F_{H_2O} = 45\,mol - (2.25\,mol) - (2)(6.75\,mol) = 29.25\,mol$$
$$F_{CO} = 2.25\,mol$$
$$F_{CO_2} = 1 + (6.75\,mol) = 7.75\,mol$$
$$F_{H_2} = 3(2.25\,mol) + (4)(6.75\,mol) = 33.75\,mol$$

So the percentage of hydrogen in the product gas is simply the amount of hydrogen relative to the total amount of gas produced.

$$\%H_2 = \frac{33.75\,mol}{29.25\,mol + 2.25\,mol + 7.75\,mol + 33.75\,mol} = 46.2\%$$

It is also useful to calculate the percent hydrogen on a "dry basis" (a dry basis is defined as the amount of gas excluding water—since water can easily be condensed and removed from the gas stream). In this case, the result is

$$\%H_2 = \frac{33.75\,mol}{2.25\,mol + 7.75\,mol + 33.75\,mol} = 77.1\%$$

6.1.3   Consecutive Reactions

In addition to the situation of parallel reactions, it is also possible for the product of the reaction to react further to additional (usually undesired) products. This is termed *reactions in series* or *consecutive reactions*. Continuing with the combustion reaction as an example, we note that complete oxidation of a fuel may actually occur by partial oxidation of C to CO, followed by oxidation of CO to CO_2. In this case, the overall reaction must be written as

$$C + \tfrac{1}{2}O_2 \longrightarrow CO \xrightarrow{+\frac{1}{2}O_2} CO_2 \qquad (6.21)$$

Although this is described as fundamentally different from the situation with parallel reactions, both types of problems can be solved by defining two extents of reaction and

TABLE 6.3. Example Stoichiometric Table for General Reactions in Series

Species	Inlet	Stoichiometric Coefficient Reaction 1	Stoichiometric Coefficient Reaction 2	Outlet
A	A_i mol	-1	0	$A_i - 4\,\xi_1$
D	0 mol	$+1$	-1	$\xi_1 - \xi_2$
U	0 mol		$+1$	ξ_2

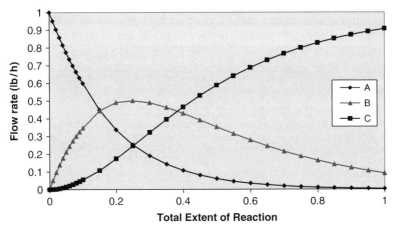

Figure 6.1. Example concentration profile for reactions in series.

combining the reactions mathematically. In order to keep things simple, let us rewrite the combustion reaction above in terms of arbitrary species A, D, and U. Thus, we have

$$A \rightarrow D \rightarrow U \qquad (6.22)$$

Assuming that A is the only component in the feed to the reactor, we obtain a stoichiometric table very similar to the one developed for the case of parallel reactions (Table 6.3).

As before, if we know two pieces of information about this system (e.g., conversion and the amount of D produced), we can calculate the composition of the product stream.

It is instructive to consider the composition within a tubular reactor, or the composition change along the reaction coordinate, for a consecutive reaction. Recall that our consecutive reactions are described as

$$A \rightarrow B \rightarrow C \qquad (6.23)$$

If we plot the concentration profile as a function of the extent of reaction, we find that the concentration of species B reaches a maximum at some intermediate extent of reaction (Figure 6.1).

Now, if species B is the desired product, then to maximize the production of B we should only run the reaction to about 15% extent, which is equivalent to about 40% conversion of reactant A. Since we also want to maximize the consumption of species A, we clearly need to recover the unreacted reactant from the exit stream and recycle it back.

Highlight 6.3 better illustrates this concept.

Highlight 6.3   Consecutive Reactions

The reaction of ethylene to produce ethylene oxide is carried out in the vapor phase over a silver catalyst. Unfortunately, the ethylene oxide can be converted into CO_2 in the same reactor. In order to minimize the amount of CO_2 produced, the feed to the process contains 2 mol/h of oxygen and 10 mol/h ethylene. At an ethylene conversion of 25%, the oxygen is completely consumed. What is the composition of the product gas?

Solution:

We will write the series reaction as two separate reaction processes:

$$C_2H_4 + \tfrac{1}{2}O_2 \rightarrow C_2H_4O \tag{1}$$

$$C_2H_4O + \tfrac{5}{2}O_2 \rightarrow 2CO_2 + 2H_2O \tag{2}$$

and relate the amount of each species to the extent of each reaction. For example, ethylene oxide is produced in reaction 1 but consumed in reaction 2. Thus, the amount of ethylene oxide in the reactor at any point can be evaluated as

$$N_{C_2H_4O} = \xi_1 - \xi_2$$

We summarize this, and similar equations for each species, in the stoichiometric table. We know both the inlet and outlet flow rates for ethylene and oxygen, so we can write the stoichiometric table (Table 6.4).

TABLE 6.4. Stoichiometric Table for Vapor Phase Oxidation of Ethylene

Species	Inlet	Stoichiometric Coefficient Reaction 1	Stoichiometric Coefficient Reaction 2	Outlet
C_2H_4	10 mol/h	-1	0	$10 - \xi_1$
O_2	2 mol/h	$-\tfrac{1}{2}$	$-\tfrac{5}{2}$	$2 - \tfrac{1}{2}\xi_1 = \tfrac{5}{2}\xi_2$
C_2H_4O	0	$+1$	-1	$\xi_1 - \xi_2$
CO_2	0	0	$+2$	$2\,\xi_2$
H_2O	0	0	$+2$	$2\,\xi_2$

Now, we need to use the conversion of ethylene (25%) and the exit flow rate of oxygen (zero, since it is completely consumed) to determine the extents of reaction. First, we know that the total conversion of ethylene is 25%, which provides

$$X = 0.25 = \frac{10 - (10 - \xi_1)}{10}$$

$$= \frac{\xi_1}{10}$$

$$2.5 = \xi_1$$

Second, we know that oxygen is completely used up in the reaction, so

$$F_{O_2,out} = 2 - \tfrac{1}{2}\xi_1 - \tfrac{5}{2}\xi_2 = 0$$

$$\xi_2 = \tfrac{2}{5}\left(2 - \tfrac{1}{2}(2.5)\right)$$

$$= 0.30$$

With these two values, it is now a relatively simple task to substitute these numbers into the equations in the last column of the stoichiometric table to determine the flow rate for each of the exiting species:

$$F_{ethylene} = 10\,mol/h - 2.5\,mol/h = 7.5\,mol/h$$

$$F_{oxygen} = 2\,mol/h - \left(\tfrac{1}{2}\right)(2.5\,mol/h) - (2.5)(0.3\,mol/h) = 0\,mol/h$$

$$F_{ethylene\ oxide} = 2.5\,mol/h - 0.3\,mol/h = 2.2\,mol/h$$

$$F_{CO_2} = F_{water} = 2(0.3\,mol/h) = 0.6\,mol/h$$

Finally, if we wanted to put these results in terms of mole fractions, we need to add up the flow rates for all of the exiting species to find $F_{total} = 10.9\,mol/h$, from which we can determine the mole fractions as

$$y_{ethylene} = \frac{7.5\,mol/h}{10.9\,mol/h} = 0.688$$

for example.

6.1.4   Chemical Equilibrium

No reaction proceeds to 100% conversion. The conversion of all reactions is controlled by equilibrium. Just as equilibrium controls the ratio of vapor and liquid and the composition of each phase, equilibrium also controls the relative amounts of products that can be converted or reactants that can be formed in the chemical reaction.

One may think of an equilibrium reaction as two reactions occurring in parallel:

$$A \rightarrow B \tag{6.24}$$

$$B \rightarrow A \tag{6.25}$$

The amount of A that exists at equilibrium is controlled by how much B is present, and vice versa. In some cases, the equilibrium condition gives extremely high conversion, which approaches 100%. In these cases, we say that the reaction goes to completion and we assume that complete conversion can actually be achieved. However, other cases exist in which less than 100% conversion may be obtained as a result of the thermodynamic limitations. It is these cases with which we are concerned in this section.

In most cases in which reversible reactions are present, we characterize the final product based on the maximum amount of conversion that we can achieve (i.e., the amount that would be present if we let the reaction run for a really long time). This can be illustrated better by looking at Figure 6.2.

If we allow the reaction to go long enough (to the right of Figure 6.2), then the conversion no longer increases and the reaction is said to be at equilibrium.

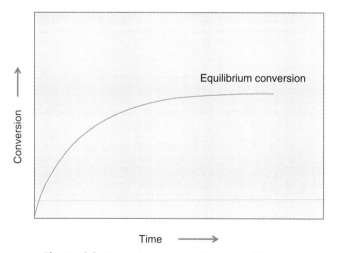

Figure 6.2. Example of conversion at equilibrium.

In terms of kinetics, a chemical equilibrium is reached when the rate of the forward reaction is equal to the rate of the reverse reaction. This is often described as a dynamic equilibrium and a double arrow ($\Leftrightarrow$) is usually drawn between reactants and products.

6.1.4.1 Example of Acid–Base Reactions at Equilibrium

In Chapter 4 (Section 4.6), we introduced the concept of acids and bases. We can now look at the concept of equilibrium associated with the dissociation of weak acids and weak bases in water. Remember that strong acids and strong bases dissociate in water, they completely dissociate and a forward reaction does not exist.

When a weak acid HA dissolves in water, it gives an H^+ to water (which becomes H_3O^+) and produces a conjugate base A^-. The ionization reaction can be summarized as follows:

$$HA\,(aq) + H_2O\,(l) \Leftrightarrow H_3O^+\,(aq) + A^-\,(aq) \qquad (6.26)$$

We can also write an equilibrium constant expression for this reaction, where the equilibrium concentrations of the products appear in the numerator and the equilibrium concentrations of the reactants appear in the denominator. The equilibrium constant expression does not involve pure solids or pure liquids (including water).

For equilibrium associated with the dissociation of weak acids, the equilibrium constant expression is called the *acid ionization constant expression* K_a:

$$K_a = [H_3O^+][A^-]/[HA] \qquad (6.27)$$

All weak acids have very small values of K_a (much smaller than 1), which means that the reaction is reactant-favored. On the opposite side, large values of K_a (more than 1) indicate a product-favored reaction.

When a weak base B dissolves in water, it accepts an H^+ from water (and becomes BH^+). In this case BH^+ is called the conjugate acid. Water loses an H^+ and produces OH^-. The ionization reaction can be summarized as follows:

$$B\,(aq) + H_2O\,(l) \Leftrightarrow BH^+\,(aq) + OH^-\,(aq) \qquad (6.28)$$

For equilibrium associated with the dissociation of weak bases, the equilibrium constant expression is called the *base ionization constant expression* K_b:

$$K_b = [BH^+][OH^-]/[B] \qquad (6.29)$$

Knowing the value of K_a or K_b, we can calculate the equilibrium concentrations of all species present, or vice versa (knowing the equilibrium concentrations of all species present, we can deduce the value of K_a or K_b). When the value for the equilibrium concentration of H_3O^+ is determined, we can then deduce what the value of the pH of the solution is since $pH = -\log[H_3O^+]$. Highlight 6.4 shows an example of calculation of pH knowing the acid ionization constant value K_a.

Highlight 6.4   Calculation of pH for Lactic Acid

Lactic acid, or $CH_3CHOHCOOH$, dissociates partially in water to produce the lactate anion, $CH_3CHOHCOO^-$, with a value of K_a of 1.4×10^{-4}. Calculate the pH of a 0.020 M solution of lactic acid at 25 °C.

Solution:

The first step is to write down the dissociation of lactic acid in water:

$$CH_3CHOHCOOH(aq) + H_2O(l) \Leftrightarrow CH_3CHOHCOO^-(aq) + H_3O^+(aq)$$

The second step is to express the acid ionization constant K_a:

$$K_a = [CH_3CHOHCOO^-][H_3O^+]/[CH_3CHOHCOOH] = 1.4 \times 10^{-4}$$

The goal is to calculate the pH of this solution, in other words to determine the concentration of H_3O^+. (*Remember*: pH $= -\log[H_3O^+]$.) To do so we need to build a table that summarizes information about initial concentrations, changes in the concentrations as the reaction proceeds, and resulting equilibrium concentrations. Chemists will sometimes call this table the ICE table (I—Initial; C—change; E—equilibrium), but we described it previously as a stoichiometric table (Table 6.5).

TABLE 6.5. Example of an ICE Table

Species	Initial Concentration (mol/L)	Stoichiometric Coefficient	Final Concentration (mol/L)
Lactic acid, $CH_3CHOHCOOH$	0.020	−1	$0.020 - \xi$
H_3O^+		+1	ξ
Lactate anion		+1	ξ

Remember since H_2O is a liquid, its concentration does not have an effect on the equilibrium.

We can substitute the concentrations by their values in the expression of K_a:

$$K_a = \frac{[CH_3CHOHCOO^-][H_3O^+]}{[CH_3CHOHCOOH]} = \frac{\xi \times \xi}{(0.020 - \xi)} = 1.4 \times 10^{-4}$$

Because K_a is very small, this implies that the reaction is reactant-favored and not much of the products will form. In other words, ξ will be very small and much smaller than 0.020 M. We can approximate that $0.020 - \xi$ will be almost equivalent to 0.020 to get

$$K_a = \frac{[CH_3CHOHCOO^-][H_3O^+]}{[CH_3CHOHCOOH]} = \frac{\xi^2}{(0.020)} = 1.4\times10^{-4}$$

We can solve for ξ:

$$\xi = \sqrt{(1.4\times10^{-4})(0.020)} = 1.7\times10^{-3}\,M = [H_3O^+]$$

Knowing the value of $[H_3O^+]$, we deduce that the pH of this lactic acid solution is

$$pH = -\log(1.7\times10^{-3}) = 2.77$$

This solution is acidic (pH < 7.0). We can also measure the percent (%) ionization by comparing the H_3O^+ concentration at equilibrium with the initial concentration of the acid:

$$\% \text{ ionization} = \frac{[H_3O^+]_{equilibrium}}{[\text{lactic acid}]_{initial}}\times100\% = \frac{1.7\times10^{-3}\,M}{0.020\,M}\times100\% = 8.5\%$$

This makes sense since lactic acid is a weak acid; its K_a value is very small and it is only 8.5% ionized.

6.1.5 Effect of Concentration on Reaction Rate

The concentration of one or more reactants influences the rate of a reaction. The relationship between the reaction rate and the concentration of a reactant is called the *rate law* and for a simple reaction such as A → products, the rate may be defined according to what is known as power law kinetics:

$$\text{Rate} = k[A]^n \qquad (6.30)$$

where k is the proportionality constant between the rate and the concentration, also known as the *rate constant*, and n is an integer called the reaction order.

If $n=0$, the reaction is zero order; in this case, the rate is independent of the concentration and is equal to the proportionality constant.

$$\text{Rate} = k[A]^n = k[A]^0 = k \qquad (6.31)$$

If $n=1$, the reaction is first order; the rate is directly proportional to the concentration of reactant:

$$\text{Rate} = k[A]^n = k[A]^1 = k[A] \tag{6.32}$$

If $n=2$, the reaction is second order; the rate is proportional to the square of the concentration of reactant:

$$\text{Rate} = k[A]^n = k[A]^2 \tag{6.33}$$

Increasing the reaction order increases the sensitivity of the rate to the concentration. For a second order reaction, a change in the concentration will affect the rate more greatly than it would for a first order reaction.

For multiple reactants, the reaction rate is affected accordingly. For the general reaction

$$aA + bB \rightarrow cC + dD \tag{6.34}$$

the rate law is expressed as follows:

$$\text{Rate} = k[A]^m[B]^n \tag{6.35}$$

with m and n being the reaction orders with respect to A and B, respectively.

One of the major challenges is to determine an appropriate kinetic model that describes the rate of the reaction. There are two common methods in use to determine kinetics, both of which can be used equally. The choice of method usually comes down to an experimental question—what reactor is available for the tests. We will look at both techniques.

6.1.5.1 Integral Method

In this process, we begin by assuming a specific form for the rate expression and then compare this form against the experimental data.

The integral method is based on data that provides concentration as a function of time. This is most easily obtained in a batch system, where the reactor can be loaded with the reactants at time zero, the reaction initiated, and samples taken on a periodic basis to determine the concentration of the various species, or the conversion of the reactant.

For a batch reactor, the change in conversion with respect to time can be related through the mass balance

$$C_{A,0} \frac{dX}{dt} = -r_A \tag{6.36}$$

Recall that $C_{A,0}$ is the initial concentration of species A in the reactor, X is the conversion (which will be measured as a function of time, t), and r_A is the rate of depletion of species A.

We can integrate this equation through separation of variables, which gives the general formula

$$C_{A,0} \int \frac{dX}{-r_A} = t \tag{6.37}$$

In some specific cases, such as the power law expression given previously, the term within the integral may be obtained analytically, providing a mathematical relationship between conversion, initial concentration, and reaction time.

In the case where $n=1$, the reaction is said to be *first order*. In this case,

$$-r_A = kC_A = kC_{A,0}(1-X) \tag{6.38}$$

which when substituted into the combined expression provides

$$\int \frac{dX}{1-X} = kt \tag{6.39}$$

This is integrated from $X=0$ to $X=X_t$ to get

$$-\ln(1-X) = kt \tag{6.40}$$

There are many chemical reactions that can be approximated through first order kinetics.

Now let's consider the case of *second order* kinetics. In this case,

$$-r_A = kC_A^2 = kC_{A,0}^2(1-X)^2 \tag{6.41}$$

which provides, after substitution into the general power law expression or integration of the combined material balance and rate expression,

$$\frac{X}{1-X} = C_{A,0}kt \tag{6.42}$$

In order to make use of the rate expression, we need to put it into a form which is amenable to linear regression, so that the data can be compared with the proposed rate model. Let's consider the second order rate expression, given above. Recall that we have data in the form of X versus t. However, if the reaction can be modeled with second order kinetics, then a plot of $X/(1-X)$ versus t should yield a straight line. Comparison of this result with the general second order rate expression reveals that the line should pass through the origin and have a slope of $C_{A,0}k$.

Highlight 6.5 will help to illustrate this method.

Highlight 6.5   Evaluating the Reaction Rate

The hydrogenation of latex polymer was accomplished using nanoparticle polymers in aqueous solution, without added solvent, and using Wilkinson's catalyst.
The use of extremely small latex particles allows high reaction rates with lower
catalyst concentrations, two green chemistry benefits. At 120 °C, 1 wt % catalyst,
and 35.3 nm particles, the data in Table 6.6 is reported.

TABLE 6.6. Reaction Conversion Data [1]

Time (min)	Conversion
1	0.10
4	0.26
5	0.34
7	0.44
9	0.55
12	0.65
15	0.73
18	0.80
28	0.91

Solution:

Let's assume that first order kinetics apply (although this may not be correct, it is
the simplest place to begin). Recall that for first order kinetics

$$-\ln(1-X) = kt$$

so that a plot of $-\ln(1-X)$ versus t should yield a straight line that passes through
the origin with a slope equal to k. To test this hypothesis, we use a spreadsheet to
convert the data to the proper form and then complete the regression analysis.

Time (min)	Conversion	$-LN(1-X)$
1	0.1	0.1054
4	0.26	0.3011
5	0.34	0.4155
7	0.44	0.5798
9	0.55	0.7985
12	0.65	1.0498
15	0.73	1.3093
18	0.8	1.6094
28	0.91	2.4079

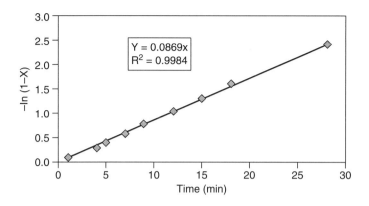

The graph, shown above is a plot of $-\ln(1-X)$ versus time, as required to satisfy first order kinetics. The trendline indicated on the graph reveals that this data is very well approximated through first order kinetics, and the slope of 0.0869 corresponds to a first order rate constant of $0.0869\,\text{min}^{-1}$.

6.1.5.2 Differential Method

Another type of reactor is the plug flow reactor, in which we often can find conversion data for a fixed residence time, but at different inlet concentrations. In this case, we can show that the rate is a function of conversion, initial concentration, and time, according to the simplified mass balance,

$$C_{A,0}\frac{dX}{d\tau} = -r_A \qquad (6.43)$$

In Equation 6.43, the variable τ represents the residence time, which is the amount of time that an element of the fluid remains within the reactor (τ is the Greek version of t, which is the symbol we use for time). The differential method is based on this form of the mass balance equation. Although it is impossible to directly measure the reaction rate, the left-hand side (LHS) (including the derivative) may be approximated by a difference equation, if the conversion is kept low (usually, the conversion is assumed to be low enough if it is less than 20%). In writing the difference equation, we substitute for the derivative

$$C_{A,0}\frac{X-0}{\tau-0} = -r_A \qquad (6.44)$$

where X is the conversion at residence time τ and there is zero conversion at the entrance to the reactor ($\tau=0$). Removing the zeros provides

$$C_{A,0}\frac{X}{\tau} = -r_A \qquad (6.45)$$

Now, if we substitute the general form of the rate expression, we find

$$C_{A,0} \frac{X}{\tau} = -kC_{A,0}^n (1-X)^n \tag{6.46}$$

Our goal is to find the values of the constants, k and n. Our data, however, is usually obtained in the form of X as a function of $C_{A,0}$ (it is often the case that these experiments are only performed at a single residence time). We would like to transform our equation into a linear form, where the slope and intercept can be used to obtain the constants. Taking logarithms of both sides provides the desired linear form.

$$-\ln\left(C_{A,0} \frac{X}{\tau} \right) = \ln k + n \ln\left(C_{A,0}(1-X) \right) \tag{6.47}$$

Now, a straight line should pass through the data when plotted as

$$\ln\left(C_{A,0} \frac{X}{\tau} \right) \text{ versus } \ln\left(C_{A,0}(1-X) \right) \tag{6.48}$$

with the slope equal to the reaction order, n, and the intercept providing the value of the rate constant, k. Notice that in this case, we do not need to make any initial assumptions about the reaction order and that it falls out as a natural consequence of the data analysis. See Highlight 6.6.

Highlight 6.6   Kinetics by the Differential Method

The catalytic performance of montmorillonite clay ion-exchanged with ionic liquids in the cycloaddition of carbon dioxide to allyl glycidyl ether (AGE) was evaluated under a range of CO_2 feed pressures [2]. Under a specific set of conditions (AGE=40 mmol, catalyst (TDAC-MMT)=0.2 g, temperature=100 °C, reaction time=2 h), the conversion of AGE was measured to determine the effect of CO_2. Estimate the reaction order in CO_2 and the rate constant.

Solution:

The data is shown in the Excel spreadsheet below. We note that there are four initial concentrations of CO_2, and the conversion increases with increasing CO_2 pressure. In order to use the differential method, we need to compare the rate of the reaction with the initial concentration of CO_2. The reaction rate is estimated from the data as

$$C_{CO_2,0} X/t$$

and the conversion of AGE provides the data for the right-hand side, as

$$C_{AGE,0}(1-X)$$

We then take the log of both sides and plot. The result is a straight line, as shown in the graph.

CO_2 pressure	Conc (mol/L)	Conversion	LHS	RHS
0.65	0.209601557	0.227	3.738499	3.431403
0.86	0.277318983	0.293	3.203317	3.342155
1.07	0.345036409	0.351	2.804222	3.256557
1.27	0.409529196	0.492	2.295171	3.011606

Age = 40 mmol, catalyst (TDAC-MMT) = 0.2 g, temperature = 100 °C, reaction time = 2 h.

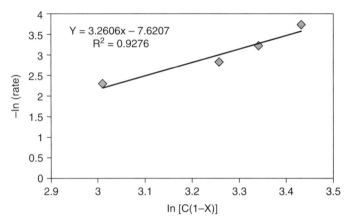

The slope on this graph should correspond to the reaction order, which in this case is found to be 3.26. That reaction order is very high and indicates that the effect of CO_2 concentration goes beyond kinetics and probably involves equilibrium and transport issues. The value of the rate constant is found by taking the exponential of the intercept:

$$k = \exp(-7.6207) = 4.90 \times 10^{-4} \, \text{mol/L·h}$$

While these results are somewhat unusual for a typical chemical reaction, this reaction involves dissolution of CO_2 into an ionic liquid, a rather unique situation. This is an excellent green chemistry example, however. We are using a benign solvent for the utilization of CO_2, addressing two very important green chemistry principles.

6.1.6 Effect of Temperature on Reaction Rate

Molecules move faster when the temperature is increased. We often read in experimental procedures that increasing the temperature hastens the reaction. Therefore, rates of reactions are sensitive to temperature. The effect of the temperature on the reaction rate is actually contained in the rate constant (which is only a constant at a specific temperature; in other words, rate constants vary with the temperature at which the

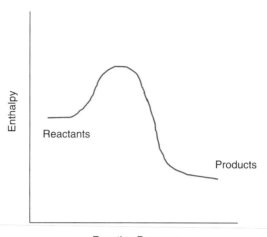

Reaction Progress

Figure 6.3. Representation of the activation energy with R=reactants and P=products.

reaction is performed). When the temperature is increased, the rate constant, k, and the reaction rate are also increased.

The relationship between rate constant k and temperature T originally defined by the Swedish chemist Swante Arrhenius is an exponential relationship:

$$k = Ae^{-E_a/RT} \quad \text{(Arrhenius equation)} \tag{6.49}$$

where A is a constant called the *frequency factor* (or the preexponential factor), E_a is the *activation energy* (or activation barrier), R is the *gas constant* (8.314 J/mol.K), and T is the temperature expressed in kelvin units.

For a reaction to take place, it is often necessary to add energy to the system. The energy of the products is lower than that of the reactants but the reactants must go over an energy barrier before they are transformed into products. When reactants reach this higher energy state level they are in a transition state (also called activated complex). As temperature increases, the number of molecules on the reactants side having enough energy to overcome the activation barrier increases.

The activation barrier is often represented in term of enthalpy versus reaction progress: see Figure 6.3.

In its modern form, the Arrhenius equation is mostly used by taking the natural log of both sides of the equation:

$$\ln k = \ln\left\{Ae^{-E_a/RT}\right\}$$
$$\ln k = \ln A + \ln e^{-E_a/RT} = \ln A - E_a/RT$$
$$\ln k = -E_a/R \times 1/T + \ln A \tag{6.50}$$

The plot of $\ln k$ versus $1/T$ generates a straight line with a slope of $-E_a/R$ and a y-intercept of $\ln A$.

Knowing the slope allows for the determination of the activation energy and knowing the y-intercept is useful to determine the frequency factor A. See Highlight 6.7.

Highlight 6.7   Evaluating the Activation Energy for a Component Decomposition

In a series of experiments, we evaluated the decomposition of dipotassium orthophthalate, a synthetic lubricant [3]. Experimental data was fit to a first order kinetic scheme, and the rate constants given in Table 6.7 were obtained as a function of temperature.

TABLE 6.7. Rate Constants for Deactivation as a Function of Temperature

T (°C)	k (h^{-1})
218	0.1357
235	0.1937
250	0.3042
400	3.633
900	548

Determine the activation energy and the preexponential factor.

Solution:

We carry out the data reduction as described in the previous discussion. The data is plotted in the prescribed form in the graph below, and the regression line is that of an exponential fit (as required according to the Arrhenius law).

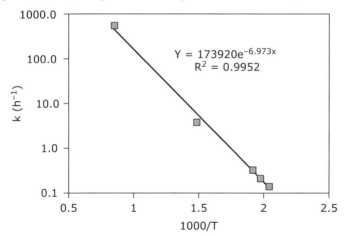

The exponential value of −6.9727 corresponds to an activation energy of

$$E_a = (6.9727)(8.314 \text{ kJ/mol})$$
$$= 57.97 \text{ kJ/mol}$$

The preexponential factor is simply the value indicated in the graph, $1.74 \times 10^5 \text{ h}^{-1}$, or 48.3 s^{-1}.

6.2   ROLE OF INDUSTRIAL AND BIOLOGICAL CATALYSTS

6.2.1   Definition of Catalysts

While it is always possible to increase the rate of a chemical reaction by increasing the concentration of the reactants or by increasing the temperature, it is not always practical to do so. It is not always possible to make highly concentrated mixtures of reactants, maybe because the reactant would promote a phase separation, it would be hazardous in the concentration required, producing the required concentration would be too costly, or many other possible reasons. Increasing the temperature requires a source of heat, which can also be costly and almost certainly leads to undesirable emissions, but may also lead to the decomposition of a reactant. So chemists and engineers require another tool that can be employed to change the reaction path in order to design an energy-efficient process leading to the desired product, with high yield and selectivity. Fortunately, many such species exist, and when employed to accelerate the rate of a chemical reaction, they are termed *catalysts*.

A catalyst increases the rate of a chemical reaction without being consumed by the reaction. The role of a catalyst is to lower the activation barrier, which leads to a faster reaction rate, a higher selectivity (a greater proportion of the desired product is formed), and a higher atom economy. A catalyst does not change the chemical equilibrium of a reaction; it only favors a chemical pathway that will normally not be possible at lower temperatures and atmospheric pressure. Figure 6.4 represents the decrease in the activation energy using a catalyst.

Choosing the most effective catalyst, especially for a commercial process, depends on a number of criteria such as the concentration of the catalyst used, the catalytic turnover, the selectivity of the catalyst to the desired product, and the ability to recover the catalyst after reaction.

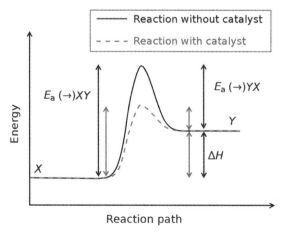

Figure 6.4. Representation of the activation energy of a reaction with and without a catalyst [4].

The catalytic turnover number (TON) is the number of moles of product per mole of catalyst; in other words, it is also the number of catalytic cycles for a given process.

The catalytic turnover frequency (TOF) is the catalytic turnover per unit time; that is, the number of moles of product per mole of catalyst per unit time.

The selectivity of a catalyst can be measured in terms of regioselectivity and chemoselectivity. Let's look at the typical transformation of alkenes into aldehydes in the presence of CO and H_2. This reaction is called hydroformylation and is presented in Scheme 6.1.

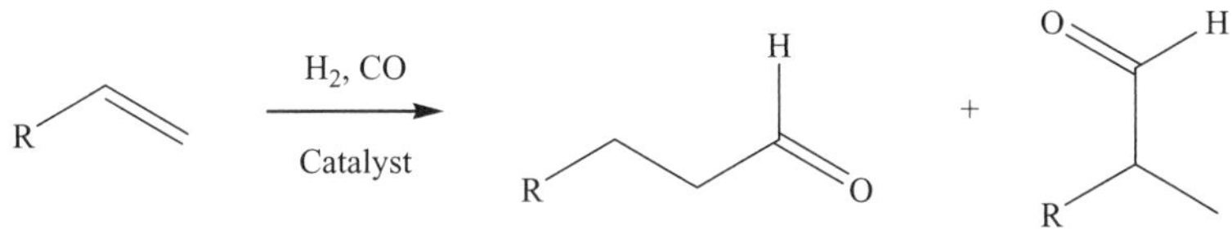

Scheme 6.1 Hydroformylation reaction.

The regioselectivity of the reaction is defined as the ratio of normal-to-branched (or linear-to-branched) aldehydes (*n:i* ratio). The chemoselectivity of the reaction is measured in terms of the ratio of different chemical species produced by the reaction. For example, side reactions can also occur at the same time as the hydroformylation reaction. The hydrogenation of aldehydes to alcohols as well as the alkene isomerization are two major competing reactions to the hydroformylation process. In this case, the chemoselectivity is defined as the aldehyde-to-alcohol ratio for a given chain.

Figure 6.5 shows the typical catalytic cycle for the hydroformylation process.

The hydroformylation or oxo process was discovered in 1938 by Otto Roelen of Ruhrchemie AG in Germany. "Oxo" means carbonyl in German and the hydroformylation (or oxonation) initially produces carbonyl compounds. Commercially both cobalt and rhodium carbonyl catalysts have been exploited since World War II. The same catalyst can be regenerated and reused through many cycles but the catalyst will eventually degrade and lose its activity or it can also be poisoned by impurities that are likely present in the feedstock.

There are different types of catalysts depending on whether the catalyst ends up in the same phase as the product or in a different phase. If the catalyst is part of the same phase as the products, it is termed a homogeneous catalyst. Such materials are often used in relatively large quantities and cannot easily be recovered and reused. A good example of homogeneous catalysts are the Lewis acids, which were discussed in Chapter 5.

When one chooses to use a catalyst in a phase that is different from the phase in which the reaction occurs, this is termed a heterogeneous catalyst. In many cases, it is possible to replace the homogeneous catalyst with a heterogeneous one. The major advantage of a heterogeneous catalyst is that separation of solid catalysts from the reaction mixture can easily be accomplished. This eliminates the need for a separate

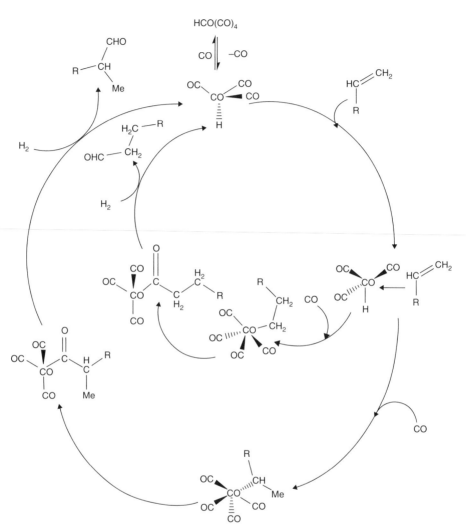

Figure 6.5. Typical hydroformylation reaction cycle

catalyst recovery step, greatly simplifying the process. And because the catalyst is often expensive, the ability to completely recover the catalyst has significant economic advantages, and thus many industrial processes use heterogeneous catalysts. Of course, there are many reasons why a homogeneous catalyst may be preferred; most homogeneously catalyzed reactions operate at milder conditions and provide higher selectivity than a comparable heterogeneously catalyzed process. In addition, mass transfer can play an important role in limiting the effectiveness of a heterogeneous catalyst.

A heterogeneously catalyzed reaction can be written as proceeding through seven consecutive rate processes, described below and shown schematically in Figure 6.6.

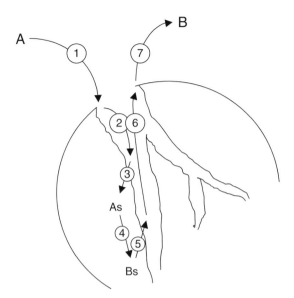

Figure 6.6. Mechanism of a heterogeneous reaction.

1. *External Mass Transfer*. Movement of the reactant from the bulk gas phase flow to the mouth of the catalyst pore.
2. *Internal Mass Transfer*. Movement of the reactant through the catalyst pores to the *active site* of the catalyst. The active site is the location that contains the catalytic material and is where the reaction actually occurs.
3. *Adsorption*. In order for the reaction to take place, a chemical bond must occur between the surface of the catalyst and the reactant. In this sense, adsorption is really a chemical reaction, and this process is frequently termed chemisorption. Chemisorption refers to the binding of the reactant to the surface in a state of minimum energy. Adsorption is an equilibrium process.
4. *Surface Reaction*. The actual chemical reaction occurs while the reactant is bound to the surface. The transformation is from one (or more) adsorbed species to another.
5. *Desorption*. The products are chemically bound to the surface following the reaction. In order for the product to leave the reactor, it must unbind from the surface, in a chemical reaction. The process of breaking the surface–product bond is termed desorption. This step is the opposite of adsorption and is also controlled by equilibrium.
6. *Internal Mass Transfer*. After leaving the surface, the product must make its way through the pores of the catalyst to the external surface of the catalyst particle.
7. *External Mass Transfer*. Once at the pore mouth, the gas phase product returns to the bulk flow of the gas in the reactor.

In order for the entire process to occur at steady state (i.e., no accumulation of species in any particular location), the rate for all of the various steps must be the same. Frequently, one of the steps will occur more slowly than the others. In order to remain at steady state, all of the other steps must slow down to the rate of the slowest step. This slowest step is termed the *rate-limiting step*.

6.2.2 Catalytic Kinetics

The presence of several rate processes occurring in series leads to the development of complex rate expressions, often described as Langmuir–Hinshelwood kinetics. Before going on, we need to separate the chemical steps from the mass transfer steps (a mass transfer step is one in which the species moves into position to react but does not undergo a chemical change). In this section, we consider only the chemical steps that lead to the development of a Langmuir–Hinshelwood rate expression; we reserve the mass transfer issues for the following section.

Because we have three consecutive reactions occurring in series (steps 3, 4, and 5 from above), the development of the overall rate expression can be fairly complex. However, the procedure provides a general expression that is well known and fairly common. Once the more detailed expression can be obtained, simplification can provide a more manageable form for use in engineering calculations.

Let's start by considering a catalytic reaction that is carried out in the gas phase over a solid catalyst. We write the overall reaction as

$$A \rightarrow B$$

According to our description above, this reaction must occur through three separate steps (adsorption, surface reaction, desorption), so we can write a chemical equation for each step,

$$A + S \leftrightarrow As \tag{6.51}$$

$$As \rightarrow Bs \tag{6.52}$$

$$Bs \leftrightarrow B + S \tag{6.53}$$

where the symbol S is used to denote an active site on the surface of the catalyst and the symbol As is used to indicate an A molecule adsorbed on the active site. Now, the rate of the reaction is given by the rate of the slowest step. Assuming that the surface reaction is the rate limiting step, we obtain

$$-r_A = kC_{As} \tag{6.54}$$

In order to use this rate expression, we require the concentration of component A adsorbed on the catalyst surface. This is obtained by assuming that the adsorption reaction is in pseudo-steady state. This means that the rate of the forward reaction

$$A + S \rightarrow As \tag{6.55}$$

is equal to the rate of the reverse reaction

$$As \rightarrow A + S \tag{6.56}$$

or

$$-r_{As} = k_f C_{As} - k_r C_A C_S = 0 \tag{6.57}$$

and then solving for C_{As} provides

$$C_{As} = \frac{k_r p_A C_S}{k_f} = k_A p_A C_S \tag{6.58}$$

After substitution, we obtain

$$-r_A = k K_A p_A C_S \tag{6.59}$$

Now, we need an expression for the concentration of vacant sites. A site balance reveals that all sites are occupied either by a species A molecule, a species B molecule, or that the site is vacant. Mathematically, this provides

$$C_t = C_S + C_{As} + C_{Bs} \tag{6.60}$$

We already have a suitable expression for C_{As} and can obtain an equivalent expression for component B:

$$C_{Bs} = K_B p_B C_S \tag{6.61}$$

Substitution provides

$$C_t = C_S(1 + K_A p_A + K_B p_B) \tag{6.62}$$

Solving Equation 6.62 for C_S and substitution into Equation 6.59 provides the Langmuir–Hinshelwood rate expression:

$$-r_A = \frac{k K_A C_t p_A}{(1 + K_A p_A + K_B p_B)} \tag{6.63}$$

This fairly complex rate expression is typical of those obtained for catalytic reactions and is valuable for further discussion of the kinetics of heterogeneously catalyzed reactions.

Heterogeneously catalyzed reactions are often accomplished in a packed bed reactor, which is essentially a tube packed with catalyst particles. In a packed bed reactor, the weight of catalyst takes the place of the residence time, and so we can show that the rate of the reaction is dependent on the flow rate of the reactant and the weight of catalyst:

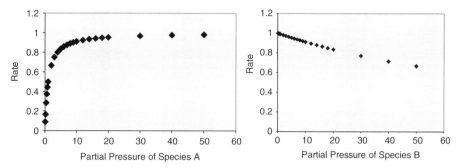

<u>Figure 6.7.</u> Illustration of the effect of species concentration on rate of reaction for complex Langmuir–Hinshelwood kinetics.

$$-r_{A} = -\frac{dF_{A}}{dW} \approx \frac{F_{A,0} - F_{A}}{W} \qquad (6.64)$$

where the last equality is approximately true for low conversions.

What can we do with this? Let's suppose that we have only a small amount of catalyst in the bed, so that the conversion of the reactant is small. Also, for a gas phase catalytic reaction we can use the partial pressure of the species in the gas phase in place of the concentration. In this case, we can assume $p_{A} \approx p_{A,0}$ and $p_{B} \approx p_{B,0}$. Also, we can replace the derivative with a differential, to provide

$$-r_{A} \approx \frac{F_{A,0} - F_{A}}{W} = \frac{k C_{t} K_{A} p_{A,0}}{(1 + K_{A} p_{A,0} + K_{B} p_{B,0})} \qquad (6.65)$$

This is a big improvement, because now we can measure the exit concentration of the reactant, convert to a molar flow rate (mol/s), and evaluate the effect of inlet concentration on the reaction rate. For the example provided, we would obtain two graphs (Figure 6.7).

In the first case, increasing the initial partial pressure of reactant A increases the rate of the reaction at low partial pressures, but appears to have essentially no effect at higher partial pressures. At high pressure, $K_{A} p_{A} \gg 1$ and the effect of the term $K_{A} p_{A}$ in the numerator cancels with the term in the denominator, making a net zero impact.

In the graph on the right, increasing partial pressure of B always decreases the rate. Here, species B is an *inhibitor*. It occupies catalytic sites, which become unavailable for the reaction because they are already in use. Unfortunately, the occupying species cannot be converted to product, so no reaction can take place on that site.

Another category of catalyst is the type present in the human body or any other biological system. Biological catalysts found in living organisms are usually large protein molecules called enzymes. Biochemical reactions operate within a relatively tight range of temperatures, pressures, and pH; efforts to operate outside the stable range of the enzyme will denature the protein and destroy its catalytic effect. Therefore, enzymes provide the means for reactions to occur in an organism at body temperature.

In enzymatic systems, the reaction often follows Michaelis–Menton kinetics. In this case, the rate of reaction is written as

$$-r_A = \frac{k[E][A]}{K + [A]} \tag{6.66}$$

where $[E]$ is the concentration of an enzyme that catalyzes the reaction, $[A]$ is the concentration of the reactant, and k and K are two different characteristic constants. This is more conventionally written as

$$-r_A = \frac{V_{max}[A]}{K_m + [A]} \tag{6.67}$$

where V_{max} is the maximum rate of reaction for a specific enzyme concentration. Operating under differential conditions, we substitute the rate expression into the material balance:

$$[A]\frac{X}{\tau} = \frac{V_{max}[A]}{K_m + [A]} \tag{6.68}$$

Now, simplifying this expression provides

$$\frac{X}{\tau} = \frac{V_{max}}{K_m + [A]} \tag{6.69}$$

We can linearize this equation by taking the inverse of both sides, and then split the RHS into two parts:

$$\frac{\tau}{X} = \frac{K_m + [A]}{V_{max}}$$
$$= \frac{K_m}{V_{max}} + \frac{1}{V_{max}}[A] \tag{6.70}$$

So we see that a plot of τ/X versus $[A]$ should yield a straight line with a slope equal to K_m/V_{max} and an intercept equal to $1/V_{max}$. This analysis for enzyme kinetics has the special name of Lineweaver–Burke plot. The reality is that the results for enzyme kinetics are entirely consistent with those of other catalyzed reactions and can be developed in a parallel way.

6.2.3   Types of Catalysts and Impact on Green Chemistry

Because of the opportunities to control selectivity and improve reaction conditions, many commercially important reactions are conducted with a catalyst. Some examples follow.

6.2.3.1 Homogeneous Catalysts

One common form of catalyst is a metal complex composed of a central metal ion surrounded by one or more groups of atoms coordinated to the central metal ion, called ligands. Homogeneous catalysts are in the same state (i.e., solid, liquid, and gas) as the reactants. One of the most well-known homogeneous catalysis reactions is the alkene hydrogenation using a catalyst called rhodium-tris(triphenylphosphine) chloride(I) $[Rh(PPh_3)_3Cl]$ created by Wilkinson in 1965. The Wilkinson catalyst is used for the oxidative addition reaction of dihydrogen to an alkene to provide an alkane (Scheme 6.2).

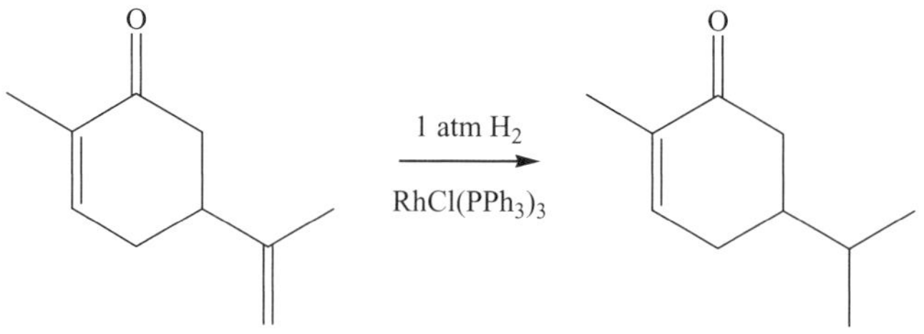

Scheme 6.2. Hydrogenation reaction

The addition of H_2 to a double bond is thermodynamically favored but the kinetic barrier is high and for the reaction to be performed at 298 K and 1 bar H_2, a catalyst is needed.

Wilkinson's catalyst can be used to produce important biological products including antibiotics, steroids, and prostaglandins. In these cases, it is particularly important that the molecule have at least one asymmetric center (all four ligands are different around the metal center) in order to produce proper biological activity. The concept of asymmetric catalysis will be covered in the Section 6.2.4 on biocatalysis.

6.2.3.2 Development of Immobilized Homogeneous Catalysts

The development of new homogeneous catalysts retaining the advantages of mild operating conditions and selectivity while overcoming the difficulty in catalyst separation has attracted considerable interest. Different strategies have been proposed to address the problem of catalyst separation.

IMMOBILIZATION ON SOLID SUPPORTS. Solid supports mainly include polymeric organic, inorganic, and dendrimeric supports. In order to attach the catalyst to a support, it is necessary to functionalize the support with a ligand that can be used to bind to the catalytic metal center. Usually the polymeric supports have a high degree of cross-linking or large surface areas or can be microporous with a low degree of cross-linking.

Polymer-supported catalysts have been used since the early 1960s but are still the subject of improvement. A common polymer is cross-linked polystyrene, which can bind the catalyst directly [5]. The other common example is the use of phosphine-functionalized surface to attach a Rh(I) catalyst (Scheme 6.3).

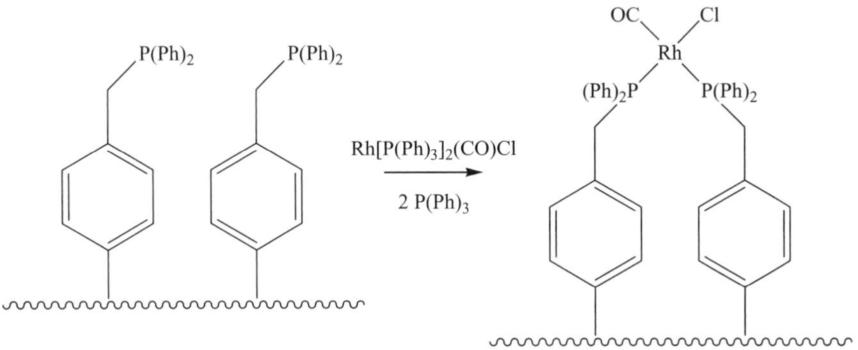

Scheme 6.3. Phosphine-functionalized surface

Even if polymers are easily functionalized, they exhibit poor heat transfer ability and poor mechanical properties and are often temperature sensitive. The choice of a reaction solvent is also important since some solvents cause the polymers to swell. Another issue is the leaching of metal from the polymer, resulting in the deactivation of the catalyst. These drawbacks can be partially eliminated if the catalyst precursors are covalently anchored on inorganic supports.

Silica, alumina, carbon nanotubes, and zeolites are common inorganic supports for catalyst precursors. For example, silica can be found in the amorphous form of a high-surface-area silica gel (200–800 m^2/g) with an average pore diameter of 2.2–2.6 nm or as low-density gel with a surface area of 300–400 m^2/g and an average pore diameter of 10–15 nm. Silica supports are usually more mechanically stable and resistant against aging, solvents, and high temperatures than the organic polymeric supports. Many metals grafted on silica and silica-supported complexes are used in hydroformylation of olefins.

Different active-carbon supports also have been tested. For example, Zhang and co-workers performed the hydroformylation of propene using carbon nanotube-supported rhodium–phosphine catalysts [6]. They found that using carbon nanotubes as supports did not improve the conversion of propene to propanal; however, the regioselectivity increased with carbon nanotubes as supports. The activity and regioselectivity were higher when using carbon nanotubes as supports versus active carbons. However, the rhodium catalyst deactivated over 30 hours due to the degradation of the ligand and/or its oxidation of PPh$_3$ to OPPh$_3$.

A considerable amount of research has been devoted to the study of zeolite-encapsulated metal complexes. Zeolites are crystalline aluminosilicates with typical pore diameters varying between 0.4 and 1.4 nm. Due to the confined space of the zeolite cavity the selectivity is improved by diffusion. It has been found that the regioselectivities improve over those obtained with typical homogeneous catalysts but the activities are lower than in homogeneous systems.

In 1992, a new family of silicate/aluminosilicate mesoporous molecular sieves M 41S with pore diameters varying between 2 and 50 nm was discovered by Beck and co-workers at Mobil Corporation [7]. The synthesis of molecular sieves is based on surfactant/silicate solution chemistry. The template is a surfactant with alkyl chain lengths varying from 8 to 16 carbon atoms. The dimensions of the carbon chain length determine the pore size of the MCM-41 (Mobil Corporation Materials).

Contrary to zeolites, MCM structures do not exhibit any acidic character and have been used in a variety of reactions ranging from olefin polymerization to esterification and hydrogenation reactions.

Dendrimeric supports are high molecular weight phosphine supports allowing precise distribution of catalytic sites and easily separated from the reaction products by precipitation and microporous membrane filtration.

BIPHASIC CATALYSTS. One means of recovering the catalyst is to perform the reaction in a two-phase system in which the catalyst naturally distributes into the nonproduct phase after reaction. These approaches have been used to achieve high activity and selectivity with water-soluble catalyst systems. Water is the only processing solvent that is naturally abundant and nontoxic. The immiscibility of water with nonpolar organic compounds creates unique opportunities for catalyst recycling. The catalyst is located in the aqueous phase and the feedstock and organic products are recovered in the organic phase. The reaction usually takes place at the organic–aqueous interface.

The more common, water-soluble, phosphine-based ligands are arylphosphines with polar functional groups ($-SO_3^-$, $-CO_2^-$, $-NR_3^+$) attached to the aryl rings [8].

Much research has been devoted to water-soluble sulfonated phosphines and, in particular, the well-known TPPTS (triphenylphosphine tris-sulfonated sodium salt) ligand used in propene hydroformylation on an industrial scale by Ruhrchemie/Rhône-Poulenc since 1982 (Figure 6.8) [9, 10].

Prior to the development of aqueous biphasic catalysis, fluorous biphasic catalysis was proposed as another alternative to overcome the low solubility of substrates. The fluorous and organic phases form a homogeneous phase at high temperatures and are immiscible at room temperature. In 1994, Horvath and Rabai modified some phosphine ligands with long-chain "fluoroponytails" to enable catalysts to be recovered in the fluorous phase of the solvent while the products would be recovered in the organic phase upon cooling at room temperature [11]. Many reactions have been performed in fluorous biphasic systems including hydroformylation of 1-alkenes, hydrogenation, hydride reduction, hydroboration, and alkene epoxidation. However, it should be noted that fluorinated compounds are persistent in the environment, and while the catalyst recovery makes the process more green, the use of the fluorous

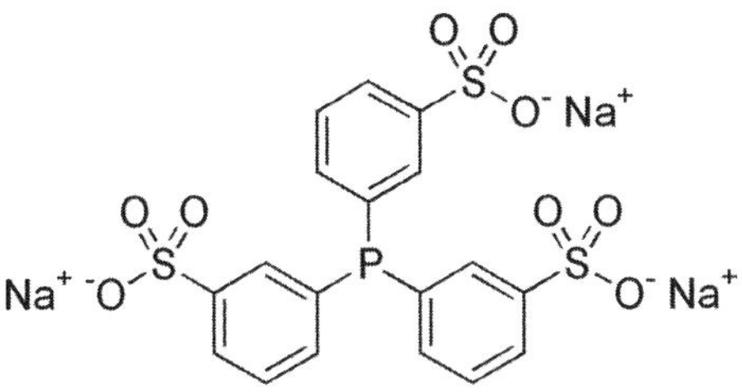

Figure 6.8. Structure of TPPTS.

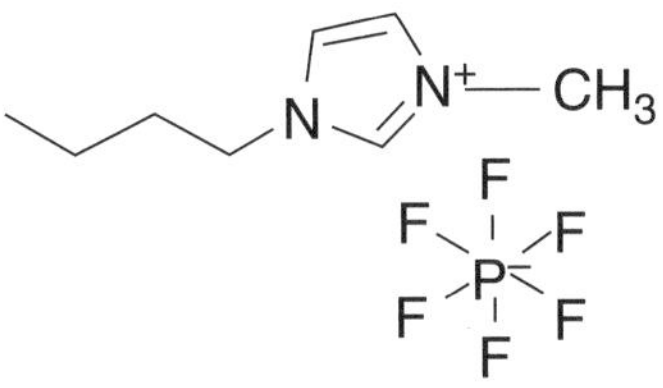

Figure 6.9. Example of ionic liquid 1-butyl-3-methylimidazolium-hexafluorophosphate (BMIM-PF$_6$).

solvent makes an overall nonsustainable process. The presence of multiple fluoroponytails on catalysts to increase solubility of organometallic complexes is often undesirable.

IONIC LIQUIDS. In earlier catalytic processes the use of organic solvents was predominant due to the low water solubility of olefins. An alternative to this problem is the use of ionic liquids introduced in Chapter 5. Ionic liquids are salts that melt at a temperature below 100 °C. Nonvolatile ionic liquids, such as those based on the 1-butyl-3-methyl-imidazolium (BMIM) cation, are used extensively to prevent the lack of thermal stability of homogeneous catalysts during the distillation recovery step and allow the catalyst to be used several times without being deactivated (Figure 6.9).

The nature of cations and anions of the ionic liquid can be adjusted to be compatible with the nature of the ligands used. The other advantage of ionic liquids is their nonvolatility and nonflammability as well as their lower energy use during processes compared to organic solvents.

6.2.3.3 Heterogeneous Catalysts

Most industrial processes use either heterogeneous catalysts or immobilized homogeneous complexes. Most catalysts can be heterogenized by covalently anchoring ligands to a support or to a polymer. This implies that ligands be modified extensively to suit the solubility of the substrate. This approach allows the reaction to be performed homogeneously while easy recovery is also practical. Recyclable solid acid catalysts have gained notoriety in the oil refining and petrochemical industries and in the manufacture of pharmaceuticals, agrochemicals, flavors, and fragrances. They constitute the ultimate replacement for the traditional Lewis acids leading to the generation of inorganic salts in aqueous waste streams. They also eliminate the need for separation at the end of the reaction as well the cost associated with the neutralization and disposal steps.

The process of heterogeneous catalysis involves the adsorption of molecules of reactants onto the catalyst surface and after the reaction is performed, the products undergo desorption. The process of adsorption can involve either weak van der Waals interactions between the surface of the catalyst and the adsorbed species (physisorption) or covalent bonds between surface atoms and the adsorbate (chemisorption). These reaction steps were explored previously when discussing the kinetics of catalytic processes.

In commercial applications diverse catalyst types can be used. For example, the production of ammonia, NH_3, from nitrogen and hydrogen through the Haber process, uses a heterogeneous catalyst based on Fe on SiO_2 or Al_2O_3 support as shown in Equation 6.71.

$$N_2(g) + 3H_2(g) \Leftrightarrow 2NH_3(g) \qquad (6.71)$$

Without a catalyst this reaction would proceed very slowly because the activation barrier for the dissociation of N_2 and H_2 into adsorbed atoms in the gas phase is very high. The adsorbates of N and H then combine to form the product NH_3, which is then desorbed.

Another example of heterogeneous catalysts is found in the polymer industry. In 1953, Ziegler discovered that the polymerization of ethene to high molecular weight polyethene can be performed at low pressures in the presence of Ti-based heterogeneous catalysts. The latest generation of catalysts in use since the 1980s involves $TiCl_4$ supported on anhydrous $MgCl_2$ with Et_3Al as cocatalyst. Karl Ziegler and Giulio Natta were awarded the 1963 Nobel Prize in Chemistry for their discovery.

We previously described zeolites as catalytic supports, but they represent another important type of catalyst used in catalytic cracking of heavy petroleum distillates. Zeolites function as Lewis acids and provide higher selectivities and reaction rates compared to those obtained with alumina/silica catalytic equivalents. The most well-known zeolites used for their robustness to withstand the conditions of the cracking process are ultrastable Y (USY) zeolites. Ultrastable Y zeolites are often coupled to ZSM-5 used as a cocatalyst due to its shape selectivity. The other advantage of using USY zeolites in catalytic cracking is the increase in gasoline octane number. The catalyst ZSM-5 is also used in the conversion of methanol-to-gasoline process (called MTG), developed by Mobil in the 1970s. ZSM-5 is able to convert methanol to a mixture of $>C_5$ alkanes, cycloalkanes, and aromatics. Zeolites have largely replaced acid catalysts in many manufacturing processes because of their ease in recovery and reduced environmental footprint.

The last example using heterogeneous catalysts is the one used in automobile catalytic converters. Catalytic converters are used to reduce the toxicity of emissions from an internal combustion engine. Chemical species found in engine exhaust are carbon monoxide (CO), unburned hydrocarbons (C_nH_{2n+2}), and nitrogen oxides (NO_x). As mentioned in Chapter 3, these products contribute to air pollution. The three-way catalytic converter simultaneously converts CO and unburned hydrocarbons to CO_2 and oxidizes NO_x to N_2, thereby reducing or eliminating most of the harmful emissions from the combustion engine: for example, the oxidation of carbon monoxide to carbon dioxide,

$$2CO(g) + O_2(g) \rightarrow 2CO_2(g) \tag{6.72}$$

the oxidation or combustion of typical unburned hydrocarbons to carbon dioxide and water,

$$C_nH_{2n+2} + 2nO_2 \rightarrow nCO_2 + 2nH_2O \tag{6.73}$$

and the reduction of NO to diatomic nitrogen,

$$2NO + 2H_2 \rightarrow N_2 + 2H_2O \tag{6.74}$$

Catalytic converters built after 1981 can simultaneously promote both oxidation and reduction and are commonly referred to as "three-way" catalytic converters.

The converter requires high temperatures to operate and is heated by the exhaust from the engine. As a result, the greatest majority of emissions from the automobile are released immediately after starting the vehicle. In the 1920s lead additives were introduced in gasoline to raise octane levels and unfortunately lead poisoned the catalyst and deactivated it. It was not until the 1970s that lead additives were eliminated from gasoline. A catalytic converter is made of three components:

1. A core or substrate, often a ceramic honeycomb structure typically made from a low porosity alumina.
2. An alumina washcoat applied to the ceramic substrate to increase the surface area available for reaction. The washcoat often contains other materials added to stabilize the alumina.
3. Fine particles of precious metals, Pt, Pd, and Rh (catalytic active species), are added to the washcoat in suspension to promote the specific reactions. New developments have reduced the amount of metal species required for successful operation of the catalytic converter.

6.2.4 Biocatalysis

Biocatalysts are enzymes used from naturally occurring organisms such as yeasts, bacteria, and plants. Biocatalysts can be applied to nearly any field involving chemical reactions. Many chemical companies are interested in this new market such as Dow Chemical, BASF, DuPont, and Cargill, just to name a few. Biocatalysis uses cleaner processes with no hazardous waste from unnecessary solvents, has high and efficient yields, has high atom economy, and achieves high chemo-, regio-, and stereoselectivity. Enzymes have a long lifespan and allow for higher yields of pure enantiomeric reactions and 50:50 racemic mixtures while operating under mild conditions. However, biocatalysts are also extremely expensive, and their selectivity limits their applicability to specific reagants, reducing the ability to apply biocatalysts across classes of compounds. We illustrate biocatalysis by looking at pharmaceutical and food industries.

6.2.4.1 Pharmaceutical Industries and Chirality

Chirality plays a major role in the development of drugs. A chiral molecule is defined as nonsuperimposable on its mirror image. Two chiral molecules commonly called enantiomers are often compared to the right and left hands. The same type and number of atoms are in the two enantiomers but the spatial arrangement of atoms is different. Usually one enantiomer is preferred over the other. In the pharmaceutical industry chiral molecules constitute a large portion of pharmaceutical sales. More than half the drugs approved worldwide are chiral. Examples include Lipitor® and Zocor®, as well as ibuprofen sold under the common brand names Motrin® and Advil®.

Lonza, a Swiss company, developed an enzyme-based process for the production of the world's best selling drug that lowers cholesterol, Lipitor. Codexis, the designer of the key chiral building block in the synthesis of the active pharmaceutical ingredient in Lipitor, received the 2006 Presidential Green Chemistry Challenge Award from the U.S. Environmental Protection Agency [12].

The key chiral building block is hydroxynitrile (HN) or ethyl (R)-4-cyano-3-hydroxybutyrate whose demand is estimated to be about 200 metric tons annually. Traditional commercial processes are based on:

- The use of chiral precursors.
- The use of hydrogen bromide to generate a bromohydrin for cyanation.
- The substitution of cyanide for halide under heated alkaline conditions, resulting in the formation of by-products.
- A high-vacuum fractional distillation to obtain a purified final product leading to a further decrease of the yield.

The three-step green process designed by Codexis is centered around the activity, selectivity, and stability of three enzymes created using cutting-edge genetic methods. This new process involves:

- Mild, neutral conditions with fewer steps.
- A high-quality product due to the activity and stability of the enzymes.
- An increase in the volumetric productivity of the first step reaction by approximately 100-fold and that of the second step by approximately 4000-fold.

- An increase in the yield, and a decrease in the formation of by-products and generation of waste.
- A reduction of solvents and purification equipment, increasing worker safety.

Merck and Co., Inc. is a global pharmaceutical company established in 1891. Merck is devoted to addressing urgent medical needs through far-reaching programs "that can make a difference in people's lives and create a healthier future" [13]. Merck puts the well-being of its patients first while committed to environmental best practices. Their "green by design" initiative through chemo- and biocatalysis generates benefits summarized by the "triple bottom line" of economics, society, and environment.

A standout example of using catalysis to reduce the number of process steps and waste, and thereby decreasing the cost of manufacturing, is Merck's Januvia™ treatment for type 2 diabetes. This case study will guide you through the understanding of how catalysis can be applied to green chemistry.

Part I: Introduction and Traditional Way to Produce Type 2 Diabetes Drugs

Insulin is a hormone secreted by the pancreas to allow the blood sugar or glucose to enter the body's cells to be used for energy. When insulin is not performing normally, too much glucose accumulates in the blood and over time can do substantial damage to eyes, heart, blood vessels, and kidneys. Type 2 diabetes is the most common form of diabetes and is a lifelong disease. One-third of all people who have diabetes are not aware of their condition [14]. Type 2 diabetes symptoms may include feeling more thirsty and hungry than usual, as well as having to urinate more frequently, losing weight without trying to, and feeling more tired.

Normally, a controlled diet and exercise should help maintain normal blood glucose levels. When these are not sufficient, medication may need to be prescribed. Medications available today either increase the insulin supply (such as sulfonylureas, secretagogues), decrease the insulin resistance, or improve its effectiveness (biguanides and thiazolidinediones) [15]. Another group of medications is called alpha-glucosidase inhibitors, which reduce the rate of glucose absorption.

Traditional oral medications such as Diabinese and Tolinase cost between $26 and $27 per month. Other medications such as Starlix and Prandin cost from $77 to $84 per month because no generic brand is available. Side effects of these medications include hypoglycemia and sometimes weight gain. Most of these medications should not be taken if a condition of heart disease is known. The other type of medication would be insulin, which is injected under the skin.

The traditional pathway to synthesize the active ingredient for type 2 diabetes drugs requires the production of aminoacid derivatives. Sitagliptin, a chiral β-amino acid derivative, is the active ingredient in Januvia, which is a recent treatment for type 2 diabetes. The process required eight steps as well as high-molecular-weight reagents that were not present in the final molecule and contributed to waste.

Part II: Environmentally Benign Pathway—Another Shot at Januvia

Merck, in collaboration with Solvias, a company specializing in catalytic hydrogenation, came up with a new method for generating β-amino acids. Merck researchers discovered that amino acid derivatives can be produced via the asymmetric catalytic hydrogenation of unprotected enamines. The catalyst is a rhodium salt of a ferrocenyl-based ligand. The hydrogenation step reduces waste by 80% and the cost of manufacture by 70%.

Merck was also able to recover and recycle over 95% of the rhodium after hydrogenation. This new synthesis has only three steps and the overall yield is increased by almost 50%. The amounts of raw materials, processing time, energy, and waste were reduced, leading to a reduction of 220 pounds of waste for each pound of sitagliptin manufactured. Because the reactive amino group of sitagliptin is created in the last step of the synthesis, there is no need for protecting groups.

Biocatalysis is a rising market within the pharmaceutical and fine chemical industry. Many companies are joining the market to help with the creation of new biocatalysts, which ultimately allow for greater specificity and activity, and therefore lower the cost of products for consumers.

6.2.4.2 Food Industry and Flavors

Heterogeneous catalysts can also be used in the food industry. University of California–Riverside researchers led by Francisco Zaera designed a platinum-based heterogeneous catalyst which favors the production of partially hydrogenated oils without making *trans* fats. The process of hydrogenation, which is the addition of hydrogen to natural oils to increase the shelf life of foods, results in the production of trans fats found in vegetable shortenings, fried foods, doughnuts, pastries, cookies, crackers, and snacks. Trans fats are known to raise the "bad" cholesterol levels (LDL) and lower the "good" ones (HDL) and are commonly associated with the development of heart disease, stroke, and type 2 diabetes.

Researchers in Zaera's lab found that by controlling the shape of the platinum particles used in the catalyst, the catalyst increased selectivity toward hydrogenated oils while minimizing the production of trans fats.

Another controversial food group is the low calorie sweeteners (also referred to as nonnutritive sweeteners, artificial sweeteners, or sugar substitutes). Artificial sweeteners provide sweetness without the addition of excessive calories. They can be found in diet soft drinks, sugar-free puddings, light yogurts, candies, and as table-top packets.

Some of the most common low-calorie sweeteners approved for use in the United States are:

- Acesulfame potassium (Ace-K; brand names: Sunett, Sweet One), which does not cause any human health problems but it is not broken down by the body. Therefore, it is eliminated in its original form by the kidneys.
- Sucralose (brand name: Splenda) is about 600 times sweeter than regular sugar. It was approved in 1998 by the FDA to be used in diabetic diet and for blood glucose control.

- Neotame (brand name: n/a) was approved in 2002 by the FDA as a general purpose sweetener, which is approximately 7000 times sweeter than sugar.
- Stevia Sweeteners (brand names: PureVia, Sun Crystals, Truvia) are highly purified steviol glycosides, found naturally in the stevia plant. They contain zero calories and are about 200–300 times sweeter than sugar.
- Saccharin (brand names: Sweet'N Low, Sweet Twin, Sugar Twin), one of the oldest discovered artificial sweeteners, was discovered in 1878. It raised a lot of concerns about thirty years ago when a study found that saccharin consumption was linked to stomach cancer. Saccharin was then banned by the FDA. However, in 2001, the National Institutes of Health (NIH) removed saccharin from the list of potential carcinogens. Saccharin is still widely used today.
- Aspartame (brand names: NutraSweet, Equal) was approved as a general purpose sweetener by the FDA in 1996. The safety of aspartame was challenged due to some concerns over its long-term possible carcinogenicity [16]. The only unsafe use would be for individuals with phenylketonuria (PKU), which is a genetic disorder causing a baby to have very low levels of the enzyme phenylalanine hydroxylase (PAH). This enzyme is necessary to convert the amino acid phenylalanine into tyrosine, another amino acid. Aspartame is not recommended for baking because it often breaks down at high temperature and loses its sweetness. Because of this issue, aspartame has lost a good share of the artificial sweeteners market to sucralose, which does not lose its sweetness when heated.

Aspartame was discovered by accident by Jim Schlatter, a chemist at G.D. Searle in 1965. Jim Schlatter was working on drugs for the treatment of gastric ulcers when he spilled some aspartyl-phenylalanine on his hand. He later licked his finger and noticed the sweet taste of the compound, which later became aspartame. Aspartame is the methyl ester of the dipeptide of the natural amino acids L-aspartic acid and L-phenylalanine. There are four possible diastereoisomers for aspartame but aspartame is the only one having sweetening properties. The taste of aspartame would not have been predictable based on its amino acid constituents.

Aspartame is made of two amino acids that are chiral. The chemical synthesis starts with a racemic mixture, which contains equal quantities of both isomers of the two amino acids. The L isomers of both phenylalanine and aspartic acid are the only desired ones. The other isomers must be removed, therefore becoming waste.

The first step in the synthesis of aspartame is the ester synthesis where L-phenylalanine reacts with methanol in the presence of hydrochloric acid. The product is the methyl ester of phenylalanine (Scheme 6.4a). The second step is the amide synthesis, which is the reaction of the methyl ester of phenylalanine with aspartic acid (Scheme 6.4b).

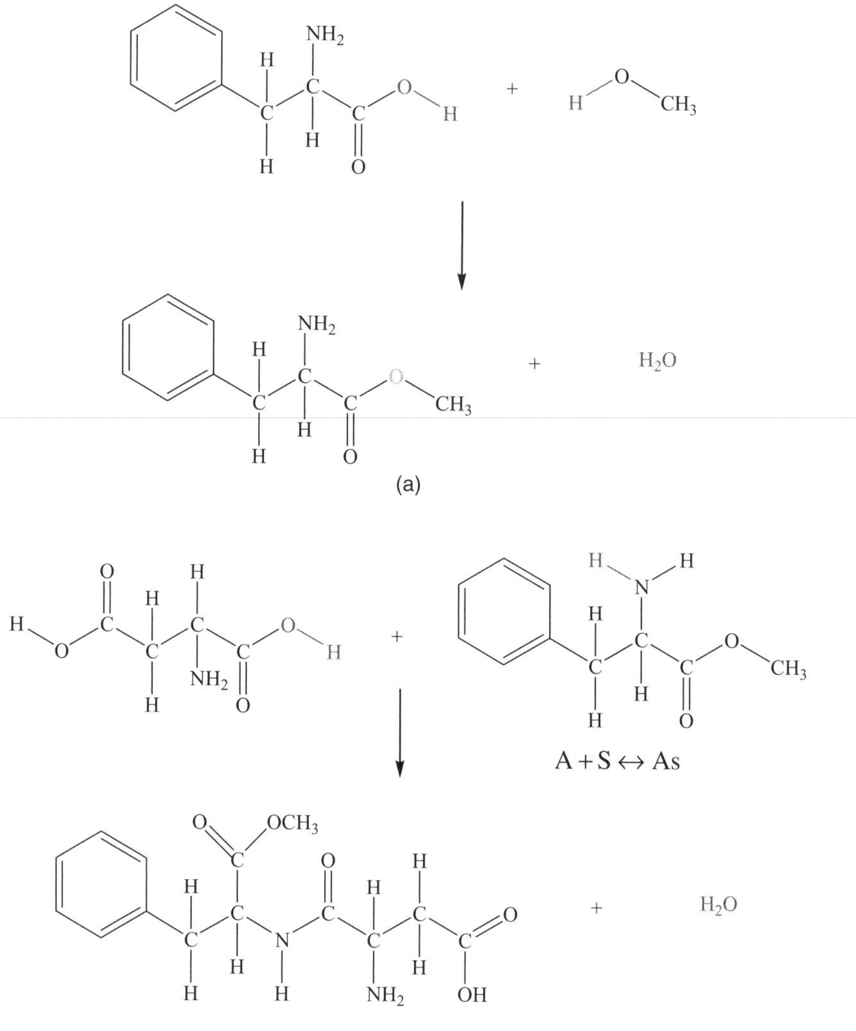

(a)

$A + S \leftrightarrow As$

(b)

Scheme 6.4. (a) First step in the synthesis of aspartame and (b) second step in the synthesis of aspartame.

Due to the necessity to resolve a racemic mixture, the traditional chemical synthesis of L-aspartic acid is very costly. An alternative would be to use a chiral catalyst that would selectively synthesize the L-enantiomer or one that would eliminate the D-enantiomer. Some of the most useful chiral catalysts provided by nature are enzymes.

There is a one-step enzymatic conversion to produce aspartic acid that consists of reacting fumaric acid with the aspartase from a bacteria, *Bacillus subtilis*, strain ASP-4, NRRL B-15536. The loss of enzyme activity can be inhibited by adding a source of ammonium ions and fumaric acid in the fermentation process [17].

6.3   REACTION ENGINEERING

Reactions are conducted in reactors, vessels designed to allow the reactants to come into contact under specific conditions. In order to determine the size of a reactor needed to carry out a specific reaction, we use a material balance equation, written in maximum generality as follows:

Flow rate entering the system + Generation or depletion due to chemical reaction
   + Accumulation or depletion = Flow rate leaving the system

The engineer who wants to determine the size of the vessel needs to use symbols in order to solve specific examples of this general mass balance. Written symbolically, we get the following very general equation:

$$F_{i,0} + \int r_j dV + \frac{dN_i}{dt} = F_i \qquad (6.75)$$

where

$F_{i,0}$ = Flow rate of component i entering the reactor (mol/time)
  F_i = Flow rate of component i leaving the reactor (mol/time)
$\int r_j dV$ = Rate of formation of component i due to reaction j (mol/ time), and r_j is the
           reaction rate (mol/vol·time)
   V = Volume of the reactor (vol)

  This is a very complicated expression that contains both a derivative and an integral. The integral term is included to account for the possibility that the reactor is not well mixed, and the rate of reaction will then be an "average" based on the concentration everywhere within the reactor. The differential term accounts for the reality that certain components may be accumulated within the reactor, depending on the design of the system.
  While it is true that this is a very difficult mathematical expression, the good news is that we very rarely need to work with the full equation to complete our basic reactor design. Through the next few sections, we consider several ideal reactors, where simplification of the general material balance equation is possible.

6.3.1   Batch Reactor

The batch reactor, shown schematically in Figure 6.10, is comparable to a batch process in that there is no flow of material into or out of the system. The components are placed in the reactor at the beginning of the process and are removed from the reactor at the end. Thus,

$$F_{j,0} = F_j = 0$$

Figure 6.10. Picture of a batch reactor [18].

Recall that F_i corresponds to the flow rate of the reactant or product. We also assume that the contents of the reactor are well mixed. This simplifies the calculation of the rate:

$$\int r_j \, dV = r_j V \tag{6.76}$$

Substituting these assumptions into the material balance provides

$$\frac{dN_j}{dt} = r_j V \tag{6.77}$$

We can rearrange this equation, and rewrite the number of moles in terms of the concentration and reactor volume, $N_j = C_j V$. Then, for a constant volume reactor, we find

$$r_j = \frac{1}{V} \frac{d}{dt}(C_j V) = \frac{dC_j}{dt} \tag{6.78}$$

We can put this equation in terms of the conversion by noting that

$$N_j = N_{j,0} - N_{j,0} X_j$$

Thus,

$$\frac{dN_j}{dt} = -N_{j,0}\frac{dX_j}{dt} \tag{6.79}$$

which provides, upon substitution,

$$N_{j,0}\frac{dX_j}{dt} = -r_j V \tag{6.80}$$

and for a constant volume process

$$C_{j,0}\frac{dX_j}{dt} = -r_j \tag{6.81}$$

In most cases, we are interested in either the time needed to achieve a particular conversion, or the conversion that will be obtained after a specific reaction time. Highlight 6.8 provides an application example related to the production of ethanol.

Highlight 6.8 *Production of Ethanol from Waste Corn Cobs*

Ethanol can be produced from waste corn cobs through the conversion of the cellulose in the cob under fermentation with yeast. At an initial concentration of 100 g/L, the rate expression is shown to be first order, with a rate constant of $0.0131\,h^{-1}$. The reaction is to be performed in a batch reactor with a reaction time of 24 h. If 1000 g of cellulose is used in the initial batch, what is the final concentration of ethanol in solution?

Assume the reaction stoichiometry as

$$C_6H_{12}O_6 \rightarrow 2C_2H_5OH + 2CO_2$$

Solution:

We start by calculating the conversion of cellulose that can be achieved in one day. This can be obtained by solving the material balance equation for the batch reactor

$$C_{A,0}\frac{dX_A}{dt} = -r_A$$

The rate is a function of conversion, so we need to separate variables and integrate. Rewriting the rate expression in terms of conversion provides

$$-r_A = kC_{A,0}(1-X)$$

Substituting into the mass balance expression and rearranging provides

$$C_{A,0} \frac{dX_A}{dt} = -r_A = kC_{A,0}(1-X)$$

or

$$\frac{dX}{1-X} = k\,dt$$

Now we can integrate both sides, noting that $X=0$ at $t=0$:

$$\int_0^{X_f} \frac{dX}{1-X} = k \int_0^{t_f} dt$$

$$-\ln(1-X_f) = kt_f$$

Since we know $t=24$ h, we rearrange to solve for X, and then substitute known values and solve:

$$X = 1 - e^{-kt}$$
$$= 1 - e^{-(0.0133\,h^{-1})(24\,h)}$$
$$= 0.27$$

We start with 1000 g of cellulose, so we convert 270 g in the allotted time. Based on the assumed molecular formula of cellulose, the molecular weight is 180 g/mol, so 270 g corresponds to

$$N_{converted} = \frac{(270\,g)}{(180\,g/mol)} = 1.5\,mol$$

From the stoichiometry of the reaction, 1.5 mol of converted cellulose produces 3 mol of ethanol. The initial concentration was 100 g/L, so 1000 g of cellulose corresponds to 10 L of solution. As a result, the final concentration of ethanol is 0.3 mol/L.

6.3.2 Continuous Stirred Tank Reactor

A continuous stirred tank reactor (CSTR) looks like a batch reactor, in that material enters the reactor, resides in the reactor for some time, and then leaves the reactor (Figure 6.11). The difference is that, in this case, the reactor is operating continuously, with new feed continually entering the reactor and products continually being removed. We assume that the contents of the reactor are well mixed, which means that the concentration everywhere within the reactor is the same. Thus, the

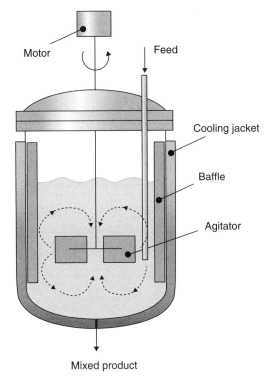

Figure 6.11. A CSTR used in a fermentation reaction [19].

material entering the reactor is immediately converted to the concentration within the reactor and the concentration of material leaving the reactor is the same as that inside.

Now, we can develop the material balance equation starting from the general balance equation given earlier, noting that there is no accumulation within the reactor (this system operates at steady state).

$$F_j - F_{j,0} = \int r_j dV \qquad (6.82)$$

Since the composition of the fluid inside the reactor is identical everywhere, the reaction rate must be the same everywhere. We replace the integral with a simple multiplication, which provides

$$F_j - F_{j,0} = r_j V \qquad (6.83)$$

Now, solving for the reactor volume provides

$$V = \frac{F_{j,0} - F_j}{-r_j} \qquad (6.84)$$

We frequently prefer to define reaction rates in terms of concentration. In order to make this transformation, we need to define the *mean residence time*, which is the average length of time that an element of fluid remains within the reactor. The residence time can be calculated from the volumetric flow rate and the volume of the reactor.

$$\tau = \frac{V}{V} \tag{6.85}$$

Now, we write the molar flow rate as $F_j = C_j\, V_0$ and the volume as $V = \tau V_0$, which provides the required residence time as a function of concentration

$$\tau = \frac{(C_{j,0} - C_j)}{-r_j} \tag{6.86}$$

Finally, we can rewrite the equation in terms of the reactant conversion

$$\tau = \frac{(C_{j,0} X_j)}{-r_{j,\text{exit}}} \tag{6.87}$$

or

$$V = \frac{F_{j,0} X_j}{-r_{j,\text{exit}}} \tag{6.88}$$

in which the use of the term $-r_{j,\text{exit}}$ reminds us that the reaction rate is constant everywhere within the reactor and equal to the rate at the reactor exit. Highlight 6.9 gives an example of how to calculate residence time.

Highlight 6.9 demonstrates that the conversion for two CSTRs in series is always greater than that of one CSTR of equal size. This is because of the way a CSTR operates. Remember, the concentration inside the tank is the same everywhere, and the same as the exit concentration. So if we break the tank into two, then the concentration within the first CSTR is higher, and the reaction rate is higher because it is proportional to the concentration. We could get even better conversion if we used three CSTRs connected in series, and likewise continue to improve as we add more (and smaller) reactors. We can generalize the result to any number of equal-sized CSTRs connected in series and obtain a single equation relating the residence time to the conversion. We find

$$X = 1 - \frac{1}{(1 + k\tau)^n} \tag{6.89}$$

if the reaction can be described by first order kinetics. While we could improve the performance using many consecutive reactors, this is not practical because of the need for many multiple vessels for the reactions and pumps needed to move the fluid from one reactor to the next.

Highlight 6.9 Oxidation of Pentane

The oxidation of pentane is occurring in a continuous stirred tank reactor. Under these conditions, the rate of the reaction is given by $-r_p(\text{mol/L·min})=0.1\,C_p$; independent of oxygen. What is the reactor residence time required to achieve 60% conversion?

Solution:

We start from the material balance equation, written in terms of residence time and conversion:

$$\tau = \frac{C_{p,0}X_p}{-r_{p,\text{exit}}}$$

and now substitute in the rate expression, also written in terms of conversion,

$$\tau = \frac{C_{P,0}X_P}{C_{P,0}(0.1\,\text{min}^{-1})(1-X_P)} = \frac{X_P}{(0.1\,\text{min}^{-1})(1-X_P)}$$

The conversion that must be obtained is 60%, so substitution in this equation provides

$$\tau = \frac{0.6}{(0.1\,\text{min}^{-1})(1-0.6)}$$
$$= 15\,\text{min}$$

Suppose we wanted to run the same reaction in two identical CSTRs connected in series. What residence time in each CSTR would be required? Would it be 7.5 min in each? In order to determine the answer, let's assume that we have two 30 L reactors connected in series, so the volumetric flow rate needs to be 4 L/min to achieve a residence time of 7.5 min in each reactor. Given this condition, let's calculate the conversion.

Previously, we showed that the conversion and the residence time in a CSTR could be related according to

$$\tau = \frac{X}{k(1-X)}$$

We can now rearrange this equation and solve for X to find

$$X = \frac{k\tau}{1+k\tau}$$

Substituting the given values of the rate constant and residence time, we find the conversion from each reactor can be obtained as

$$X = \frac{k\tau}{1+k\tau}$$

$$= \frac{(0.1\,\text{min}^{-1})(7.5\,\text{min})}{1+(0.1\,\text{min}^{-1})(7.5\,\text{min})} = 0.429$$

So how much material is remaining in the stream leaving reactor 1? If we started with 1 mol/L of the reactant, then 0.571 mol/L remains at the end. This is also the amount entering reactor 2. So 42.9% conversion from this reactor then provides an exit composition of

$$\begin{aligned} C_A &= C_{A,0}(1-X) \\ &= (0.571\,\text{mol/L})(1-0.429) \\ &= 0.326\,\text{mol/L} \end{aligned}$$

which corresponds to an overall conversion of 0.674. This is higher than the conversion obtained from a single CSTR with twice the residence time.

6.3.3 Plug Flow Reactor (PFR)

A reaction can also occur as a fluid moves along a length of heated pipe. We assume that the fluid moves in a plug and that the plug of fluid moves without mixing with any other fluid elements. As the plug moves along the tube, reaction occurs. Fresh feed continuously enters at one end of the tube, and product is continuously removed from the other end of the tube (Figure 6.12).

Now we apply the general mole balance, assuming that the reactor operates at steady state,

$$F_j = F_{j,0} + \int_V r_j dV \tag{6.90}$$

or, in terms of a differential reactor element,

$$-r_j dV = F_j - (F_j + dF_j) \tag{6.91}$$

which simplifies to give

$$\frac{dF_j}{dV} = r_j \tag{6.92}$$

Figure 6.12. A tubular (plug flow) reactor inside a tube furnace [20]. (Courtesy of Robert Hesketh, Rowan University, Glassboro, NJ.)

Now, we write the molar flow rate as $F_j = C_j V$ and the volume as $V = \tau V$, which is substituted into the equation to give

$$\frac{d(C_j V)}{d(\tau V)} = \frac{dC_j}{d\tau} = r_j \tag{6.93}$$

where the last expression requires that the volumetric flowrate V be constant along the length of the reactor. Note that this last result is identical to the result obtained for the batch reactor, except that the time is now expressed as a residence time, rather than process time.

Because of the similarity between the PFR and the batch reactor, we can by analogy develop an expression of the material balance, which is based on conversion:

$$C_{j,0} \frac{dX_j}{d\tau} = -r_j \tag{6.94}$$

or in integral form, and in terms of the reactor volume,

$$F_{j,0} \int_0^X \frac{dX}{-r_j} = V \tag{6.95}$$

Highlight 6.10 provides another example about residence time required for conversion of pentane.

If we compare Highlights 6.9 and 6.10, we see that we have an identical reaction with an identical reaction rate. However, the residence time required in the

Highlight 6.10   Oxidation of Pentane

The oxidation of pentane is occurring in a plug flow reactor. Under these conditions, the rate of the reaction is related to the composition according to $-r_P=0.1$ C_P (mol/L·min)—independent of oxygen. Calculate the residence time required to achieve 60% conversion (the half-life on the reactant).

Solution:

We start from the material balance equation and substitute the rate expression given by the problem statement:

$$\tau = C_{P,0} \int_{0}^{X=0.6} \frac{dX_P}{0.1C_P}$$

and recognize that $C_P = C_{P,0}(1-X_P)$, which gives, after substitution,

$$\tau = \int_{0}^{X=0.6} \frac{dX_P}{0.1(1-X_P)}$$

Integration provides

$$\tau = \frac{-\ln(1-X_P)}{0.1\,\mathrm{min}^{-1}}$$

and after substitution of $X_P=0.6$ provides the answer, $\tau=9.16\,\mathrm{min}$.

plug flow reactor is less than that required in the CSTR, or even in two identical CSTRs connected in series. The reason is again the decreasing concentration as the reaction proceeds, leading to a continually decreasing reaction rate. At all points in the reactor, the concentration is the highest, and thus the reaction rate is maximized.

Although we looked at only a few specific types of reactor systems, the green chemist and green engineer have many different variations on these ideal systems that can be used to maximize the performance of the reactor. For example, it is possible to use microwave energy to add heat to a reactor, and because of the penetration of the microwaves, this delivers energy directly to the reactants rather than the vessel. High shear systems can add large amounts of energy and provide good mixing in viscous systems. There are many vessel types that allow for contacting between species in a gas phase with those in a liquid phase. In all cases, the purpose is to maximize the concentration of the reacting species in the phase in which the reaction will occur, and to deliver heat energy to the location of the reaction in an efficient manner. Analysis of the reactor system often follows as illustrated in the three ideal situations considered above.

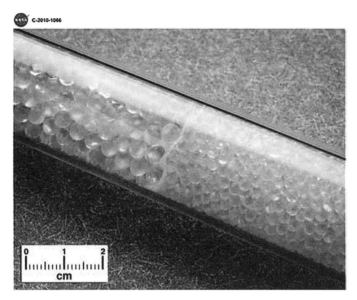

Figure 6.13. Photograph of a packed bed reactor, in which fluid enters at the top and flows down through the catalyst bed [21].

6.3.4 Multiphase Reactor Design

There are numerous other types of reactors that can be used to carry out a chemical reaction. The choice of reactor depends on the physical situation, and the economics of the process. In most cases, design of the reactor can be completed through a variation of that used for one of the "basic" reactor profiles.

First, let's briefly describe the packed bed reactor—the analog of the PFR for heterogeneously catalyzed reactions. The packed bed is essentially a tube in which catalyst particles are packed (Figure 6.13).

As gas flows across the catalyst particles, reactant is converted into product. In the plug flow reactor, the concentration of the reactant varied along the length of the reactor—here, the length (or volume) is replaced with the weight of the catalyst. As a result, the material balance for a packed bed reactor provides

$$\frac{dF_j}{dW} = r_j \tag{6.96}$$

or in integral form,

$$F_{j,0} \int_0^X \frac{dX}{-r_j} = W \tag{6.97}$$

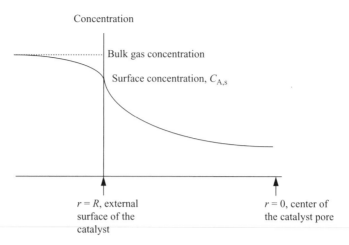

Concentration

Bulk gas concentration

Surface concentration, $C_{A,s}$

$r = R$, external
surface of the
catalyst

$r = 0$, center of
the catalyst pore

Figure 6.14. Schematic diagram of the concentration profile in a catalytic system.

This equation is essentially identical to that for the plug flow reactor, except the volume of the reactor is now replaced with the catalyst weight. If the reaction rate is defined in terms of catalyst weight (as would be common for a heterogeneously catalyzed reaction), then all of the units cancel and the design of the packed bed reactor is completely analogous to that of the PFR.

Slurry reactors are three-phase reactors used for contacting a gas, a liquid, and a solid. They are often used during hydrocarbon processing, such as in hydrogenation. The gaseous reactant (hydrogen) must diffuse through the liquid layer (usually the hydrocarbon) to the surface of the solid catalyst, where the reaction can occur. The reactor can be arranged so that the liquid and gas pass in a continuous flow arrangement, maintaining the catalyst particles within the reactor through the use of a screen or filter. In this case, the reactor is designed using the basic equations of a CSTR. These reactors provide good temperature control and are useful for cases in which the catalyst cannot be palletized. However, the need to separate the catalyst from the liquid phase can cause problems of plugging, sometimes making these types of systems less desirable.

Whenever we consider a reaction between species in different phases, we must confront the issue of mass transfer. Because of the presence of the phase boundary, the concentration of the species at the point where the reaction occurs is different from the concentration of the species in the bulk fluid. The concentration profile is shown schematically in Figure 6.14. Since the rate of reaction is dependent on the concentration of the reacting species, we see that the effect of mass transfer is to decrease the rate below that which would be expected based on the bulk fluid concentration.

The concentration profile outside the catalyst particle is controlled by the principles of convective mass transfer (or in a packed bed, interparticle mass transfer). Although the details of convective transport are beyond the scope of this course, it is

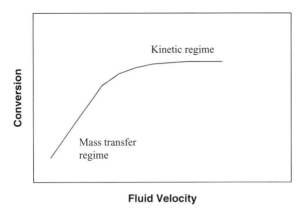

Fluid Velocity

Figure 6.15. Schematic description of the effect of external mass transfer on conversion.

sufficient to know that increasing the fluid velocity around the outside of the particle will increase the rate of mass transfer.

Usually, we choose to run the reaction in such a way that the external mass transfer rate is sufficiently high and the effect of external mass transfer may be neglected. The standard test for determining if external mass transfer is important is to vary the flow rate through the reactor, at a constant residence time (it is necessary to keep the residence time constant so that the kinetically controlled conversion will remain fixed). This is achieved by varying the flow rate and the weight of catalyst simultaneously. If the change in flow rate does not affect the measured conversion, then external mass transfer can be neglected. A typical graph of the experimental result is shown in Figure 6.15.

Internal mass transfer is more difficult to work with, since the concentration of the reactant decreases as you move further within the catalyst pore. Although it is possible to calculate the concentration at every point within the pore, the mathematics quickly becomes very difficult. What is normally done is to define a term that accounts for the inhibition of the rate due to internal mass transfer. This is termed the effectiveness factor, which is defined as

$$\eta = \frac{\text{actual rate of the reaction}}{\begin{array}{l}\text{rate of the reaction that would result if the}\\ \text{entire pellet were exposed to the reactant}\\ \text{at the surface concentration}\end{array}} \qquad (6.98)$$

Based on this definition, we write, for a first order reaction,

$$-r_{\text{app}} = \eta k_1 C_A \qquad (6.99)$$

which provides the apparent rate of reaction within the particular catalyst pellet.

To determine if internal mass transfer is important, one performs the reaction using several different sized catalyst pellets, all with the same surface area per unit volume. As the catalyst particle size increases, the importance of internal mass transfer also increases. If the particles are small enough, internal mass transfer will become unimportant. As shown in Figure 6.14, a graph is obtained which is similar to that obtained for external mass transfer, except the kinetic regime is now found to occur at small catalyst particles.

6.4 SUMMARY

The green chemist has many tools available to decrease the environmental footprint of a selected reaction. By adjusting the reaction conditions, it is possible to reduce the amount of undesired product. Adding a catalyst can reduce the temperature of the reaction or improve its selectivity. There is a choice of homogeneous or heterogeneous catalyst, and there are many types of reactors that can be used. The successful green chemist will put all of these tools together to create a reaction system that can be scaled for production while keeping attention on the waste produced.

REFERENCES

1. Wang, H.; Yang, L.; Scott, S.; Pan, Q.; Rempel, G.L. *J. Polym. Sci. Part A: Polym. Chem.*, **2012**, *50*, 4612–4627.

2. Kim, M.-I.; Kim, D.-K.; Bineesh, K.V.; Kim, D.-W.; Selvaraj, M.; Park, D. W. *Catalysis Today*, **2013**, *200*, 24–29.

3. Natarajan, S.; Olson, W. W.; Abraham, M. A. *Ind. Eng. Chem. Res.*, **1999**, *39*(8), 2837.

4. http://en.wikipedia.org/wiki/File:Activation_energy.svg.

5. Ajjou, A. N.; Alper, H. *J. Am. Chem. Soc.*, **1998**, *120*(7), 1466–1468.

6. Zhang, Y.; Zhang, H. B.; Lin, G. D.; Chen, P.; Yuan, Y. Z.; Tsai, K. R. *Appl. Catal., A*, **1999**, *187*(2), 213–224.

7. Beck, J. S.; Vartuli, J. C.; Roth, W. J.; Leonowicz, M. E.; Kresge, C. T.; Schmitt, K. D.; Chu, C. T. W.; Olson, D. H.; Sheppard, E. W.; McCullen, S. B.; Higgins, J. B.; Schlenker, J. L. *J. Am. Chem. Soc.*, **1992**, *114*(27), 10834–10843.

8. Hanson, B. E. *Coord. Chem. Rev.*, **1999**, *185–186*, 795–807.

9. Herrmann, W. A.; Kohlpainter, C. W. *Angew. Chem. Int. Ed. Engl.*, **1993**, *32*(11), 1524–1544.

10. Herrmann, W. A.; Kohlpainter, C. W.; Bahrmann, H.; Konkol, W. *J. Mol. Catal.*, **1992**, *73*(2), 191–201.

11. Fish, R. H. *Chem.—Eur. J.*, **1999**, *5*(6), 1677–1680.

12. The Presidential Green Chemistry Challenge Awards Program award entries and recipients summary, **2006**, p. 6.

13. www.merck.com.

14. http://www.diabetes.org/about-diabetes.jsp.

15. http://www.diabetic-lifestyle.com/articles/nov02_whats_1.htm.

16. http://www.ific.org/publications/factsheets/lcsfs.cfm.

17. www.freepatentsonline.com/4569911.html.

18. http://en.wikipedia.org/wiki/File:Batch_reactor.2.jpg.

19. http://en.wikipedia.org/wiki/File:Agitated_vessel.svg.

20. http://encyclopedia.che.engin.umich.edu/Pages/Reactors/PFR/PFR.html.

21. http://encyclopedia.che.engin.umich.edu/Pages/Reactors/PBR/PBR.html.

7

THERMODYNAMICS, SEPARATIONS, AND EQUILIBRIUM

Thermodynamics is the branch of science that deals with the movement of energy between materials, very frequently fluids (liquids and gases). There are two essential elements: (1) the state of the material, in other words, its physical characteristics such as temperature or specific voume; and (2) the processes that the materials that make up the system may undergo as energy is exchanged.

A critical component of thermodynamics is the definition of the system. The system is usually the materials contained within a physical space, such as a unit of equipment. However, it can also be a subset of the materials within that physical space. The system will exchange energy with the surroundings, all of the material located outside the system.

7.1 IDEAL GASES

In 1662, Robert Boyle discovered that if he placed an ideal gas into a container, as shown in Figure 7.1, and was very careful to maintain a constant temperature on the system, the pressure on the gas (p) would increase inversely to the volume of the cylinder (V). In other words, moving the piston such that the volume would be

Green Chemistry and Engineering: A Pathway to Sustainability,
Anne E. Marteel-Parrish and Martin A. Abraham.
© 2014 American Institute of Chemical Engineers, Inc. Published 2014 by John Wiley & Sons, Inc.

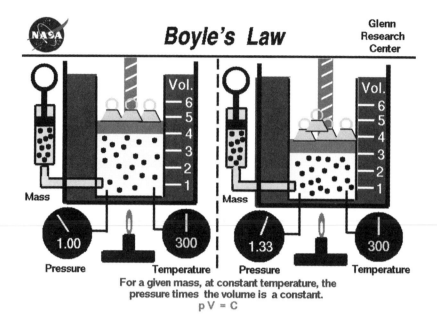

Figure 7.1. A pictorial description of Boyle's law [1].

reduced to half of the original volume would result in a doubling of the pressure. One can envision that the molecules of the ideal gas, which are constantly moving within the cylinder, would strike the wall of the cylinder more frequently. The number of strikes of the molecules on the wall of the cylinder corresponds to the pressure of the gas. The operation described in Figure 7.1, and making up the experiment that leads to Boyle's law, would be termed an isothermal compression.

Mathematically, Boyle's law is best described as

$$p_1 V_1 = p_2 V_2 \qquad (7.1)$$

A similar relationship was found by Jacques Charles in the 1780s, when he determined that the volume of a gas (V) increased as the temperature of the gas (T) also increased. When evaluated using an appropriate thermodynamic temperature scale (which did not exist at the time of Charles' discovery), it was determined that the temperature and volume increased proportionally, as shown in Figure 7.2.

We can write Charles' law as

$$\frac{V_1}{T_1} = \frac{V_2}{T_2} \qquad (7.2)$$

By combining Boyle's law with Charles' law, we can develop a relationship that describes the properties of the ideal gas through changes in the pressure (p), volume (V), and temperature (T):

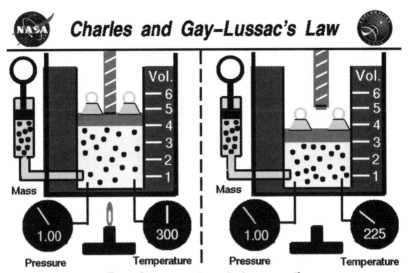

Figure 7.2. A pictorial description of Charles' law [2].

$$p_1 \frac{V_1}{T_1} = p_2 \frac{V_2}{T_2} \qquad (7.3)$$

This combined relationship serves to relate the properties of a fluid in a closed system, but does not permit the absolute calculation of any specific property, since the amount of the substance present is not known. For that, we require Avogadro's law, developed in 1811 by Amedeo Avogadro, which states that the number of molecules (n) in a specific volume of fluid (V) at a given pressure (P) and temperature (T) is always the same. Together, this provides the ideal gas law, which can be used to relate the properties of the ideal gas under any conditions.

$$PV = nRT \qquad (7.4)$$

The ideal gas law holds under conditions in which the molecules behave as independent, randomly moving particles—that is, there are no molecular interactions. Many gases can be modeled as an ideal gas, particularly at high temperature and low pressure. R is the universal gas constant, which identifies the internal kinetic energy of a mole of an ideal gas at the indicated temperature, and is given in a variety of units in Table 7.1 (J = joules; K = kelvin; Pa = pascal; atm = atmosphere; cal = calorie; lbmol = the number of moles on an English scale; °R = absolute temperature scale in English units; BTU = British thermal unit; psia = pounds per square inches, absolute).

One place where the ideal gas law is found to be useful in processes is in the definition of flow rates identified at *standard conditions* (or sometimes, standard

TABLE 7.1. Values of R Constants

7.314 J/mol K	
7.314 m3 Pa/mol K	0.7302 ft3 atm/lbmol °R
82.05 cm3 atm/mol K	10.73 ft3 psia/lbmol °R
1.987 cal/mol K	1.987 BTU/lbmol °R

Highlight 7.1   Calculation of Flow Rate

Air enters a pipe at a rate of $10\,\text{ft}^3/\text{min}$, at $200\,°\text{F}$ and $2\,\text{atm}$. The air leaves the pipe at standard conditions ($60\,°\text{F}$ and $1\,\text{atm}$). Determine the flow rate of the exiting air.

Solution:

We apply the ideal gas law twice—first to determine the molar flow rate of the entering air, and then a second time to calculate the volumetric flow rate at standard conditions.

$$\text{Molar flow rate: } n = \frac{PV}{RT} = \frac{(2\,\text{atm})(10\,\text{ft}^3/\text{min})}{(0.7302\,\text{ft}^3\text{atm}/\text{lbmol}\,°\text{R})(660\,°\text{R})} = 0.041\,\text{lbmol}/\text{min}$$

Volumetric flow rate at standard conditions ($1\,\text{atm}$, $60\,°\text{F}$) is calculated as

$$V = \frac{nRT}{P} = \frac{(0.7302\,\text{ft}^3\text{atm}/\text{lbmol}\,°\text{R})(520\,°\text{R})(0.041\,\text{lbmol}/\text{min})}{(1\,\text{atm})} = 15.6\,\text{scfm}$$

temperature and pressure, STP). These are defined as $273\,\text{K}$ (or $492\,°\text{R}$) and $1\,\text{atm}$. There is also an "industry standard" referred to as normal temperature at $60\,°\text{F}$ (F = Fahrenheit). With this definition, we can define volumetric flow rates as standard liters per minute or standard cubic feet per minute (scfm) or standard cubic centimeters per minute (sccm). A slpm is a liter per minute at STP. The ideal gas law provides a mechanism for converting actual flow rates to standard flow rates, as described in Highlight 7.1.

The ideal gas law applies to both pure gases and gas mixtures. In order to apply the ideal gas law to a particular species in a gas mixture, we need to multiply both sides of the equation by the mole fraction (y_i) of the particular component, which provides

$$(y_i P)V = (y_i n)RT \tag{7.5}$$

Using this equation, we can define the partial pressure (p_i) as

$$p_i = y_i P \tag{7.6}$$

and see that the ideal gas law provides, for an individual species in a gas mixture,

$$p_i V = n_i RT \qquad (7.7)$$

If we sum over all of the species in the mixture, we obtain

$$\sum p_i = \sum \frac{n_i RT}{V} = \sum n_i \frac{RT}{V} = \frac{nRT}{V} = P \qquad (7.8)$$

Thus, we see that the sum of all of the species partial pressures provides the total pressure of the gas. This is often termed Dalton's law.

7.2 THE FIRST LAW OF THERMODYNAMICS

The first law of thermodynamics derives from the fundamental physical principle that energy can be neither created nor destroyed. We can move energy around in a system, but we must always be able to account for where it is and where it goes. Thus, if we are looking at any system, we can do an energy balance to relate the energy flows entering and leaving the system. The general energy balance equation is

$$\begin{array}{ccc} \text{Energy entering} - & \text{Energy leaving the} & = \text{Accumulation in} \\ \text{the system} & \text{system} & \text{the system} \end{array}$$

Figure 7.3 gives a general view of energy flows entering and leaving a system.

Notice that energy can enter a system either through a flowing stream, or through work being done on the system, or heat being added to the system. Let's look at each of the energy terms in more detail.

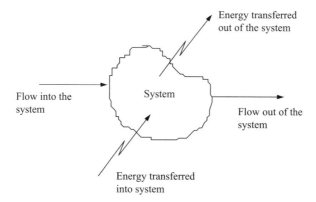

Figure 7.3. General view of energy flows.

1. **Flow into or out of the system.** This is the energy that is brought into the system by the flow of material, or that leaves the system by the flow of material out of the system. This consists of three separate parts:

- *Kinetic energy* is the energy associated with motion and is given by

$$E_k = \frac{mv^2}{2g_c} \qquad (7.9)$$

where v is the velocity of the material and m is its mass. The constant g_c is included in this equation to allow for the conversion between the mass unit pounds (lb_m) and the use of pounds as a unit of force (lb_f). In English units $g_c = 32.2$ (lb_m ft/s²)/lb_f. In metric units, $g_c = 1$ (N m/s²)/J.

- *Potential energy* is the energy associated with the system not being in a state of mechanical equilibrium (e.g., a ball sitting at the top of a hill will tend to roll down the hill). The potential energy is given by

$$E_p = m\frac{g}{g_c}z \qquad (7.10)$$

where z is the height of the system above some defined reference point. Here, g is the acceleration due to gravity, which is 32.2. ft/s² or 9.81 m/s² in metric units. Note that g and g_c are very different and need to be used carefully to account for units.

- *Internal energy* is the latent energy of the species and is due to molecular motion. Internal energy is given the symbol U and is calculated relative to a defined standard state.

For most of the systems that a chemist is concerned with, the changes in internal energy will far exceed those of kinetic or potential energy. While those terms can be important under special circumstances, we'll restrict further discussion to the internal energy and neglect the contributions of kinetic and potential energy components.

2. **Heat.** Heat is the flow of energy resulting from a temperature difference. Energy always transfers from a system at high temperature to a system at lower temperature. Heat is given the symbol Q and is defined as positive when heat is transferred into the system.

3. **Work.** Work is energy transferred through some physical process, such as mechanical agitation. Work is given the symbol W and is defined as positive when work is done on the system. The fundamental definition of work (from solid mechanics) being done on an object is the force on the object multiplied by the distance over which the force acts. In thermodynamics, we are interested in pressure (force per unit area) and volume (distance times area) so the work done by the system becomes

$$W = -\int P\,dV \qquad (7.11)$$

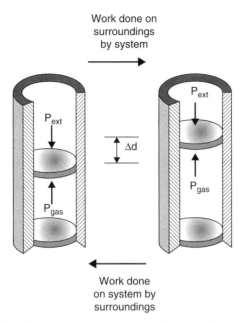

Figure 7.4. Description of work as evaluated through changes in pressure and volume [3]. (Courtesy of Carl Hepburn.)

Figure 7.4 describes the relationship of work to the pressure and volume of the fluid.

Now that we have identified all of the possible energy terms, we can write the energy balance equation appropriate for the chosen system.

7.2.1 Closed System

Let's start with the closed system, which is defined as one in which there is no movement of fluid. Heat can transfer into the system from the surroundings, and it's possible to do work on the system, but energy cannot enter or leave the system with the fluid, since there is no movement of fluid. Thus, any energy entering or leaving the system must be related to heat (Q) or work (W), and the energy balance is written as

$$U_{final} - U_{initial} = Q + W \tag{7.12}$$

In other words, the change in the internal energy of the system is strictly related to the amount of energy that is added to the system by heat transfer or mechanical work. Often, one will use the symbol Δ to describe the difference between two points, in which case the first law of thermodynamics for a closed system is written as

$$\Delta U = Q + W \tag{7.13}$$

7.2.2 Open System

In working with an open system, the problem is somewhat more complex. First, let's assume that the system is operating at steady state. Then, we can derive an equation nearly identical to the above equation for a closed system, except that we need to account for all of the flowing streams. This gives

$$\underbrace{\sum m_j \hat{U}_j}_{\substack{\text{output} \\ \text{streams}}} - \underbrace{\sum m_j \hat{U}_j}_{\substack{\text{input} \\ \text{streams}}} = Q + W \tag{7.14}$$

where $\hat{U}_j$ is the internal energy per unit mass of the flowing stream. One difficulty in solving the balance for an open system is that the fluid entering the system is doing work on the system, while the fluid leaving the system is having work done on it by the system. Thus, we need to account for these flow work terms. Following our previous definition for work, we will now define two different types of work—shaft work (W_s) and flow work (W_f). Then the total work is

$$W = W_f + W_s \tag{7.15}$$

According to the previous definition of work as being associated with the force exerted on an object, we see that the flow work term is given by

$$W_f = P_{in} V_{in} - P_{out} V_{out} \tag{7.16}$$

which accounts for the net work done on the system. Substitution into the energy balance provides

$$\underbrace{\sum m_j (\hat{U}_j + P_j V_j)}_{\substack{\text{output} \\ \text{streams}}} - \underbrace{\sum m_j (\hat{U}_j + P_j V_j)}_{\substack{\text{input} \\ \text{streams}}} = Q + W_s \tag{7.17}$$

Now, in thermodynamics, we define the *enthalpy* as

$$H = U + PV \tag{7.18}$$

which is substituted into the energy balance to provide

$$\underbrace{\sum m_j \hat{H}_j}_{\substack{\text{output} \\ \text{streams}}} - \underbrace{\sum m_j \hat{H}_j}_{\substack{\text{input} \\ \text{streams}}} = Q + W_s \tag{7.19}$$

Finally, if we use the symbol Δ to stand for the difference between total output and total input, we have

$$\Delta H = Q + W_s \tag{7.20}$$

which is the first law of thermodynamics for open systems.

7.3 IDEAL GAS CALCULATIONS

While the energy balance (or first law) is required to evaluate the energy changes of the system, the ideal gas law describes the relationship between the gas properties. For a closed system, the energy balance was written as

$$\Delta U = \delta q + \delta W \tag{7.21}$$

where δ is used to indicate a small change in the property. If we know the conditions under which the process change is implemented, we can calculate the amount of work required, or the heat added, to complete the change. We take advantage of the fact that the properties of the gas are independent of the process used to get this gas to the indicated conditions. In other words, the properties of the gas depend only on the state of the gas. However, the amount of energy required to accomplish the change (or the work) depends on the process used. We can calculate the energy requirements if we know the process, by substituting appropriate forms for work, heat, and internal energy, into our energy balance.

Let's suppose we are interested in evaluating the work associated with an isothermal compression. Since an ideal gas has no molecular interactions, the internal energy is a function of temperature only, and thus the energy balance reduces to

$$\Delta U = 0$$

or

$$\delta q = -\delta W \tag{7.22}$$

Next, we substitute our known definition for the work associated with the change in volume:

$$W = -\int P \, dV \tag{7.23}$$

For 1 mole of an ideal gas, $PV = RT$, and substitution of this relationship into the definition of work provides a relationship that can be integrated. We find

$$W = -\int P \, dV = -RT \int \frac{dv}{v} = -RT \ln \frac{V_2}{V_1} \tag{7.24}$$

Substituting the ideal gas relationship into this last expression, we can also obtain the work requirement in terms of the pressure change:

$$W = RT \ln \frac{P_2}{P_1} \qquad (7.25)$$

We can complete similar calculations to determine the work required for processes that occur under other conditions. The following example helps to illustrate this behavior.

What happens if the temperature is not constant? Our first law equation doesn't include the temperature explicitly, but we've said several times that the energy of the fluid is related to its temperature. What is that relationship? We define another property of the material, termed the heat capacity (C), which describes the change in the temperature (ΔT) of the fluid when it is heated. In mathematical terms, this provides

$$C = \frac{Q}{\Delta T} \qquad (7.26)$$

The units of the heat capacity are J/°C. The heat capacity is normally described based on the amount of the material. For example, chemists would use the specific heat capacity defined as

$$C = \frac{Q}{\Delta T \times m} \qquad (7.27)$$

In this case the units of the specific heat capacity are $J\,°C^{-1}\,g^{-1}$.

The molar heat capacity (C_m) is defined similarly by using

$$C_m = C \times M \quad (\text{with } M = \text{molecular weight}) \qquad (7.28)$$

The units of the molar heat capacity are $J\,°C^{-1}\,mol^{-1}$.

At the beginning of this section, we stated that the energy change depended on the process used to effect the change. As a result, there are two different types of heat capacity that can be defined. One assumes that the heating of the fluid occurs in a constant volume system, such as a closed vessel. Recall the first law for a closed system (Equation 7.13):

$$\Delta U = Q + W$$

If there is no work being done, then

$$\Delta U = Q \qquad (7.29)$$

From the definition of the heat capacity above, substitution provides the definition of the heat capacity at constant volume:

$$\Delta U = C_V \Delta T \qquad (7.30)$$

We can complete a similar analysis for an open system, one in which the pressure remains constant. For an open system, the first law is written as

$$\Delta H = Q + W \qquad (7.20)$$

which upon substitution provides the definition for the heat capacity at constant pressure:

$$\Delta H = C_p \Delta T \qquad (7.31)$$

Thermodynamic relationships reveal that the heat capacities at constant volume and constant pressure must be related, and it turns out that

$$C_P = C_V + R \qquad (7.32)$$

where R is once again the gas constant, usually expressed in units of J/mol K.

Using these relationships for an ideal gas and the first law of thermodynamics expressed for the correct type of system, it is possible to determine changes in pressure, temperature, and volume for a range of important processes (Highlight 7.2).

Highlight 7.2   Calculating the Energy Effects of Processes for Ideal Gases

Under many conditions, air can be considered as an ideal gas, with a heat capacity $C_p = \frac{7}{2}R$ (where R is again the universal gas constant, in this case, 1.986 BTU/lbmol °R) A compression operation is used to compress 100 lb of air from an initial condition of 1 atm and 70 °F to a final state of 3 atm and 70 °F through a reversible process in a closed system. Determine the work required and the heat transferred for each case:

(a) Isothermal compression.
(b) Heating at constant volume followed by cooling at constant pressure.
(c) Adiabatic compression followed by cooling at constant pressure.

Solution:

We will convert the mass quantity to a molar value (since the ideal gas law is in molar units):

$$n = \frac{100\,\text{lb}}{29\,\text{lb/lbmol}} = 3.45\,\text{lbmol}$$

The value of the gas constant, 1.986 BTU/lbmol °R, is used to determine

$$C_p = 3.5(1.986\,\text{BTU/lbmol}°R) = 6.95\,\text{BTU/lbmol}°R$$

(a) We derived the equations describing the work associated with isothermal compression previously, so we can simply plug in the data into the formula to calculate the work required.

$$W = nRT \ln\frac{P_2}{P_1}$$

$$= (3.45\,\text{lbmol})(1.986\,\text{BTU/lbmol}°R)(530°R)\ln\left(\frac{3\,\text{atm}}{1\,\text{atm}}\right)$$

$$= 4000\,\text{BTU}$$

(b) We consider the steps independently. For the first process (heating at constant volume), we find the final temperature using the ideal gas law:

$$\frac{T_2}{T_1} = \frac{P_2}{P_1}$$
$$T_2 = (530°R)(3\,\text{atm}/1\,\text{atm}) = 1590°R$$

For a constant volume process, $\Delta V = 0$; therefore, $W = 0$, so $\Delta U = Q$.

$$Q = \Delta U$$
$$= nC_V\Delta T = (3.45\,\text{lbmol})(4.97\,\text{BTU/lbmol}°R)(1590°R - 530°R)$$
$$= 18,150\,\text{BTU}$$

The second step involves cooling at constant pressure. In this case,

$$Q = \Delta H$$
$$= nC_p\Delta T = (3.45\,\text{lbmol})(6.95\,\text{BTU/lbmol}°R)(530°R - 1590°R)$$
$$= -25,420\,\text{BTU}$$

For the two-step process,

$$Q = Q_1 + Q_2 = 18,150\,\text{BTU} - 25,420,\text{BTU} = -7270\text{BTU}$$

Since $\Delta U = 0$ (for the process), $W = -Q = 7270\text{BTU}$.

(d) For adiabatic compression, $Q=0$, so the energy balance provides

$$dU = \delta W$$

which after substitution provides

$$C_V dT = -P dV$$

Now we substitute the ideal gas law and rearrange the expression to provide a solution that allows calculation of T_2 as a function of V_2:

$$\frac{T_2}{T_1} = \left(\frac{V_1}{V_2}\right)^{R/C_V}$$

And then substitute the pressure ratio in place of the volume ratio,

$$\frac{T_2}{T_1} = \left(\frac{P_2}{P_1}\right)^{\gamma-1/\gamma}$$

where we have made use of a shorthand to describe the ratio of the gas constants:

$$\gamma = C_P / C_V$$

This allows us to insert the pressure ratio, from which we can calculate the temperature obtained by the adiabatic compression:

$$\frac{T_2}{T_1} = \left(\frac{P_2}{P_1}\right)^{\gamma-1/\gamma}$$

$$T_2 = (590\,°R)\left(\frac{3\,\text{atm}}{1\,\text{atm}}\right)^{1.4-1/1.4} = 807.6\,°R$$

To calculate the work, we return to the energy balance for a closed system

$$dU = Q + W$$

and with $Q=0$, we see that the work of this process is simply the change in internal energy. Recall that $dU = C_V dT$ and evaluate the integral

$$\Delta U = C_V (T_2 - T_1)$$

from which we find

$$W = \Delta U$$
$$= nC_V \Delta T = (3.452 \, \text{lbmol})(4.97 \, \text{BTU/lbmol} \, ^\circ \text{R})(807.6 \, ^\circ \text{R} - 530 \, ^\circ \text{R})$$
$$= 4760 \, \text{BTU}$$

In order to determine the work on cooling, we will need to know the volume at both high and low temperature conditions. The ideal gas law provides

$$V = \frac{nRT}{P} = \frac{3.45 \, \text{lbmol}(0.7302^{\text{ft}^3} \, \text{atm/lbmol} \, ^\circ \text{R})(807.6 \, ^\circ \text{R})}{3 \, \text{atm}} = 678.2 \, \text{ft}^3$$

and

$$V = \frac{nRT}{P} = \frac{3.45 \, \text{lbmol}(0.7302^{\text{ft}^3} \, \text{atm/lbmol} \, ^\circ \text{R})(590 \, ^\circ \text{R})}{3 \, \text{atm}} = 495.4 \, \text{ft}^3$$

Finally, we calculate the work of cooling

$$W = -P\Delta V$$
$$= -(3 \, \text{atm})(495.4 \, \text{ft}^3 - 678.2 \, \text{ft}^3)\left(\frac{1 \, \text{BTU}}{0.3676 \, \text{ft}^3 \, \text{atm}}\right) = -1491.6 \, \text{BTU}$$

So that

$$W = W_1 + W_2 = 4760 \, \text{BTU} - 1492 \, \text{BTU} = 3268 \, \text{BTU}$$

It is helpful to summarize these results through the following table. Recall that, in all cases, $\Delta U = 0$ for the process, so $W = -Q$.

Process	Work (BTU)
(a) Isothermal compression	4000
(b) Heating ($\Delta V = 0$), cooling ($\Delta P = 0$)	7270
(c) Adiabatic compression, cooling ($\Delta P = 0$)	3268

7.4   ENTROPY AND THE SECOND LAW OF THERMODYNAMICS

The energy balance, also termed the first law of thermodynamics, describes how energy can be exchanged within or between processes, but it makes no statement about whether an exchange of energy is possible. For example, consider the heat exchange operation in Figure 7.5. The energy balance for this system provides

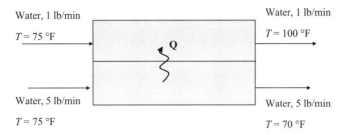

Water, 1 lb/min

$T = 75$ °F

Q

Water, 5 lb/min

$T = 75$ °F

Water, 1 lb/min

$T = 100$ °F

Water, 5 lb/min

$T = 70$ °F

Figure 7.5. An impossible heat exchange operation.

$$\Delta H = (mC_p\Delta T)1 + (mC_p\Delta T)_2 = 0 \qquad (7.33)$$

Substituting numbers, we find

$$\Delta H = (1\text{lb}/\text{min})\,(1\text{cal}/\text{lb °F})\,(25 \text{ °F}) + (5\text{lb}/\text{min})\,(1\text{cal}/\text{lb °F})\,(-5 \text{ °F})$$
$$= 0$$

So the energy balance is satisfied. But we know that the situation drawn in Figure 7.5 is impossible. What have we missed?

The second law of thermodynamics describes whether a particular energy exchange is possible. Specifically, recall that we previously indicated energy always flows from a higher energy state to a lower energy state. An alternate description would be that energy must always act in a way to increase the microscopic randomness of the system. We use this concept to define a new thermodynamic property called *entropy*:

> There exists a property called entropy, S, which is an intrinsic property of the system, functionally related to the measurable conditions of the system.

Now, noting that energy must always flow from higher states to lower states, we define the second law of thermodynamics in terms of our definition of entropy:

> The entropy change of any system, and its surroundings, considered together, is positive and approaches zero as the process more closely approximates a reversible process.

This form of the second law can be translated into a mathematical expression describing the entropy change of a process:

$$\Delta S \geq 0 \qquad (7.34)$$

What is meant by a reversible process? Let's look at a simple example. Suppose we heat a 1 lb block of copper ($C_p = 94$ kcal/lb °F) from 50 °F to 100 °F. In order to raise the temperature of the copper, the energy input, according to the first law, is

$$\Delta H = mC_p \Delta T = (1\text{lb})(94\text{ kcal/lb °F})(50\text{ °F}) = -4700\text{ kcal}$$

If we were to cool that same block of copper from 100 °F to 50 °F, the first law would then provide

$$\Delta H = mC_p \Delta T = (1\text{lb})(94\text{ kcal/lb °F})(-50\text{ °F}) = -4700\text{ kcal}$$

So the first law indicates that the energy change of heating is the same as the energy change of cooling. However, for each of these processes individually, the entropy must either increase or (if the process is reversible) remain the same. So, in order to recover the full 4700 kcal that was stored in the copper block when it was heated, the heat exchange processes must be reversible. A *reversible process* implies a situation in which the process can be run either forward or backward with the same energy effects.

In a real system with dissipative losses, it requires more than 4700 kcal of energy to increase the temperature of this copper block by 50 °F. Likewise, we cannot recover the full 4700 kcal of energy when the block is cooled back to 50 °F. This irreversibility results in energy losses that cannot be recovered.

For a reversible process, changes in entropy may be calculated as

$$dS = \frac{\delta Q\text{rev}}{T} \tag{7.35}$$

The use of entropy in solving problems requires a mathematical relationship for entropy as a function of temperature and pressure. We return to the first law,

$$dU = Q + W \tag{7.13}$$

and substitute for $Q = T\, dS$ and $W = -P\, dV$ to provide

$$dU = T\, dS - P\, dV \tag{7.36}$$

which is solved for dS:

$$dS = \frac{dU}{T} - \frac{P}{T}\, dV \tag{7.37}$$

Finally, noting that $dU = C_V\, dT$, and for an ideal gas, $P/T = R/V$ provides

$$dS = C_V \frac{dT}{T} - R \frac{dV}{V} \tag{7.38}$$

For an *isentropic process* (i.e., a process in which there is no change in entropy), direct integration of this expression provides

$$\frac{T}{T0} = \left(\frac{V_0}{V}\right)^{\gamma-1} \tag{7.39}$$

We can develop a similar expression relating the temperature change to a known pressure change. We again start with the energy balance

$$dU = Q + W \tag{7.13}$$

This time, we note that $dU = dH - d(PV)$, and again we use $Q = T\,dS$ and $W = -P\,dV$. This provides

$$dH - P\,dV - V\,dP = T\,dS - P\,dV \tag{7.40}$$

Dividing by T, simplifying, and rearranging provides

$$dS = C_P\frac{dT}{T} - R\frac{dP}{P} \tag{7.41}$$

When $dS = 0$, integration provides

$$\frac{T}{T0} = \left(\frac{P}{P0}\right)^{\gamma-1/\gamma} \tag{7.42}$$

Note that this result is the same as we obtained for adiabatic compression.

We will also need to obtain a relationship for the work of an isentropic compression (Highlight 7.3). Using the prior result from adiabatic compression, we obtain

$$W = \frac{RT_1}{\gamma-1}\left[\left(\frac{P_2}{P_1}\right)^{\gamma-1/\gamma} - 1\right] \tag{7.43}$$

Highlight 7.3 Operation of an Air Compressor

Air is compressed isentropically from 11.5 bar and 30 °C to 18 bar. What is the temperature for the air at the end of this process? What is the required work of compression? Assume that $\gamma_{air} = 1.4$.

Solution:

First, we calculate the compression ratio

$$R = \frac{18\,\text{bar}}{11.5\,\text{bar}} = 1.57$$

We will assume that air is an ideal gas, so we can calculate the exit temperature using the isentropic compression relationship

$$\frac{T}{T_0} = \left(\frac{P}{P_0}\right)^{\gamma - 1/\gamma}$$

$$T = (303\text{K})\left(\frac{18\,\text{bar}}{11.5\,\text{bar}}\right)^{1.4 - 1/1.4} = 344.4\text{K}$$

Next, we can calculate the work of compression

$$W = \frac{RT_1}{\gamma - 1}\left[\left(\frac{P_2}{P_1}\right)^{\gamma - 1/\gamma} - 1\right]$$

$$W = \frac{(8.314\,\text{J}/\text{mol K})(303\,\text{K})}{1.4 - 1}\left[\left(\frac{18\,\text{bar}}{11.5\,\text{bar}}\right)^{1.4 - 1/1.4} - 1\right]$$

$$= 860\,\text{J}/\text{mol}$$

7.5 REAL GAS PROPERTIES

We know from experience that a rapid decrease in pressure will lead to a decrease in the temperature of the fluid. This is why frost is formed on the exhaust from a gas relief pipe.

Can we describe this process using our knowledge of thermodynamics? We rewrite the first law of thermodynamics for an open system and, recognizing that there is no work being done on the fluid and no heat being added to the system, we find that

$$\Delta H = Q + W$$
$$= 0$$

Recall that the change in the temperature is simply a function of the change in the enthalpy, and for an ideal gas, $dH = C_p dT$. Thus, there will be no temperature change when an ideal gas is throttled. But we know from experience that there is a significant temperature change when a gas passes through a valve and into the surroundings. Clearly, there must be more to the properties of gases than can be described through the ideal gas law.

For real fluids, the enthalpy can also be a function of pressure. Under these circumstances, we need a new method of evaluating the thermodynamic properties of the fluid. There are several alternatives:

1. Assume ideal gas behavior (not very helpful in many cases).
2. Use an alternate equation of state that adequately models the real gas properties. One choice is the virial expansion,

$$Z = \frac{PV}{RT} = 1 + \frac{B}{V} + \frac{C}{V^2} + \cdots \tag{7.44}$$

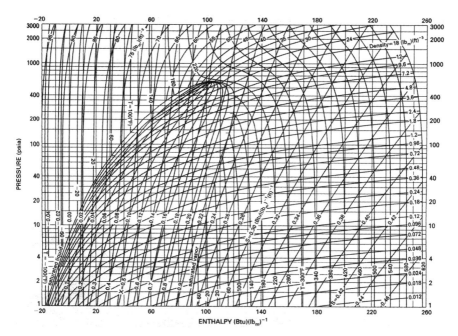

<u>Figure 7.6.</u> The thermodynamic properties of R134a [4]. (Courtesy of M. Huber and M. McLinden.)

An alternative is to use a cubic equation of state, such as the van der Waals equation (this and similar cubic equations of state are often used by chemists)

$$P = \frac{RT}{V-b} = -\frac{a}{V^2} \qquad (7.45)$$

3. Finally, we can make use of tabulated data (where available). One example is the steam tables. For other common thermodynamic fluids, such as refrigerants, data is provided in graphical form. Figure 7.6 thermodynamic data for Refrigerant 134a (1,1,1,2-tetrafluoroethane) in pressure–enthalpy coordinates.

This thermodynamic diagram provides descriptions of fluid properties as a function of pressure and enthalpy. One can follow lines of constant temperature, density, or entropy. The (mostly) horizontal curves represent constant density, and the (mostly) vertical curves are lines of constant entropy.

The large region in the middle of the diagram (shaded) is the two-phase region, the region in which the fluid can coexist as both a liquid and a vapor. Note that, in this region, the lines of constant temperature (isotherms) are horizontal, indicating that a pure substance boils at a constant pressure and temperature.

In order to see how to use the data presented in the diagram, we consider a throttling process using a real gas in Highlight 7.4.

Highlight 7.4 Expansion Through a Valve

One particularly important function of throttling is in cooling operations, such as in an air conditioner or refrigerator. A working fluid is expanded through a valve to create cooling, and then later in the cycle, that fluid is used to absorb heat from inside the refrigerator to keep it cool, and transfer that heat to the environment (i.e., your kitchen). In this specific example, R134a is expanded, at constant enthalpy, through a valve (this is termed a Joule–Thompson expansion) from 1000 psia and 220 °F to 100 psia. Determine the exit temperature and the quality of the fluid (percent liquid). Incidentally, R134a is termed a hydrofluorocarbon and was the first generation replacement for fluorocarbons that were used in refrigeration systems prior to being taken out of service per compliance with the Kyoto Protocol.

Solution:

We can begin by drawing a flowchart of the process:

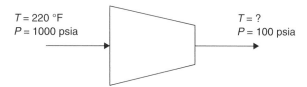

Based on the data, we can identify the thermodynamic condition of the fluid at the inlet condition using the thermodynamic diagram. This provides $H = 90\,\text{BTU/lbm}$.

The expansion takes place at constant enthalpy. On the thermodynamic diagram, this means that we draw a vertical line, until we reach the point at 100 psia and $H = 90\,\text{BTU/lbm}$. We can read the temperature from the chart as 80 °F, as indicated by the figure below.

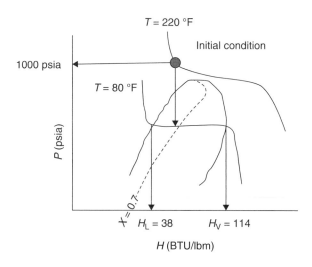

Quality refers to the percentage of the mixture that is a liquid. In order to determine the quality, we can read this value from the chart, following the curve marked x. This is the fraction that is vapor, so the quality is $1-x$, or 0.3. Alternatively, the overall enthalpy is equal to the fraction liquid times the liquid enthalpy plus the fraction vapor times its enthalpy. In other words,

$$H = xH_L + (1-x)H_V$$

From the graph, we find the liquid enthalpy is 38 BTU/lbm, and the vapor enthalpy is 114 BTU/lbm. We require the overall enthalpy to be 90 BTU/lbm. Substituting and solving provides

$$x = \frac{H - H_V}{H_L - H_V} = \frac{90 - 114}{38 - 114} = 0.32$$

We see that the two solutions are the same, within the limits of our ability to read the chart.

7.6 THE PHASE DIAGRAM

In order to understand how thermodynamic data is tabulated, it is helpful to look at a phase diagram. A simple one-component diagram is shown in Figure 7.7. Here, you can see the curves marking the phase boundaries, or two-phase lines, where, for

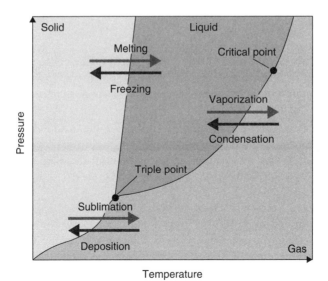

Figure 7.7. Phase diagram for a pure species [5]. (Courtesy of Carl Hepburn.)

example, a liquid is in equilibrium with a vapor. We will consider the specific case of steam and water, but for now, just consider the generic diagram shown here. The phase boundary is defined by a curve; thus, knowing the pressure defines the temperature at equilibrium and thus all of the thermodynamic properties of the fluid. However, each phase is defined by a region, and thus you need to specify both the temperature and pressure to know precisely where you are on the diagram.

We begin by looking at the phase diagram for a pure component. Pure fluids may be present in any one of three phases: solid, liquid, or gas. The phase boundaries occur along well-defined equilibrium curves, which determine where two phases can exist simultaneously. Figure 7.7 is a typical phase diagram (*PVT* diagram), which describes the phase behavior of some arbitrary fluid.

There are three distinct lines shown on this diagram.

1. The *sublimation curve*, which separates the solid phase from the vapor phase. At low temperature and low pressure a material can go directly from the solid to the vapor phase. A good example of this is CO_2, which is sold as dry ice (a solid at low temperature and atmospheric pressure).

2. The *melting* or *fusion* or *freezing curve* separates the solid from the liquid. As you are aware, water freezes at $0\,°C$ and 1 atm. However, if we increase the pressure, the freezing point temperature goes down (water is a unique case, in most cases, as we increase the pressure, the freezing point temperature will also increase, as indicated in Figure 7.7).

3. The *boiling* or *vapor pressure* or *condensation curve*, which separated the liquid and vapor phases. As the pressure is increased, the boiling point temperature of the fluid increases. As a practical example, water boils at a temperature below $100\,°C$ at high elevations (for example, in Denver, Colorado), requiring one to boil water for a longer period of time to cook spaghetti noodles in Denver than in Toledo, Ohio.

There are two additional important points plotted on the phase diagram.

1. *Triple Point*: This is the one point for which a single pure species may coexist as three separate phases.

2. *Critical Point*: This marks the end of the vapor–liquid equilibrium line. Above this temperature and pressure, the fluid will pass from properties that are liquid-like to properties that are gas-like without undergoing a phase change. A fluid above its critical point is said to be supercritical.

A phase change typically experienced in chemical processing occurs when we adjust the conditions of the fluid to cross a phase boundary. For example, consider heating a liquid at constant pressure, a horizontal line on the diagram. We will eventually cross the vapor pressure curve, at which point the liquid will start boiling and a vapor will be produced. As we continue to add more energy into the system, more of the liquid will be converted to the vapor phase. Eventually, all of the fluid

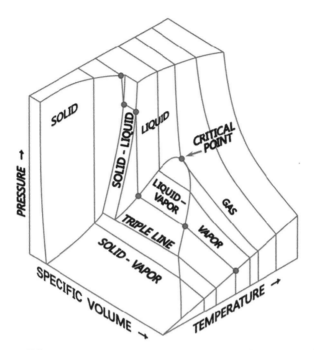

Figure 7.8. A full *PVT* projection showing the two-phase regions [6].

will exist as a vapor, and if we continue to add energy, the temperature of the vapor will increase.

A more complete phase diagram is provided by Figure 7.8, in which the pressure is plotted as a function of volume and temperature. Here we see that depending on the temperature, we may either pass through the two-phase region or, if the fluid is above its critical temperature, we avoid a phase separation. The point at the top of the liquid–vapor region is the critical point of the fluid. The region marked "LIQUID–VAPOR" depicts the two-phase envelope—below this curve two phases exist in equilibrium.

Figure 7.8 shows clearly that the vapor and liquid phases have a different specific volume (density), although the pressure and temperature of the system is the same. Here we see that the molar volume of the liquid and the vapor are connected in the phase diagram by a *tie-line*. The tie-line graphically reveals the properties of the two phases that are in equilibrium. The pressure at which the two phases coexist is termed the *saturation pressure*.

The system is said to be in phase equilibrium when it can exist simultaneously as both a gas and a liquid. The pressure at which a fluid is in equilibrium is said to be the saturation pressure and is a function of the temperature of the system. The data in the steam tables includes the saturation pressure for water, as a function of temperature. Likewise, the saturation pressure for other species can sometimes be found in graphical form, as in the *PVT* diagram shown in Figure 7.8.

For species in which tabulated data is not available, the most common technique for estimating the vapor (or saturation) pressure is to use Antoine's equation, which is written as

$$\log P^{\text{sat}} = A - \frac{B}{C+t} \qquad (7.46)$$

where P^{sat} is in mm Hg, and t is in °C. The constants are obtained from experimental measurement and are tabulated in tables. Antoine's equation can be used to estimate the vapor (or saturation) pressure if the temperature is known or, conversely, can be used to estimate the saturation temperature if the pressure is known.

Chemists will often be faced with multicomponent mixtures, and thus it is important to be able to extend the discussion from pure species. We have already seen that in a system containing only one component there is only one point at which a pure component can coexist as three phases. Also, we have seen that if we specify the temperature at which two phases coexist then we have fully specified the system. In other words, by stating that water is in equilibrium with water vapor at 100 °C, we know the pressure is 1 atm. However, if I have pure liquid water at 60 °C completely filling a container, I know nothing about the pressure. Thus, we would need to specify two variables (P and T) to define the system.

Now, let's suppose that we have two components. In this case, the two-phase line becomes a two-phase region (i.e., we can vary the temperature and pressure independently and remain within the two-phase region). This is because the composition of the system is an additional variable. Within the two-phase region, we need to specify the temperature, the pressure, and the overall composition of the mixture. Figure 7.9 shows two very common ways of describing the equilibrium of a two-component mixture. In the figure on the left, the temperature is plotted on the y-axis and the composition is plotted across the x-axis. The region above the two-phase region is the gas phase, and below the two-phase region is the liquid. In the figure on the right, the pressure is plotted on the y-axis, while both the liquid and vapor compositions are plotted on the x-axis. In the case, the region above the two-phase region is the liquid phase. Note that two phases can exist over the entire range of compositions. At a specific pressure (or temperature), the composition of the liquid in equilibrium with the vapor is defined by a horizontal line.

Many important chemical processes involve the transition between the single-phase and two- phase regions. Consider the situation in the figure on the right in Figure 7.9. If we start from a pure liquid state and decrease the pressure on the system, then at some point the liquid will begin to boil. This is termed the *bubble point* and is the location where the first drop of vapor is formed. If instead we initially have a vapor phase mixture and we increase the pressure (or decrease the temperature), then liquid will form at the *dew point*. The dew point is defined as the location where the first drop of liquid is formed.

When we generalize this result to many species and many phases we obtain the Gibbs phase rule, which provides

$$F = C + 2 - \pi \qquad (7.47)$$

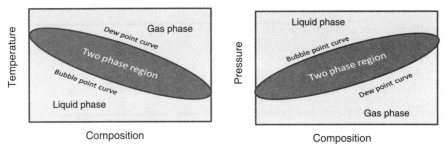

Figure 7.9. Illustration of the two-phase region for a two component system.

where the degrees of freedom F is the number of intensive variables that can be independently specified, C is the number of components, and π is the number of phases. In the example above, with 2 components and 2 phases we find F=2. This means that we can independently vary the pressure and the composition and still remain within the two-phase region.

7.7 EQUILIBRIUM

In many common chemical operations, the liquid phase is in contact with another vapor, and a portion of the liquid will vaporize and mix with the vapor phase. We frequently want to know how much liquid can be contained in the vapor stream, or the saturation condition for the vapor. We define a gas as being saturated when it contains as much of a liquid substance as possible. For example, water evaporating from a pond increases the moisture content in the air above the pond. During the winter, this creates lake effect snow, as saturated air is carried back across the land, experiences a temperature decrease, and loses excess moisture in the form of snow.

In 1882, French Chemist François-Marie Raoult determined that the pressure generated by a solution was a function of the mole fraction of the species in the solution and the vapor pressure of the species. Starting from a simple solution in which only one species has a significant vapor pressure allows a simple expression of *Raoult's law*

$$p_V = y_V P = p_V^{sat}(T) \tag{7.48}$$

where y_V is the mole fraction of the volatile species in the vapor phase. Thus, the saturation condition occurs when the partial pressure (p_V) is equivalent to the vapor pressure. Raoult's law assumes that there are no molecular interactions in the liquid or vapor phases, which is the assumption of an ideal solution. Solutions of like fluids (such as a mixture of two alkanes) may often be assumed to be ideal; however, dissimilar fluids may form nonideal mixtures, which will complicate the equilibrium.

If the vapor stream contains as much of the liquid as it can possibly hold, then it is saturated. We also define a percent saturation as

$$\text{Percent saturation} = 100\frac{Y}{Y_s} = 100\left(\frac{y_a}{1-y_a}\right)\left(\frac{1-y_a^{sat}}{y_a^{sat}}\right) = 100\left(\frac{p_a}{p-p_a}\right)\left(\frac{P-p_a^{sat}}{p_a^{sat}}\right) \tag{7.49}$$

where Y and Y_s represent the molar ratio of the condensable species to the non-condensable species and the subscript s represents the saturation conditions. Similarly, y_a and y_a^{sat} are the mole fraction of the condensable species and its mole fraction at saturation, while p_a and p_a^{sat} are the partial pressures. P is the total pressure of the system.

The percent saturation should be distinguished from the percent relative saturation, which is defined as the partial pressure of the condensable component present divided by the partial pressure present at the saturation condition. Thus,

$$\text{Relative saturation} = 100\frac{y_a}{y_a^{sat}} = 100\frac{p_a}{p_a^{sat}} \tag{7.50}$$

Water holds a unique place in green chemistry, most notably because it is a benign species. Historically, water is important because it is freely available and easy to manipulate. As a result, we often use the term humidity when describing the saturation condition of water.

For a multicomponent system, Raoult's law applies to all species in a solution, so we need to take into account the mole fraction of the species. In this case, the full expression of Raoult's law provides

$$y_i P = x_i p_i^{sat} \tag{7.51}$$

where x_i is the mole fraction of the species in the liquid and y_i is the mole fraction of the species in the vapor.

One may also define a distribution or equilibrium coefficient as

$$K = \frac{y_i}{x_i} \tag{7.52}$$

When Raoult's law is valid, the distribution coefficient can be equated to the ratio of the vapor pressure to the total pressure

$$K = \frac{p_i^{sat}}{P} \tag{7.53}$$

For an ideal multicomponent system, Raoult's law must hold for all of the species in the mixture. Thus, if we have a binary system consisting of benzene and toluene, Raoult's law provides two equilibrium relations:

$$y_{ben}P = x_{ben}p_{ben}^{sat}$$
$$y_{tol}P = x_{tol}p_{tol}^{sat} \tag{7.54}$$

Note that the total pressure of the system is the same in both cases. Then, summing over all of the species provides

$$\sum x_i p_i^{sat} = \sum(y_i P)$$
$$= \left(\sum y_i\right)P \tag{7.55}$$
$$= P$$

since

$$\sum y_i = 1$$

For a two-component system, there are four variables that can be varied—temperature, pressure, and liquid and vapor compositions. In order to calculate the properties of the system, we must know two of these properties, from which we can calculate the other two. Since these are not completely independent, we end up with five different types of equilibrium problems that can be solved. These are:

Problem Type	Variables Specified	Calculate
1. Bubble point temperature	P, x_i	T, y_i
2. Bubble point pressure	T, x_i	P, y_i
3. Dew point temperature	P, y_i	T, x_i
4. Dew point pressure	T, y_i	P, x_i
5. Flash calculation	T, P	y_i, x_i

In order to solve these problems, we need to derive the thermodynamic relationships between the vapor and liquid phase compositions, and then combine this information with the appropriate material balance.

Earlier, it was stated that we could solve problems by combining the equilibrium relationships with the mass balance relationships, and then solve for the unknown variables. For example, in the case of the bubble point pressure problem, the liquid compositions and the system pressure are known. Thus, we combine the equilibrium relations with the mass balance and find

$$\sum x_i p_i^{sat} = P^{bubl} \tag{7.56}$$

as described previously. Once the bubble point pressure is known, we can calculate the composition of the vapor phase using Raoult's law:

$$y_i = \frac{x_i p_i^{sat}}{P^{bubl}} \tag{7.57}$$

Similar solutions can be obtained for all of the bubble and dew point calculations.

We are now ready to consider several examples. See Highlight 7.5.

Highlight 7.5 Composition of Vapor Phase

In 2004, DuPont developed a greener process to produce 1,3-propanediol (PDO) through fermentation of glycerol. Following the reaction, the PDO needs to be recovered from the fermentation broth, and vacuum distillation is used. We want to determine the composition of the vapor phase in equilibrium with a liquid containing 40% PDO and glycerol at 180 °C.

Solution:

We start by looking up constants for Antoine's equation for these species, so that we can apply Raoult's law. The NIST Chemical Webbook reports Antoine's data in the following form:

$$\log_{10}(P) = A - (B / (T + C))$$

where P = vapor pressure (bar) and T = temperature (K), and provides the appropriate constants:

	A	B	C
1,3-Propanediol	6.29523	3105.018	6.101
Glycerol	3.93737	1411.531	−200.566

We first calculate the vapor pressure for each species in the mixture at the indicated temperature by substituting the appropriate constants into Antoine's equation:

$$\log_{10} P_{PDO} = 6.29523 - \frac{3105.018}{(180 + 273.15) + 6.101}$$

which provides

$$P_{PDO} = 0.342\,bar = 260.0\,mm\,Hg$$

Similarly for glycerol, we find

$$\log_{10} P_{G} = 3.93737 - \frac{1411.531}{(180 + 273.15) - 200.566}$$

which provides

$$P_{G} = 0.022\,bar = 17.0\,mm\,Hg$$

Once we know the vapor pressure of each species, we can find the equilibrium pressure for the mixture as

$$P = 0.4(260.0 \, \text{mm Hg}) + 0.6(17.0 \, \text{mm Hg}) = 114.2 \, \text{mm Hg}$$

Once we know the equilibrium pressure, we can calculate the amount of each species in the vapor phase:

$$y_i = \frac{x_i p_i^{\text{sat}}}{P}$$

which gives for PDO

$$y_{\text{PDO}} = \frac{0.4 \, (260.0 \, \text{mm Hg})}{114.2 \, \text{mm Hg}} = 0.911.$$

From the sum of mole fractions equal to 1, we can quickly determine that $y_G = 0.089$.

For a dew point problem, it is the vapor phase composition that is known, while the liquid phase composition is unknown. Rearranging the combined equilibrium and material balance equations as before will not provide any benefit in this case. However, let's return to Raoult's law,

$$y_i P = x_i p_i^{\text{sat}} \tag{7.51}$$

and now solve for the liquid phase mole fraction:

$$x_i = \frac{y_i P}{p_i^{\text{sat}}} \tag{7.58}$$

As before, the sum of the mole fractions must equal 1, so we sum the above equation over all the species in the mixture to get

$$1 = \sum \frac{y_i P}{p_i^{\text{sat}}} \tag{7.59}$$

Here, P is the total pressure that is the same in all cases, so we can rearrange this equation and solve for P as

$$P = \frac{1}{\sum \left(y_i / p_i^{\text{sat}} \right)} \tag{7.60}$$

Now we can solve for the dew point pressure directly. See Highlight 7.6.

Highlight 7.6 Liquid Composition at Dew Point

Styrene is an important monomer of the plastics industry, generally produced from benzene (a known carcinogen) and ethylene (produced from thermal cracking of hydrocarbons). A more desirable process would be to convert toluene to ethylbenzene through a reaction with methane, the primary component of natural gas. In a 2009 BASF patent, they report a new catalyst that allows such a process to occur. Our goal is to identify the temperature at which the unreacted toluene can be recovered as a liquid when the system is operating at 100 psia. The composition of the vapor product contains 80% ethylbenzene and the remainder toluene. What is the liquid composition at the dew point?

Solution:

This is the dew point temperature problem, which can most easily be solved through the use of iterative calculations. The process is relatively straightforward and similar to the previous example, except that we don't know the initial temperature. So we complete the following steps:

Step 1. Guess a temperature.

Step 2. Calculate P_i^{sat}.

Step 3. Calculate the equilibrium pressure for the assumed temperature.

Step 4. If $P_{guess} < P_{actual}$, choose a higher temperature and repeat from step 2. If $P_{guess} > P_{actual}$, choose a lower temperature and repeat from step 2.

Step 5. Continue until $P_{guess} = P_{actual}$.

We start by looking up Antoine's parameters:

Species	y_i	A	B	C
Ethylbenzene	0.8	4.4054	1695.026	−23.7
Toluene	0.2	4.5444	1737.123	0.394

If we assume a temperature of 200 °C, then we can calculate the vapor pressure for ethylbenzene just as we did in the previous example:

$$\log_{10} P_{EB} = 4.4054 - \frac{1695.026}{(200 + 273.15) - 23.7}$$

From which we find $P_{EB} = 62.29$ psia.

A similar calculation for toluene reveals a vapor pressure of $P_T = 109.96$ psia.

Since we know the composition of the vapor, we can calculate the dew point pressure for this mixture at 200 °C as

$$P = \frac{1}{0.8/62.29 + 0.2/109.96} = 69.16 \, \text{psia}$$

The operating pressure is 100 psia, so we need to go to a higher temperature. A series of iterations reveals that the actual dew point temperature is 220.13 K.

Once we know that, then we can easily calculate the composition of the liquid at the dew point,

$$x_{EB} = \frac{y_{EB}P}{P_{EB}^{sat}} = \frac{(0.8)(100\,\text{psia})}{(91.83\,\text{psia})} = 0.871$$

By difference, the amount of toluene in the liquid is 0.129 mole fraction.

7.7.1 The Flash Calculation

A flash operation occurs when a mixture is expanded through a valve from a region of relatively high pressure to a region of relatively low pressure. Because of the change in temperature and pressure, a phase separation can be induced. In this type of process, the temperature and pressure at the final state are known, and while the compositions of the equilibrium vapor and liquid phases are unknown, the composition of the fluid entering the valve is known.

In order to solve this type of problem, we require an overall mass balance to relate the amount of vapor (V) and liquid (L) to the total amount of material (F)

$$F = V + L \tag{7.61}$$

Similar equations can be written for each species in the mixture. Defining the mole fraction of the species in the vapor as y_i, the mole faction of the species in the liquid as x_i, and the mole fraction of the species in the original fluid as z_i, provides the species mass balance

$$z_i F = y_i V + x_i L \tag{7.62}$$

Substituting for $V = F - L$ in the component mass balance provides

$$z_i F = y_i (F - L) + x_i L \tag{7.63}$$

We incorporate the equilibrium data using the distribution coefficient (which can be obtained from Antoine's equation if necessary)

$$y_i = K_i x_i \tag{7.64}$$

which can be substituted to provide

$$\begin{aligned} z_i F &= K_i x_i (F - L) + x_i L \\ &= x_i \left[K_i (F - L) + L \right] \end{aligned} \tag{7.65}$$

This equation contains two unknowns, L and x_i. We would like to solve for x_i and then sum over all of the species, since $\Sigma x_i = 1$. Skipping several steps of mathematics, we find

$$x_i = \frac{z_i}{L/F + K_i(1 - L/F)} \tag{7.66}$$

and after summation, we find

$$\sum x_i = \sum \frac{z_i}{L/F + K_i(1-L/F)} = 1 \qquad (7.67)$$

This equation contains only one unknown, the ratio L/F. An iterative solution is required. Once the ratio is determined, the various liquid compositions can be found using Equation (7.66), and the vapor phase composition can be obtained using the equilibrium relation, expressed by Equation (7.67).

Now let's look at an example in Highlight 7.7.

Highlight 7.7   The Case of Ethyl Acetate

Ethyl acetate is a green solvent that can be produced from bio-derived ethanol. After the reaction, unreacted ethanol is recovered from the ethyl acetate through pressure swing distillation. Let's assume that we have a product stream leaving the reactor that is cooled to ensure that it is full liquid at 25 °C at 0.1 atm, and containing 50% ethyl acetate and the remainder ethanol. What are the compositions of the vapor and liquid phases at this condition?

Solution:

We again start by finding Antoine's constants for these species:

Species	x_i	A	B	C
Ethanol	0.7	5.24677	1597.673	−46.424
Ethyl acetate	0.3	4.2281	1245.702	−55.19

From this information, we can calculate the vapor pressure of each species at this temperature.

$$\log_{10} P_{EtOH} = 4.9253 - \frac{1432.526}{(25+273.15)-61.82}$$

From which we find $P_{EtOH} = 0.079$ atm. We want to use a distribution coefficient, and since the total pressure is 0.1 atm, then $K_{EtOH} = 0.79$.

A similar calculation for ethanol reveals a vapor pressure of $P_{EA} = 0.126$ atm and a distribution coefficient of $K_{EA} = 1.26$.

Application of the formula provides

$$\frac{0.5}{L/F + 0.79(1-L/F)} + \frac{0.5}{L/F + 1.26(1-L/F)} = 1$$

Solving for L/F by iteration provides a solution when $L/F = 0.555$. Then, upon substitution, we can calculate the mole fraction of ethyl acetate in the liquid as

$$x_{EA} = \frac{0.5}{0.555 + 1.26(1 - 0.555)} = 0.448$$

And finally the mole fraction of ethyl acetate in the vapor is

$$y_{EA} = (1.26)(0.448) = 0.565$$

7.8 SOLUBILITY OF A GAS IN A LIQUID

In many examples of interest, a single component from a gas stream is selectively dissolved into a liquid solvent. This situation was seen previously in Chapter 6, in the case of gas-forming reactions. In that situation, a species that could form a gas was produced by the reaction, and the product would leave the liquid solution to form a separate gas phase. In a similar way, a gas can be dissolved into a liquid and be present as a reactant.

For liquid–vapor equilibrium, we used Raoult's law to estimate the amount of the liquid that could transfer into the vapor phase. The key parameter was the saturation pressure, which could be estimated from Antoine's equation. However, for gases that do not normally condense, such as oxygen or SO_2, there is no vapor pressure and we need an alternate measure. In addition, the species transferring between the phases is usually very dilute in the liquid phase. Under the conditions of a dilute organic species in a liquid phase, the equilibrium is written using Henry's law

$$p_i = H_i x_i \qquad (7.68)$$

where p_i is the partial pressure of the organic species in the vapor phase, x_i is the mole fraction of the organic species in the liquid phase, and H_i is termed the Henry's law constant, usually given in units of atm. Values for H_i are available in many tabulated sources, although the specific form of Henry's law varies depending on the reference. See Highlight 7.8.

Highlight 7.8 Concentrations of Oxygen

Estimate the concentration (mol/L) of oxygen in water at atmospheric pressure and 68 °F.

Solution:

Air is 21% oxygen, so the partial pressure of oxygen in air at 1 atm is 0.21 atm. Literature reports indicate that $H = 40,100$ atm/mol fraction at the indicated temperature. Thus, we find

$$x = P_i / H_i = 0.21\,atm/(40100\,atm/mol frac)$$
$$= 5.24 \times 10^{-6}\,mol frac$$

Now we need to convert this value from mole fraction to concentration. The concentration of water is 55.6 mol/L, so the concentration of oxygen in water can be calculated as

$$C_{O_2} = (5.24 \times 10^{-6} \text{ mol frac})(55.6 \text{ mol} / \text{L})$$
$$= 2.91 \times 10^{-4} \text{ mol} / \text{L}$$

If we wanted to use gaseous oxygen to promote an oxidation reaction in water, this would be the concentration available for the reaction in solution. It is this value that is critical in biological systems, since aerobic microorganisms rely on dissolved oxygen for their metabolism.

7.9 SOLUBILITY OF A SOLID IN A LIQUID

Solid liquid equilibrium is observed when a solid is dissolved into a liquid solvent, such as a salt in water. In such cases in the inorganic chemical industry, crystallization is employed as a separation process, particularly where salts are recovered from aqueous media. The feed to a crystallization system consists of a solution from which a solute is crystallized (or precipitated), and the solids are then removed. High recovery of the refined solute is generally the desired design objective.

For an overall process, one must consider three phases of operation—vaporized solvent that contains no solute, liquid solvent containing residual solute, and the produced solid crystals. Solvent is often removed so that the liquid solution is sufficiently concentrated to permit precipitation of solid crystals.

The concentration of solute that remains in the liquid solution is dependent on solid–liquid equilibrium. Phase diagrams can be of many types, depending on the system. The maximum amount of solute that can be dissolved in a given volume of solvent at equilibrium and at a given temperature is called solubility. While it is typical for solubility to increase with increasing temperature, that is not always the case; for example, see $Ce_2(SO_4)_3$, as seen in Figure 7.10.

The equilibrium diagram is used to determine the amount of solute that can be recovered by cooling a saturated solution from an elevated temperature to a lower temperature. For example, if one has a saturated solution of potassium chloride (KCl) at 80 °C, the solubility diagram indicates that solution contains roughly 51 g of KCl per 100 g of water. If we now decrease the temperature to 40 °C, the solubility is reduced to 40 g KCl per 100 g water. Assuming that we started with 1 kg of water, this means that decreasing the temperature resulted in the production of 110 g of KCl salt crystals on the bottom of the beaker.

The addition of a salt (generally called *solute*) into a liquid (generally called *solvent*) can be used to change the boiling and freezing point properties of that liquid;

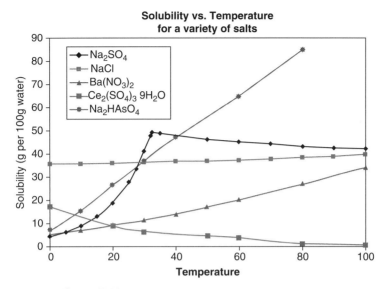

Figure 7.10. Solubility of salts versus temperature [7].

for example, adding a salt into water results in a depression of the freezing point. This is an example of a colligative property (property that depends only on the concentration of the solute in the solution), and the amount of the change in the freezing point depends on the amount of solid added to the solution. It is this chemical phenomenon that we take advantage of when we put salt on the roads in the winter to melt ice (or keep the water from freezing).

When the solution can be considered an ideal solution, then the freezing point depression can be obtained as a simple linear relationship of the molality of the solution, according to

$$\Delta T_{\mathrm{f}} = K_{\mathrm{f}} C_{\mathrm{s}} i \tag{7.69}$$

where K_{f} is termed the *cryoscopic constant* and is a function of the solvent only, C_{s} is the molality (g of solute per kg of solvent) of the solute (*not to confuse with the concentration or molarity in mol solute/L solution*), and i is the number of ion molecules generated per molecule of solute. Highlight 7.9 introduces how to calculate freezing point of a solution.

If the system is sufficiently ideal, it is possible to estimate the composition of the solution in equilibrium with the pure solute using thermodynamic data theory. After sufficient simplification, we can use the van't Hoff relationship,

$$\ln x_2 = \frac{\Delta H_{\mathrm{f}}}{RT}\left(\frac{T}{T_{\mathrm{m}}} - 1\right) \tag{7.70}$$

Highlight 7.9   Calculation of Freezing Point

Determine the freezing point (in °C) of a solution containing 1 g of KCl and 100 g of water.

Solution:

For water, the cryoscopic constant K is 1.853 K kg/mol. KCl will dissociate to form 2 ions per mole of salt (K^+ and Cl^-), and the concentration is found to be 10 g KCl/kg water, which corresponds to 0.134 mol KCl/kg water. Substituting provides

$$\Delta T_f = (1.853 \text{ K kg /mol})(0.134 \text{ mol KCl / kg water})(2 \text{ ion / mol})$$
$$= 0.497 \text{K}$$

Thus, the freezing point of this solution will be roughly −0.5 °C.

Highlight 7.10   Calculation of Solubility

Ethyl acetate is often considered a cleaner solvent than some harsher organic solvents and may be recovered by extraction with water. Determine the equilibrium solubility of ethyl acetate in water at 25 °C.

Solution:

This is a simple application of the van't Hoff equation. We can find the heat of fusion for the solute ethyl acetate as $\Delta H_{f, EA} = 10.48$ kJ/mol at the melting temperature of 83.8 °C. Converting to absolute temperatures and substituting provides

$$\ln x_2 = \frac{\Delta H_f}{RT}\left(\frac{T}{T_m} - 1\right)$$
$$= \frac{10,480 \text{ J / mol}}{(8.314 \text{ J / mol K})(298 \text{K})}\left(\frac{298 \text{K}}{356.95 \text{K}} - 1\right)$$
$$= -0.698$$
$$x_2 = 0.498 \text{ mol frac}$$

This is the maximum concentration of ethyl acetate that can be found in a solution with water at 25 °C.

where ΔH_f is the heat of fusion at the melting temperature, T_m. The van't Hoff equation provides a simple method for calculating the mole fraction of solute in the liquid at a given temperature using only the properties of the solute, without regard to solvent properties. An example is given in Highlight 7.10.

7.10 SUMMARY

Since reactions involve collisions between molecules, reactions can only occur between two species in a common phase. But many systems will have multiple phases. It is important for the chemist to be able to understand the phase behavior of these different types of systems in order to be able to predict the reactions that may occur.

Furthermore, green chemists can use the phase behavior of a system to control the reaction pathways, or to recover a solvent that may be used. By manipulating the temperature and pressure, the phase behavior can be modified and improved reaction performance may be obtained.

REFERENCES

1. http://www.grc.nasa.gov/WWW/K-12/airplane/boyle.html.

2. http://exploration.grc.nasa.gov/education/rocket/glussac.html.

3. http://www.splung.com/content/sid/6/page/work.

4. Huber, M. L.; McLinden, M. O. "Thermodynamic Properties of R134a (1,1,1,2-tetrafluoroethane)." *International Refrigeration and Air Conditioning Conference*, 1992. Paper 184.

5. http://www.splung.com/content/sid/6/page/phasesofmatter.

6. http://en.wikipedia.org/wiki/File:PVT_3D_diagram.png.

7. http://en.wikipedia.org/wiki/File:SolubilityVsTemperature.png.

8

RENEWABLE MATERIALS

8.1 INTRODUCTION

Resources can be classified as finite (with limited existence) or infinite (endless supply); or as renewable (capable of being replaced naturally and replenished on a reasonable timescale), nonrenewable (only capable of being replenished at a very slow rate), and perpetual (such as sunlight and wind). Nonrenewable resources (such as petroleum, coal, and minerals) are generally extracted from the earth. On the other hand, renewable resources (such as biomass) are grown.

One of the central tenants of green chemistry is that raw materials should be derived from renewable resources whenever possible. Effectively the seventh principle of green chemistry states that "a raw material or feedstock should be renewable rather than depleting whenever technically and economically practicable" [1, p. 45]. Petrochemical resources are clearly limited, although there remains discussion as to how long these nonrenewable resources can last. Thus, for processes to be truly sustainable, raw materials must be derived from plants or microbial sources and,

Green Chemistry and Engineering: A Pathway to Sustainability,
Anne E. Marteel-Parrish and Martin A. Abraham.
© 2014 American Institute of Chemical Engineers, Inc. Published 2014 by John Wiley & Sons, Inc.

because fossil resources can be used in the production of renewable materials, should be produced with noninvasive chemical processing.

This chapter explores the sources of renewable feedstocks, in particular, carbohydrates, lignin, lipid oils, and proteins, followed by the production of chemicals from renewable resources, and finally some current applications of renewable materials.

8.2   RENEWABLE FEEDSTOCKS

8.2.1   Role of Biomass and Components

Biomass is often defined as "a renewable energy source, living or recently living biological material" such as wood, waste, and alcohol fuels [2]. Biomass is often associated with plant-based material but it also applies to both animal- and vegetable-derived material.

The chemical structure of biomass is comprised of a mixture of elements found in organic molecules, including carbon, oxygen, hydrogen, and nitrogen atoms, as well as other atoms such as alkali, alkaline earth, and heavy metals. But in general, biomass is a polymeric material without a regular structure, and thus there is no such thing as a biomass molecule.

According to the Biomass Energy Center in the United Kingdom, there are five categories of biomass materials including "virgin wood from forestry, arboricultural activities, or from wood processing; energy crops such as high yield crops grown specifically for energy applications; agricultural residues; food waste from food and drink manufacture, preparation and processing, and post-consumer waste; and industrial waste and co-products from manufacturing and industrial processes." Common sources of biomass include wood materials, such as bark, sawdust, and mill scrap; agricultural wastes such as corn stalks and straw, manure from cattle, and waste from poultry; municipal waste such as paper and yard clippings; and energy crops such as corn (starch), switchgrass, soybean (oil), and alfafa (pasture and hay crop) [3].

In order to evaluate the environmental impact of biomass, one needs to evaluate the carbon cycle of the process. Carbon from the atmosphere is converted into biological matter through photosynthesis. If the biomass is used to produce energy through combustion, the final carbon product is carbon dioxide, a significant greenhouse gas. However, the carbon originated from atmospheric CO_2, and thus combustion merely returns the carbon back into the atmosphere and there should be no net gain in atmospheric CO_2. However, if the biomass is not regenerated through new planting, this cycle is disturbed and biomass combustion and use can contribute to a larger extent to global warming. This happens in the case of deforestation and intense urbanization. In Figure 8.1, the full cycle is shown.

Figure 8.1 shows four steps in the complete carbon cycle.

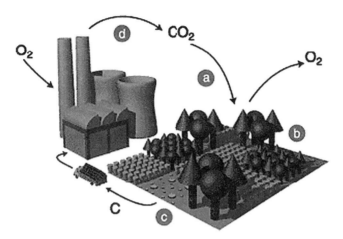

Figure 8.1. Absorption and emission of carbon dioxide through combustion of biomass [4]. (R. Mathews for IEA Bioenergy Task 38.)

(a) Trees in the energy plantation grow, absorbing carbon dioxide from the atmosphere.

(b) During photosynthesis, the trees store carbon in their woody tissue and oxygen is released back to the atmosphere.

(c) The biomass is harvested and then transported from the plantation to the power plant. Unless the vehicle is fueled from bioderived materials, there is a net consumption of nonrenewable materials and a contribution to atmospheric CO_2.

(d) The wood is burned in the power plant, generating carbon dioxide through the combustion of carbon stored in the woody tissue. The CO_2 is released back to the atmosphere in the exhaust gases.

Under perfect conditions, the absorption of CO_2 through plant growth would be perfectly balanced with the emission of CO_2 through combustion. However, the process of growing biomass is facilitated through the use of fertilizers and pesticides, and machinery is used to perform planting, harvesting, processing, and transport, consuming fossil resources. Thus, biomass is often referred to as "a low carbon fuel."

Fossil fuels such as coal, oil, and gas are also derived from biological materials. However, the biological materials from which fossil fuels derive absorbed carbon dioxide from the atmosphere many millions of years ago. When their combustion products are released into the atmosphere, there is no process occurring at a comparable rate that consumes this additional CO_2. Unless they are captured and stored, the combustion of fossil resources contributes to increased levels of CO_2 in the atmosphere.

Biomass is a complex substance made up of many organic species combined into a heterogeneous polymeric material. Although the actual structure of biomass cannot be defined, portions of it can be related to the structure of known organic compounds. These structures include carbohydrates (cellulose and hemicelluloses, approximately

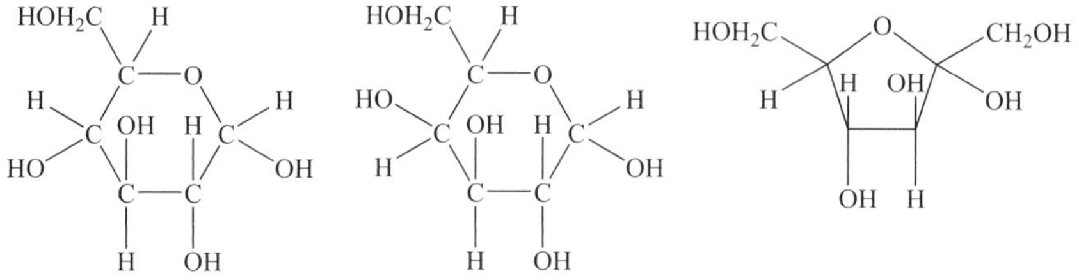

Figure 8.2. Basic carbohydrates of (a) glucose, (b) galactose, and (c) fructose.

75%, dry weight) followed by lignin, then fats and proteins, and small amounts of minerals, such as sodium, phosphorus, calcium, and iron.

8.2.1.1 Carbohydrates

Carbohydrates, or saccharides, make most of the organic matter on earth. They hold numerous roles such as the storage and transport of energy in living things (e.g., starch and glycogen) and are part of the structural framework of RNA (ribonucleic acid) and DNA (deoxyribonucleic acid) (e.g., ribose and deoxyribose, a component of DNA) and of plant cell walls and plant tissues (e.g., polysaccharides such as cellulose).

Carbohydrates are related to simple organic compounds, such as aldehydes and ketones with added hydroxyl groups. Monosaccharides, such as glucose (necessary "blood sugar" for cellular respiration), galactose (sugar found in milk and yogurts), and fructose (sugar found in honey), are the basic carbohydrates. They are structural isomers: they have identical molecular formulas $C_6H_{12}O_6$ but different structural formulas (Figure 8.2).

When monosaccharides are linked together, they form polysaccharides (or oligo-saccharides) such as lactose, which is a disaccharide occurring naturally in mammalian milk. Polysaccharides, such as starch and cellulose, are an important class of biological polymers.

Starch, a polymer of glucose, is used as an energy storage polysaccharide in all green plants and is present in a lot of food sources for humans (breads, pastas, potatoes, rice, some cereals). There are two types of starch molecules: amylose and amylopectin (Figures 8.3 and 8.4).

"Animal starch," also known as glycogen, is related to glucose in an analogous way as amylopectin is related to amylose; glycogen provides a convenient way for animals to store energy as it gets metabolized quickly in active animals.

There is around 60 million tons of raw starch produced annually for food and nonfood uses. The major sources of starch include rice, wheat, potatoes, cassava (tapioca), and corn, which accounts for over 80% of the world production. Corn is used in the United States as a fermentation feedstock for bioethanol production, discussed later in this chapter. Starch is produced by steeping the raw material in hot water, which releases the starch from the binders also present in the biomass.

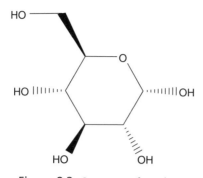

Figure 8.3. Structure of amylase.

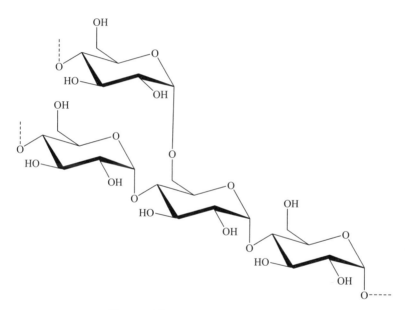

Figure 8.4. Structure of amylopectin.

The water is then evaporated, and the starch can be separated from the oil and other components released through the hot water treatment.

Cellulose is another important carbohydrate consisting of linked D-glucose units found in the primary cell wall of green plants and other organisms such as algae and water molds.

Cellulose, a long-chain polymer with the molecular formula $(C_6H_{10}O_5)_n$, is the most common organic compound on earth (Figure 8.5). Commercial cellulose is mainly obtained from wood pulp and cotton. The list of commercial products made from cellulose is extensive, including primarily paper, cardboard, and cardstock, and textiles made from cotton, linen, and other plant fibers. Cellulose can also be converted into cellophane, a thin transparent film, or used in the manufacture of nitrocellulose

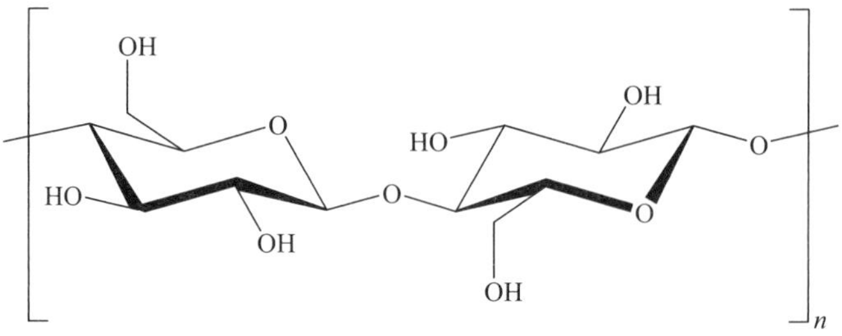

Figure 8.5. Structure of cellulose.

(cellulose nitrate) employed historically in smokeless gunpowder. Cellulose esters such as cellulose acetate are used in a wide range of products including cigarette filters, films, and filtration membranes in the pharmaceutical sector. Alkyls of cellulose esters are also used in the cosmetic and pharmaceutical industries.

Cellulose is generally produced through wood pulping, a process not too dissimilar from the steeping process used to produce starch. In this case, high temperatures are required to break down the lignin that binds the cellulose fibers together. Additional oxidizers may also be added to enhance the process. Once the lignin is removed, the cellulose fibers can be recovered. Cellulose can also be recovered from recycled paper using a similar process. Because of the nature of the secondary processing, recycled cellulose may be used as a material for building insulation, which is environmentally preferable over the often petrochemical-based insulation types including foam and fiberglass.

8.2.1.2 Lignin

Following carbohydrates (starch and cellulose), the next most abundant species in biomass is the biopolymer, lignin, found in the secondary cell walls of plants and some algae. Lignin is a noncarbohydrate macromolecule with a complex, highly aromatic and indeterminate primary structure. Its main function is to fill void spaces in the cell wall and therefore to provide mechanical strength as well as prevent enzymatic and chemical degradation of cellulose. Wood with high lignin content is durable and is consequently used in a lot of applications.

Lignin is present in high quantity in the wood pulp used to make paper, and its presence contributes to the yellowing of newsprint with age. In order to produce high quality bleached paper, lignin needs to be removed from the wood pulp as sulfonates, usually via the Kraft process known as sulfate pulping. Sulfonates are the conjugate bases of sulfonic acids and are anionic species of the type RSO_3^-.

After removal of the lignin, it can be dewatered and then burned as a fuel, providing more energy than cellulose and enough energy to run a typical paper mill. The sulfonates of lignin can be used as dispersants in paints, in cement applications, as additives in pesticides and cleaning agents, and as environmentally conscious dust suppression agents.

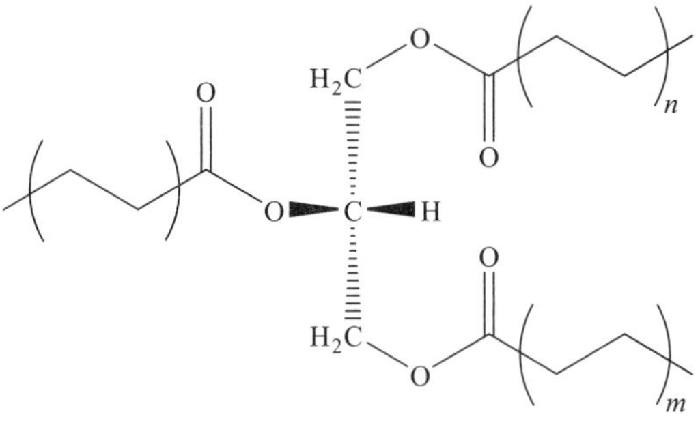

Figure 8.6. General structure of a triglyceride molecule (*m*, *n*, and *p* are whole numbers).

8.2.1.3 Oils/Fats

The most efficient oil-producing crops are tropical crops such as coconut and palm, followed by sunflower and soybean. Other fruit and nut oils are also extracted for food use but the need for high content of fatty acids or fatty acid derivatives tends to limit their industrial use. Vegetable oils, animal fats, or recycled greases are used as feedstocks for the production of biodiesel, with soybean oil as the most commonly used oil in the United States.

The most important component of plant-derived oils is triglycerides or triacylglycerides, which constitute 93–98% by weight of the oil. Each plant or vegetable species will have a different mixture of triglyceride molecules. A triglyceride molecule contains a molecule of glycerol commonly called glycerin or glycerine combined with three fatty acids (Figure 8.6).

The carbon chains of the fatty acids in the natural fats vary in length. For example, in cocoa butter, three fatty acids—palmitic, oleic, and stearic acids—have a low melting point of about 34 – 38 °C. This explains why chocolate is a solid at room temperature but melts once in the mouth.

Vegetable-based oils find many applications in industry such as for lubricants (increased biodegradability of vegetable oils is an advantage when used in marine or forest environments), for detergent, soap, and cleaning formulations, for solvents, and for plastics manufacturing.

8.2.1.4 Proteins and Amino Acids

Seeds of legumes are important sources of protein and are commonly used commercially as animal feed supplements or may be used for cosmetic purposes. Unfortunately, extracting proteins in high quantities without modifying their structural and chemical properties is a challenge.

The proteins in plants are comprised of amino acids, small molecules that contain both amine and acidic groups (Figure 8.7).

The length of the chain and any additional functionality in the chain provide the amino acid with its specific chemical properties and determine what types of further

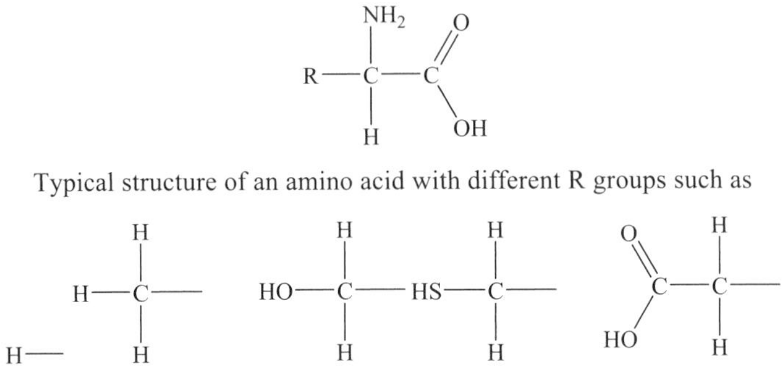

Typical structure of an amino acid with different R groups such as

Figure 8.7. Amino acids (in this order: glycine, alanine, serine, cysteine, and aspartate).

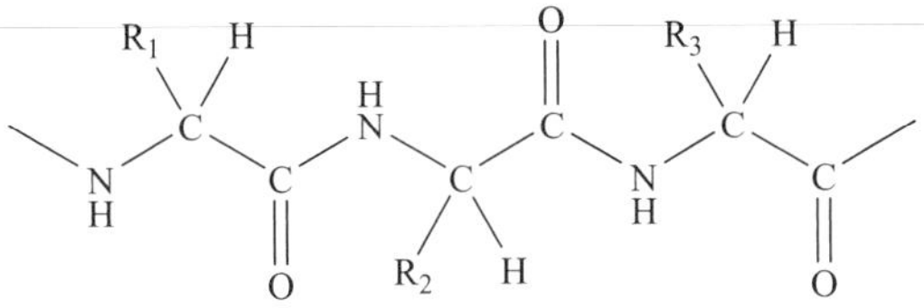

Figure 8.8. Primary structure of a protein (R_1, R_2, R_3 are alkyl groups).

processing might be appropriate. The protein is developed as the amino acids join together through peptide bonds (Figure 8.8).

The nature of the protein is then determined by the exact sequence of amino acids from which it is comprised.

8.2.2 Production of Chemicals from Renewable Resources

The chemical industry is heavily dependent on oil, which is the major feedstock for the production of chemicals, and a significant source of energy, particularly for vehicle applications. More than 90% of all organic compounds are derived from petroleum. However, the increased world demand for petroleum-based products and energy, as well as the finite reserve of crude oil, pose an enormous sustainability challenge. Combustion of fossil resources produces CO_2, which leads to climate change. And there is no doubt that fossil resources are in limited supply. Thus, strategies based on renewable sources are warranted.

Biomass resources can be used directly as a fuel, generally through combustion. But to be able to derive a chemical industry based on renewable resources, it is necessary to convert the structures within the biomass into useful chemicals that can replace chemicals currently produced from petroleum-derived compounds. This remains a significant challenge and the basis of a great deal of ongoing research.

Recognizing the efficiencies of the petroleum refinery in producing a large variety of products simultaneously, conventional thought is that the most effective plan for conversion of bioresources will be through a biorefinery process. In the petrorefinery process based on a "build-up" approach, feedstocks such as crude oil

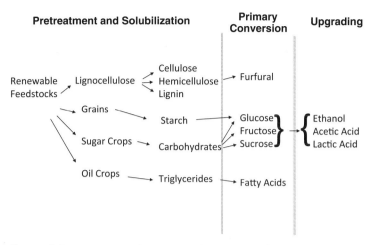

Figure 8.9. Conversion of renewable feedstocks to fuels and chemicals.

and natural gas are converted into products such as energy (fuels), specialty and commodity chemicals, and materials (plastics). In the integrated biorefinery approach the feedstock is biomass (trees, crops, plants, grass, waste, etc.), which is changed into energy, chemicals, and materials plus food and feed. However, one crucial difference between the two processes is that the plant materials need to be broken down into simpler building block molecules first, before they can be converted into more complex products. Thus, a typical biorefinery might involve the steps shown in Figure 8.9 as the basis by which fuels and chemicals can be produced from renewable resources.

In general, the reactions needed for the conversion process required for biomass resources are different from those associated with fossil resources. Fossil fuels contain primarily carbon and hydrogen, and the conversion processes are based on oxidation of the derived molecules. Most of the species are not soluble in water and remain relatively stable at elevated temperatures. Bio-derived molecules are very different. They often contain significant quantities of oxygen and must be reduced to give the desired functionalities. They decompose at elevated temperatures and are soluble in water. So while the same range of chemistry is available for the processing of bio-derived molecules as would be the case for fossil resources, such processes are not usually economically practical, and new processing techniques are required.

There are two primary mechanisms available for conversion of biomass into useful chemicals, as shown in Figure 8.10 [5]. One is through thermochemical treatment, which essentially takes advantage of traditional chemical processing techniques to convert the biomass into a material that can be accommodated through conventional technologies used for processing of fossil resources. A second route focuses on biological processes using enzymatic methods and processing in water, usually at relatively low temperature, but often at rates significantly slower than traditional chemical processes. Either route can provide a viable pathway to desired chemical products and should be evaluated on a life cycle basis to determine the

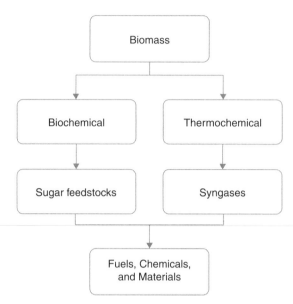

Figure 8.10. The biorefinery concept.

sustainability of the process route. Eventually, the biorefinery will need to take advantage of both platforms through integrated processes that also allow for interchange of materials between the two routes and the use of thermal energy produced as a part of the processing.

8.2.2.1 Conversion of Biomass Through Thermochemical Processes

Gasification is a thermochemical technique for converting biomass into a product conventionally called synthesis gas (or syngas), essentially CO and H_2. The process is conducted at elevated temperatures (typically greater than $700\,°C$), with only a small amount (or no) oxygen present. By carefully controlling the oxygen content in the feed gas, the composition of the product can be controlled. Typically, methane and CO_2 are also produced during gasification, as well as some liquid oils and solid char. Oils produced during gasification can be separately recovered and recycled for further gasification or converted to other materials. However, the char is essentially a pure carbon waste product that has very limited commercial value and is thus an undesirable side product of the reaction. By adjusting the process conditions, one can tune the composition of the products to achieve a desirable gas mixture that can be used for later processes. A typical thermochemical process is pictured in Figure 8.11.

Syngas is a platform product that allows biomass to be converted to traditional chemical products, most typically through a process known as Fischer–Tropsch (F-T). The F-T technology is not new and has been employed for the conversion of

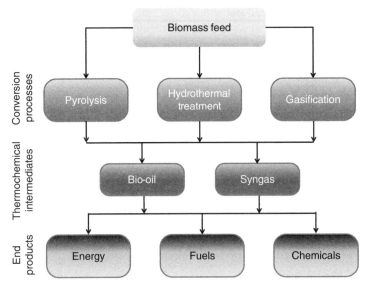

Figure 8.11. A typical thermochemical process and its products.

coal and natural gas to chemicals since World War I, when the Germans used this process for the production of gasoline. More recently, South Africa relied on this technology to produce hydrocarbon products from coal throughout the apartheid era. In all of these cases, the economics of the process relative to other choices for obtaining the desired product drove the development of the technology. The current interest for conversion of biomass to traditional hydrocarbon products represents a similar economic opportunity, since the efficiency of other conversion processes has not yet been demonstrated on a large scale.

The first step of the gasification process is through combustion, in which the carbon and hydrogen are converted to CO_2 and water. Because the reactions are exothermic, temperatures in excess of $1000\,°C$ can be achieved, depending on how much residual water is retained within the fuel. The extent of combustion is controlled by the amount of oxygen present in the fuel. After all of the oxygen is consumed, a series of reduction reactions occur that convert the CO_2 to CO, either by disproportionation with residual carbon

$$C + CO_2 \rightarrow 2CO \tag{8.1}$$

or through reaction of residual carbon with water

$$C + H_2O \rightarrow CO + H_2 \tag{8.2}$$

or through the water gas shift reaction

$$CO_2 + H_2 \leftrightarrow CO + H_2O \tag{8.3}$$

When the reaction is carried out using air as the source of oxygen, a product is formed that is roughly equal parts of syngas, containing roughly equal portions of CO, H_2, and CO_2, and the remainder being diluent nitrogen. Pure oxygen can be used in place of air, which substantially increases the energy content of the syngas, but at increased cost required for the separation of oxygen from air.

Once the syngas is produced, the Fischer–Tropsch process can be used to produce primarily alkanes, through what is essentially a polymerization reaction

$$(2n+1)H_2 + nCO \rightarrow CnH_{2n+2} + nH_2O \qquad (8.4)$$

where n is a positive integer. A range of alkanes are produced, generally following the Anderson–Shultz–Flory distribution, described as

$$W_n / n = (1-\alpha)^2 \alpha^{n-1} \qquad (8.5)$$

where W_n is the weight fraction of hydrocarbon molecules containing n carbon atoms and α is the chain growth probability. The selection of the catalyst and process conditions determines the value of α, which in turn determines the relative fraction of small or large hydrocarbon molecules in the product. The reaction also produces alkenes and alcohols as side products.

Two different catalysts can be used, depending on the desired products. A highly active cobalt-based catalyst can be used for low-temperature conversion at 150–200 °C, whereas an iron-based material is used at elevated temperatures of 300–350 °C. During the low-temperature synthesis, the catalyst is dispersed in a liquid slurry, whereas the high-temperature process is normally carried out in the vapor phase. The high-temperature process yields a greater preponderance of oxygenates, whereas the low-temperature process yields a product that more closely mimics traditional diesel fuel. The low-temperature F-T process is considered the more efficient, because the reactor cost is less, lower catalyst loading is possible, higher conversions can be achieved, and the catalyst has greater longevity. Regardless of the F-T process used, the product must then be upgraded to create a viable alternative fuel, or further refined to create the desired chemical product.

8.2.2.2 Conversion of Glucose Through Chemical Processes

Naturally occurring sugars such as glucose and fructose are gaining popularity as one of the most used feedstocks for the production of many commodity materials. Glucose, the building block of cellulose, can be converted to 5-hydroxymethylfurfural (HMF), a "platform chemical" in the production of compounds currently derived from petroleum such as solvents, fuels, and monomers [6]. HMF is produced through dehydration of glucose, by conversion to fructose through isomerization in the presence of alkali, and then through dehydration in the presence of an acid catalyst, or with titanium dioxide TiO_2. In either case, the reaction is carried out in hot compressed water at about 473 °C (Figure 8.12).

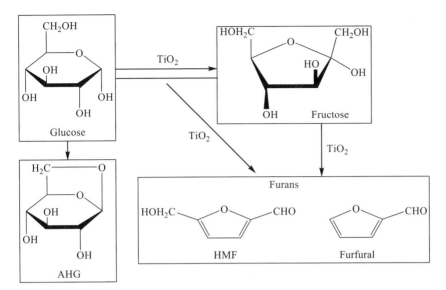

Figure 8.12. Conversion of glucose to HMF.

HMF can then be converted to 2,5-diformylfuran through oxidation, from which it can be used as a monomer for polymer formation, or as an intermediate for pharmaceuticals and other chemical intermediates [7].

Numerous chemicals can also be derived through fermentation of glucose and sucrose. A wide range of microorganisms can be used to break down the sugars into smaller molecules, under various reaction conditions and with variable efficiencies. However, these reactions are generally not particularly selective, as the microorganism is supporting the reaction in order to produce energy for its own sustenance. Thus, the products of the reaction are often dilute and in low yield, and substantial separation and purification steps are required to achieve a useful chemical product. Regardless, the ability to make chemicals directly from biomass continues to increase in interest, as the costs and environmental impacts of consuming nonrenewable resources become more definite.

Fermentation of glucose yields several primary acid products, as shown in Figure 8.13 [8]. Each of these products can then be converted to other useful chemicals.

Approximately 350,000 tons/year of lactic acid is produced from glucose. The lactic acid can be converted through traditional chemical routes to yield well-known chemical products such as acrylic acid and propanediol. Lactic acid is also the building block for polylactic acid (PLA), a biodegradable polymer that has the potential to replace several existing polymer products in high use (Figure 8.14) [9].

The other primary fermentation products of glucose are equally valuable as raw materials for chemical production. Succinic acid can be converted through hydrogenation to butanediol, and then through dehydration to tetrahydrofuran. 3-Hydroxypropionic acid (3HPA) has bifunctionality that allows for reduction of the acid group to alcohol, esters, and amides, dehydration of the alcohol function to

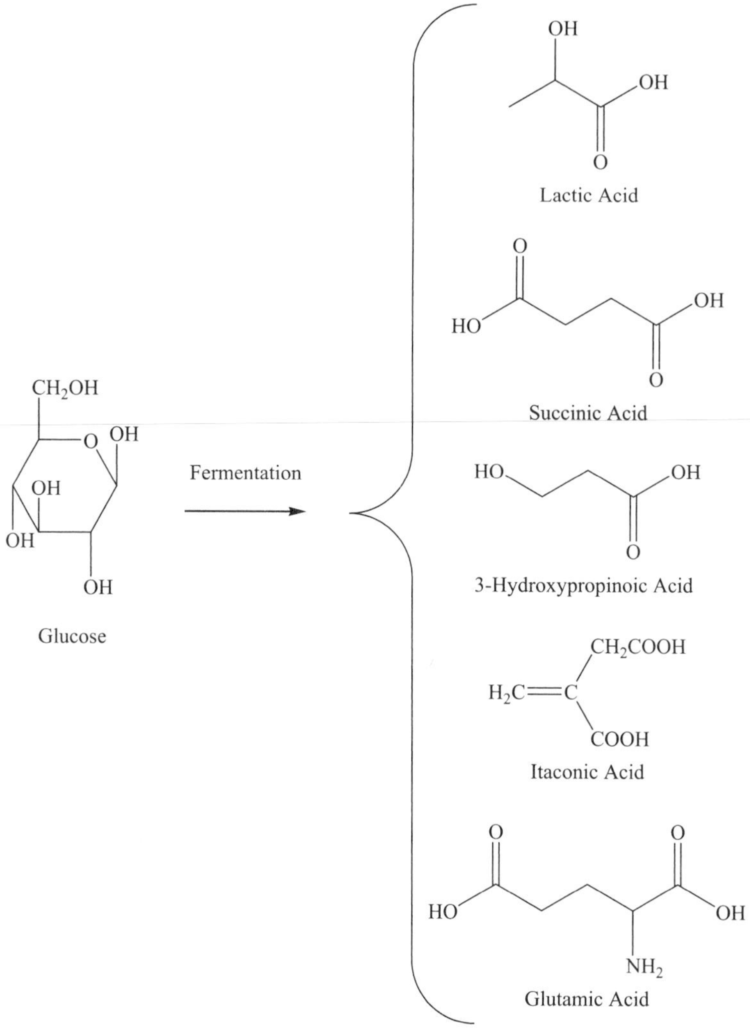

Figure 8.13. Products of fermentation of glucose.

produce unsaturated compounds, polymerization to polyesters and oligomers, and cyclization to propiolactone and lactides.

While it is clear that conversion of glucose into chemical building blocks through fermentation is a valuable route to chemicals production, glucose can be a valuable raw material in its own right. One of the important commercial synthetic routes to the production of vitamin C, ascorbic acid, is based on glucose [10]. Several process improvements have been made since the basic synthetic route developed in the 1930s. The bacterial process now uses *Acetobacter suboxydans*, a bacterium more resistant to the nickel catalyst used in the hydrogenation step. This nickel catalyst is the subject

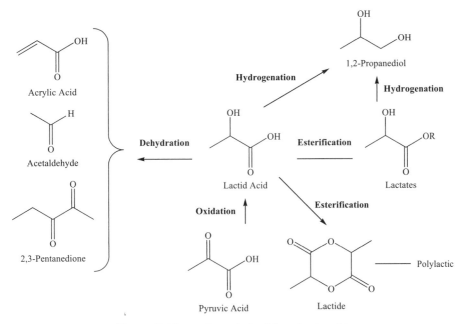

Figure 8.14. Products derived from lactic acid.

of environmental concern due to its toxicity. More recently, a one-step synthesis of L-ascorbic acid from D-glucose was presented [10]. According to the U.S. Patent #5998634 this mixture is "catalytically oxidized in aqueous solution by hypochlorous acid. L-Ascorbic acid can be separated from the aqueous solution, and the unconverted reactants recycled for greater conversion."

Glucose is also an excellent replacement for benzene, a known carcinogen, in the synthesis of many aromatic compounds such as hydroquinone [$C_6H_4(OH)_2$], catechol [$C_6H_4(OH)_2$], and adipic acid [$(CH_2)_4(COOH)_2$] represented in Figure 8.15 [1, p. 94].

The use of glucose instead of benzene allows the substitution of organic solvents by water. The standard process for the production of adipic acid is based on the nitric acid oxidation of a cyclohexanol/cyclohexanone mixture derived from cyclohexane, which is in turn derived from benzene. Note that this process produces N_2O as a by-product of the reaction (Scheme 8.1).

Alternatively, adipic acid can be produced through biological conversion of benzoic acid or other aromatics derived from glucose to *cis*-muconic acid, which can then be converted to adipic acid by hydrogenation. Life cycle analysis of these two processes shows that the biological route can reduce the CO_2 equivalent emissions from 17.4 to 14.0 ton CO_2 eq/ton adipic acid, even beyond the environmental benefits associated with the elimination of the use of benzene derived from fossil resources [11].

Another example of the direct use of a renewable chemical is the biocatalytic conversion of D-glucose into vanillin used as a flavoring agent in food and beverages [12]. The use of a recombinant *Escherichia coli* biocatalyst in fermentation offers many advantages over the synthetic vanillin manufacture based on the use of

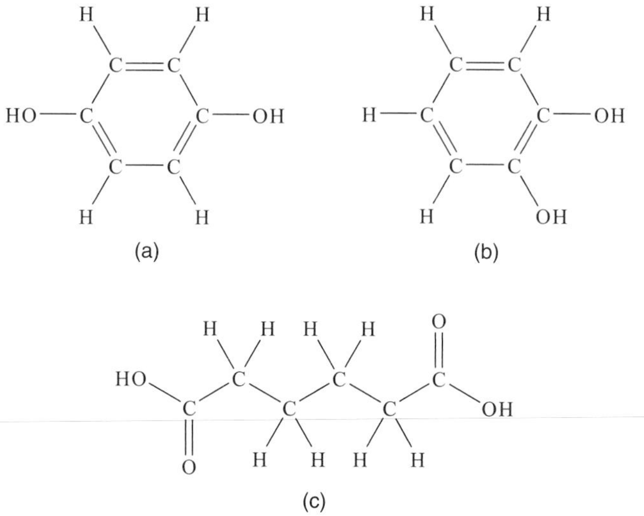

Figure 8.15. Structural formulas of (a) hydroquinone, (b) catechol, and (c) adipic acid.

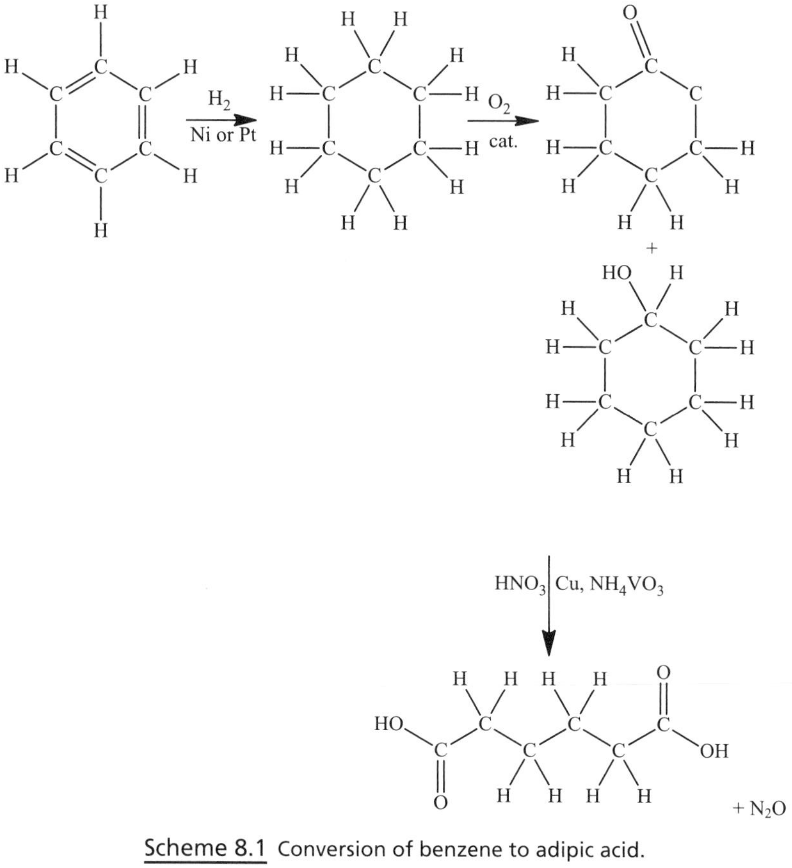

Scheme 8.1 Conversion of benzene to adipic acid.

nonrenewable petroleum. The feedstock commonly used in the synthesis of vanillin is phenol derived from carcinogenic benzene; phenol gets eliminated and replaced by glucose derived from renewable starch.

8.3 APPLICATIONS OF RENEWABLE MATERIALS

8.3.1 The Case of Biodegradable Plastics

In almost every product we buy, there is some plastic! Packaging and containers is the largest market for plastics. They are cheap to produce, ensure excellent protection to the product, and last forever. Nevertheless, this third property is not ideal from an environmental point of view since traditional plastics are manufactured from nonrenewable and fossil-fuel-based resources such as oil, coal, and natural gas.

Designing for biodegradability is part of the intrinsic nature of green chemistry. The tenth principle of green chemistry states that "chemical products should be designed so that at the end of their function they do not persist in the environment and break down into innocuous degradation products" [1, p. 52].

A common definition of "biodegradable" is "capable of being broken down especially into innocuous products by the action of living things (microorganisms)" according to the Merriam–Webster online dictionary. The biodegradation of substances in the environment is helped by the action of bacteria, fungi, and protozoa. The goal of biodegradation is to reduce and, if possible, eliminate the toxicity of a substance by modifying it. Biodegradable plastics are typically derived from naturally occurring materials such as lactic acid and can be broken down by naturally occurring organisms. Highlight 8.1 provides options on what to do with plastic bottles.

Highlight 8.1 What Can We Do with Plastic Bottles?

Even if recycling plastic bottles seems very common these days, only 23% of water plastic bottles that Americans use are recycled. This leads to 38 billion water bottles ending up in landfills, where they can remain in their manufactured form for 1000 years before they are fully decomposed. When water bottles are burned in an incinerator, they can release toxic fumes and other gases involved in the destruction of the ozone layer.

Recycled water bottles made of polyethylene terephthalate (PET) have found many uses, especially in the clothing industry. Some of the companies using recycled PET fabric include Billabong's Eco-Supreme Suede, Wellman Inc.'s Eco-fi, and Reware's Rewoven.

The most-often recycled plastic is polyethylene, designated with the number 2 on bottles, which can be used in applications such as roadside curbs, trash receptacles, and benches.

However, the bottle's cap is not made of the same type of polymer as the bottle itself and therefore needs to be separated from the PET bottle during the recycling process.

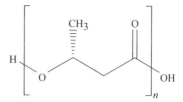

Figure 8.16 . Structure of poly-3-hydroxybutyrate (PHB).

Polyhydroxyalkanoates or PHA are thermoplastics produced by bacterial fermentation of sugars or lipids. Maurice Lemoigne, Director of the Fermentation Laboratory of the Pasteur Institute in Lille, France, first isolated and characterized the simplest and most commonly occurring polyester of PHA, poly-3-hydroxybutyrate, PHB or poly(3HB) represented in Figure 8.16, in the 1920s [13].

The production of PHA is triggered by a microorganism such as *Alcaligenes eutrophus* under certain conditions such as lack of oxygen and excess of carbon supply. The polyester is then extracted and purified from the bacteria during the fermentation of carbohydrates such as sugar and glucose, or vegetable oil or glycerine from biodiesel production.

Polylactic acid or PLA is a plastic made from corn, an annually renewable resource, leading to 100% biodegradability and compostability. Plastic products made from corn are in all common household products all around the world. The NatureWorks® PLA is derived from naturally occurring plant starch. According to the NatureWorks LLC website, the different steps of the transformation of plant sugar into NatureWorks PLA are the following [14]:

1. *Photosynthesis* (use of sunlight) used in conjunction with carbon dioxide and water create glucose (sugar) and oxygen. The plant uses the sugar as a fuel and the unused sugar gets stored as starch in the kernel of the corn.
2. After the *harvest of field corn*, it is cooked and then ground and screened to isolate the starch.
3. The starch is converted into sugar and microorganisms convert the starch into lactic acid through fermentation.
4. The lactic acid molecules form the monomer lactic acid, which then is polymerized to form the polylactide polymer. During the polymerization, tens of thousands of monomers of lactic acid are linked together.
5. The polymer PLA is then shaped into pellets, which are sold under the brandname NatureWorks PLA, which is often found in packaging and fiber products.

The PLA polymer can be converted into a number of common products beyond simple plastic sheet. One of the most commonly seen products is compostable plastic tableware (cutlery, plates, etc.), which can result in 50% less nonrenewable energy consumption and 60% fewer greenhouse gas emissions. Other applications include apparel, fresh food packaging, and durable goods such as computers and cell phones.

In 2000, 3 million tons of polystyrene, commonly known as Styrofoam, was made for use in disposable coffee cups and food packaging, and 2.3 million tons was thrown away. Polystyrene waste does not break down quickly and is difficult to recycle. A new strategy based on the recycling of Styrofoam into PHA was developed by a team of microbiologists at University College Dublin [15]. These scientists found that a soil bacterium, based on *Pseudomonas putida*, thrives on dirty styrene oil (C_8H_8) derived from the pyrolysis of polystyrene. In this type of reaction, the styrene is the source of carbon, which is then stored as PHA (Figure 8.17).

Highlight 8.2 focuses on a type of biodegradable plastic bottles named ENSO.

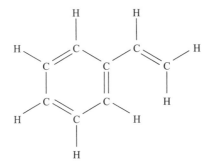

Figure 8.17. Molecular structure of styrene.

Highlight 8.2   Biodegradable Plastic Bottles: ENSO Bottles

A company named ENSO Bottles LLC and launched in 2008 is making biodegradable plastic bottles. ENSO bottles are not made of PLA or corn-based plastic. They are made of everyday plastic and an undisclosed organic additive. They have similar durability and shelf life properties as regular plastic bottles.

When these ENSO bottles are not recycled and end up in an anaerobic landfill environment, the microbes are attracted to the organic additive and eat through the plastic for food and energy. These microbes break down the ENSO bottles into biogases and inert humus. The plastic bottle making process is the same as with any other plastic with the exception of the addition of the organic compound. Thus, there is no need to change the existing process and therefore ENSO bottles are part of the natural cycle of life.

8.3.2 The Case of Compostable Chemicals

In 2010, SunChips® bags became the first fully compostable chip bag made of PLA film [16]. Researchers have found that under aerobic compost conditions (in the presence of oxygen) at >55 °C the film structure completely breaks down in 12 to 16 weeks. These conditions are very similar to those used in a typical industrial compost facility. However, under anaerobic conditions commonly found in a landfill, the film would not decompose. In a home composting process under adiabatic conditions (in the absence of heat transfer) and in the presence of an ideal mixture of greens, leaves, and grass clippings, the SunChips bags completely decomposed in about 12 to 16 weeks. If the conditions are less than ideal (poor mixture of greens and browns with lower reaction temperatures) the decomposition would still take place but at a slower pace.

Highlight 8.3 focuses on composting and compostable available materials.

8.3.3 Production of Ethanol from Biomass

The production of ethanol from cellulosic biomass for automobile fuel is not a new idea. The ancient Greek society was already using agricultural products such as grains to produce ethanol. Ethanol is currently used as an alternative to gasoline in flex-fuel light vehicles in Brazil and as an oxygenate for gasoline in the United States. Most cars in the United States can function with a mixture of 10% ethanol mixed with gasoline. The main focus for the production of bio-ethanol is to use fermentation of starch or sugar present in a wide variety of crops such as sugarcane,

---

Highlight 8.3   Old Plastics, Fresh Dirt

It is commonly believed that waste simply biodegrades or decomposes in a landfill. However, since most landfills are kept dry and deprived of oxygen, this is not true. Waste is not supposed to biodegrade in a landfill and some biodegradable products may only be partially biodegradable. When the latter happens, groundwater pollution and gas emissions can lead to more serious problems.

Since biodegradability may not always be a favorable option, scientists have studied how to control biodegradation outside of a landfill through composting. Many compostable materials, such as bags, films, resins, food ware, and packaging items, can naturally be converted to useful products for farming or gardening.

While we all know that food, leaves, grass clippings, and garden wastes can all go into the compost pile, it is less common to throw a cup or a lid on that compost pile. This is definitely possible through the purchase of GreenGood PLA goods such as cups and cutlery that are 100% biodegradable and compostable.

sugar beet, potato, sunflower, wheat, straw, cotton, switchgrass, manioc, corn, and any other types of cellulose-containing biomass or waste. However, growing crops for fuel raises several other sustainability questions:

1. The production of crops requires substantial consumption of fertilizer, most of which is produced from nonrenewable sources such as gas. The life cycle analysis indicates that ethanol production from crops is only marginally beneficial for the reduction of consumption of nonrenewables.

2. Social justice aspects of sustainability require that one considers the food versus fuel aspects of producing ethanol from food crops. When arable land is used for the production of fuel, less food is produced, leading to an increase in food prices and an increase in worldwide hunger.

While there are clear concerns with using food crops as a source of ethanol, other crops, such as wood-type material and specifically fibrous cellulose as a source of biomass, are likely to play a bigger role in the future. Cellulose, one of the carbohydrate polymers present in woody biomass, can serve as a potential raw material for the production of fuel ethanol.

Before ethanol can be produced from woody biomass, the cellulose needs to be released from the biomass superstructure. This can be accomplished chemically through acid pretreatment or steam explosion. Afterwards, the liberated cellulose chains are broken down through hydrolysis to produce starch and sugars. The conversion of cellulose into sugar may also be achieved through the use of enzymes and yeasts. Cellulase is usually used to hydrolyze the cellulose into glucose before the fermentation step.

Glucose is decomposed into ethanol and carbon dioxide through fermentation, per the following general reaction pathway:

$$C_6H_{12}O_6\,(\text{glucose}) \rightarrow 2C_2H_5OH\,(\text{ethanol}) + 2CO_2\,(\text{carbon dioxide}) + \text{heat} \quad (8.6)$$

Many bacteria may be used to catalyze the fermentation reaction, and depending on the choice of microorganism and the process conditions, products other than ethanol may be obtained. Microorganisms such as *Saccharomyces cerevisae*, *Zymomonas* sp., and *Candida* sp. are well-known fermentation agents. However, *Saccharomyces cerevisae*, commonly known as baker's yeast, can be used either in its natural form or as a genetically modified form.

There is also interest in converting biomass to butanol, since butanol is more similar to gasoline and less soluble in water. *Clostridium acetobutylicum* is the typical microorganism for the process, which is similar to the process used for production of ethanol.

It is essential to keep in mind that the selection of the appropriate enzyme and process conditions is specific to the source of biomass and the desired end product. High performance yeast strains are now selected and commercialized for dry grind corn ethanol production using batch fermentation [17]. The commercial yeast formats and fermentation conditions should be optimized. Stress factors affecting yeast metabolism include high temperatures, high osmotic pressure, high sodium and other ions concentrations, and high concentrations of organic acids. Prevention of bacterial growth is also essential to successful ethanol production.

Fermentation may be accomplished in either a batch or continuous mode. Regardless, the rate of the reaction determines the ethanol yield and is described through traditional Michaelis–Menton kinetics. In this case, the rate of reaction is written as

$$-r_A = \frac{k[E][A]}{K+[A]} \tag{8.7}$$

where [E] is the concentration of an enzyme that catalyzes the reaction, [A] is the concentration of the reactant, and k and K are two different characteristic constants. This is more conventionally written as

$$-r_A = \frac{V_{max}[A]}{K_m+[A]} \tag{8.8}$$

where V_{max} is the maximum rate of reaction for a specific enzyme concentration. Operating under differential conditions, we substitute the rate expression into the material balance

$$[A]\frac{X}{\tau} = \frac{V_{max}[A]}{K_m+[A]} \tag{8.9}$$

This equation is also sometimes written as Monod's equation, except the reaction velocity is replaced with the specific growth rate (μ_{max}) for the microorganism

$$\mu = \frac{\mu_{max}[S]}{K_m+[S]} \tag{8.10}$$

When performed in batch mode, the cell density varies with time, which impacts the overall formation of ethanol product. A typical growth curve is described in Figure 8.18.

8.3.4 The Case of Flex-Fuel Vehicles

While most automotive fuel sold in the United States contains 10% ethanol, it is also possible to power the vehicle on a blend that is primarily ethanol. The optimum blend sold in both the U.S. and European markets for flexible-fuel vehicles is E85 (mixture of gasoline containing 85% of ethanol). In Brazil, flex-fuel vehicles can operate with ethanol mixtures up to E100. The Brazilian 2008 Honda Civic flex-fuel car is built with a secondary reservoir gasoline tank. In March 2009 the first Polo E-Flex was launched in Brazil without the auxiliary gasoline tank. Ethanol-based engines can be used to power automobiles, tractors, city buses, distribution trucks, and waste collectors.

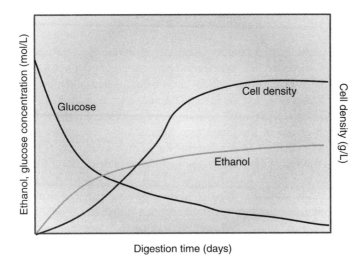

Figure 8.18. Representation of the production of cell density, and the change of glucose and ethanol concentration through digestion under anaerobic conditions.

As with any other combustion reaction, ethanol reacts with oxygen to form carbon dioxide and water, releasing significant amounts of heat:

$$C_2H_5OH + 3O_2 \rightarrow 2CO_2 + 3H_2O + heat \qquad (8.11)$$

The high temperature that is generated from combustion and the large volume of combustion gases cause substantial pressure accumulation in the engine cylinder chamber, which pushes back on the piston through the transfer of work.

In order to start the combustion reaction, there must be sufficient concentrations of air and fuel in the vapor phase, and a spark to ignite the mixture. During cold weather, the vapor pressure of ethanol is insufficient to generate enough ethanol in the vapor to spark the ignition. This problem can be alleviated by including a small secondary gasoline reservoir, which allows the car to start on pure gasoline at low temperatures, as is typical of Brazilian flex-fuel vehicles.

The United States and Brazil are the world's top ethanol fuel producers accounting together for 70% of the world's production and nearly 90% of ethanol used for fuel. The production of ethanol in Brazil is more than 30 years old and is stimulated through low-interest loans for the building of ethanol distilleries as well as tax incentives provided to the buyers of ethanol vehicles. There are no pure gasoline run vehicles in Brazil. It is now mandatory for light vehicles to run on a blend of ethanol (20%) with gasoline. The development of flex-fuel technology is now spreading across the motorcycle industry.

8.3.5 Production of Biodiesel

Biodiesel is a nontoxic and biodegradable diesel fuel equivalent made from plant oils, chemically converted to alkyl esters. The most common feedstock is vegetable oil such as soybean or corn oil (either virgin or used), but other sources including animal fats such as tallow, lard, or chicken fat may also be used. It is also possible to produce bio-oil from algae used to treat sewage waste. Production of biodiesel requires the conversion of the long-chain fatty acids in the oil into a triglyceride through chemical reaction. Depending on the source of the oil, different process conditions are required and different yields can be obtained.

Biodiesel can be used in its pure form (B100) or blended with petroleum diesel (a common blend is known as B20, a mixture of 80% petroleum diesel and 20% biodiesel). Biodiesel is not raw vegetable oil, but rather is a product derived from the oil. Biodiesel has a slightly lower fuel value than petroleum diesel, but has greater lubricity than current low sulfur diesel fuel. Biodiesel also has a higher flash point, and will gel at low temperatures of about −10°C, making use in cold weather conditions difficult. Biodiesel is compatible with most parts in commercial diesel engine systems, and thus could almost be used as a drop-in replacement in trucks, buses, boats, and construction equipment. However, the difficulties associated with low-temperature operation and the potential degradation of plastic parts in the fuel system have made the B20 blend the typical fuel associated with biodiesel use.

The process of esterification used for the production of biodiesel involves a chemical reaction of a fat or a vegetable oil with an alcohol, usually methanol, in the presence of a catalyst like sodium or potassium hydroxide (KOH). This is the preferred process, because it operates at low temperature and pressure and achieves high conversion (98%). The products are glycerine and fatty-acid methyl esters or biodiesel consisting of straight saturated and unsaturated hydrocarbon chains. The methanol is usually in excess to provide quick conversion but is recovered for reuse. In this first route nothing is wasted [18]. The inputs are alcohol (12%), catalyst (1%), and oil (87%). The outputs are alcohol (4%), fertilizer (1%), glycerine (9%), and methyl ester (86%). It is also possible to produce biodiesel through acid-catalyzed esterification of the oil with methanol, or by conversion of the oil to fatty acids and then to alkyl esters using acid catalysis.

Instead of purified vegetable oil, recycled grease can also be used as a starting material for biodiesel production. In this case, an acidic esterification comprising sulfuric acid and methanol is the first step before the transesterification can occur. The basic technology is summarized in Figure 8.19.

Biodiesel can also be produced from algae grown on waste biomass. The requirements for algae growth are just a few: carbon dioxide, sun, and water. Algae can grow under any conditions: high or low temperatures, acidic or alkaline environments, and any level of salinity. Algae-growing facilities can be built any-where and the agricultural land is not depleted to the extent of crops grown for fuel production. The production of biofuels from biomass, such as agricultural waste, is appealing from multiple points of view:

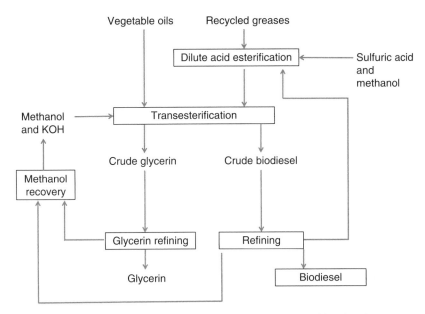

Figure 8.19. Basic technology for the production of biodiesel.

1. The feedstock is not dependent on the availability of fossil fuels.
2. Burning biofuels decreases the net amount of carbon dioxide rejected into the atmosphere since the amount of carbon dioxide given off in the process should be similar to the amount originally absorbed by the biomass.
3. The use of genetically engineered organisms can produce biofuels more efficiently and lead to a cost-competitive energy production.
4. Waste biomass can be used as the food on which algae are grown, and this growth process does not compete with the need to produce food for animal or human consumption.

The key to successful growth of algae is the correct identification of strains with the highest oil content as well as the development of cost-effective growing and harvesting methods. Algae can be grown either in ponds or lakes or in closed, translucent tubes or containers called photobioreactors. One of the challenges with converting algae to fuel is harvesting the algae from the growth medium, which tends to be easier in closed systems. The use of microalgae with high growth rates, high lipid production, and relative ease of handling in conventional reactor systems has increased the availability of biodiesel and other bio-based fuel resources.

Biodiesel 2020, a study exploring the opportunities for biodiesel fuel through the next decade, projects a rapid growth in the production of biodiesel from algae [19]. Some growth areas include:

1. Capture the carbon dioxide from the smokestacks of a power plant and use the carbon to produce algae for low-emissions biofuels for transport.

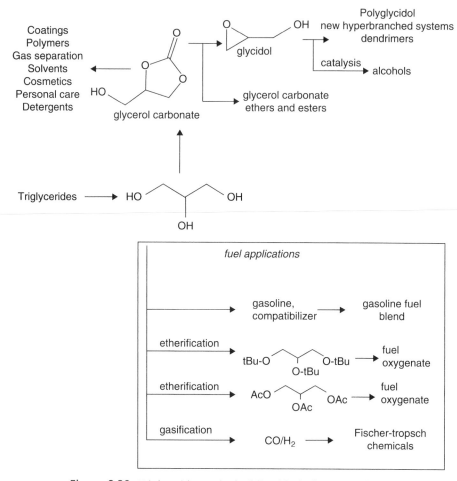

Figure 8.20. Triglycerides as the building blocks for a biorefinery.

2. Use the entire algae plant mass and convert it via a biomass-to-liquid (BTL) thermochemical process to create bio-oil or bio-crude oil, which can then be reprocessed into diesel.
3. Use waste water from municipalities to grow algae.

Some of these opportunities are already present today. The company GreenFuels uses algae to capture and convert the carbon from coal-power generation sources into biodiesel or ethanol for transport fuel. Other private ventures, such as PetroAlgae, are finding ways to expand the production of algae oil using large open pond technology. A public–private venture between UOP (a division of Honeywell) and the U.S. Defense Advanced Research Project Agency is working toward the conversion of algae into military grade jet fuel.

While the oils associated with vegetable matter and those pressed from algae used in waste treatment processes are most typically converted to biodiesel fuel, it is also possible to produce a broad range of products from vegetable oil and animal fats. These products can then once again serve as the basis for the production of chemicals from renewable resources (Figure 8.20). As with all bio-based products, it is important to complete a full economic and environmental life cycle analysis to determine the most viable routes for desired products.

8.4 CONCLUSION

The use of renewable materials for chemicals and energy production is becoming increasingly attractive as realization grows that fossil resources are limited. While there are other choices for the production of energy, which will be covered in Chapter 9, the conversion of biomass represents a real alternative for production of chemicals. Thus, the growth and expansion of the integrated biorefinery is of significant interest for the modern chemical company.

The Office of Biomass Energy in the Department of Energy has a significant research program focused on better technologies for using biomass resources, and while it is primarily focused on energy production, it also addresses the conversion of renewables for chemicals. Its program encompasses activities well beyond the traditional scope of the chemist and the chemical engineer, from achieving greater yield during crop growth, to the use of new fertilizers and pesticides, conversion of biomass into energy and chemicals, and ultimately waste disposal.

Several issues remain to be resolved. For one, it is not typically cost effective for large-scale production from renewable resources. Sourcing is a concern, and it is unclear whether there are sufficient biomass resources available to meet societal demands. The intermittency of biomass availability is also an issue. All of these aspects need to be considered in the evaluation of what may ultimately become the major source of chemicals and energy for the future.

REFERENCES

1. Anastas, P. T.; Warner, J. C. *Green Chemistry: Theory and Practice*, Oxford University Press, New York, 1998.

2. http://en.wikipedia.org/wiki/Biomass.

3. http://www.biomassenergycentre.org.uk/portal/page?_pageid= 76,15049&_dad=portal&_schema=PORTAL.

4. http://www.biomassenergycentre.org.uk/portal/page?_ pageid=76,15068&_dad=portal&_schema=PORTAL or http://www.sustainableheatingsolutions.com/explained_biomass.php.

5. http://www.nrel.gov/biomass/biorefinery.html.

6. *Chemical and Engineering News*, July 6, 2009, p. 26–28.

7. Cao, S.; Yeung, K. L.; Kwan, J. K. C.; To, P. M. T; Yu, S. C. T. *Applied Catalysis B: Environmental*, **2009**, *86*(3–4), 127–136.

8. De Almeida, R. F.; Nederlof, C.; Li, J.; Mouijn, J. A.; O'Connor, P.; Makkee, M. *Chem. Rev.*, **2007**, *107*(6), 2411–2502.

9. Lancaster, M. *Green Chemistry: An Introductory Text*, The Royal Society of Chemistry, London, 2007, p. 269.

10. Murphy, A. P.; Henthorne, L. R. *One-step synthesis of vitamin-C (L-ascorbic acid)*, December 7, 1999, U. S. Patent #5998634.

11. Van Duuren, J. B. J. H; Brehmer, B.; Mars, A. E.; Eggink, G.; Martins dos Santos, V. A. P., Sanders, J. P. M. *Biotechnol. Bioeng.*, **2011**, *108*, 1298–1306.

12. Li, K.; Frost, J. W. *J. Am. Chem. Soc.*, **1998**, *120*, 10545–10546.

13. Lenz, R. W.; Marchessault, R. H. *Biomacromolecules*, **2005**, *6*(1), 1–8.

14. www.natureworksllc.com.

15. Ward, P. G.; Goff, M.; Donner, M.; Kaminsky, W.; O'Connor, K. E. *Environ. Sci. Technol.*, **2006**, *40*(7), 2433–2437.

16. http://www.sunchips.com/resources/pdf/SunChips_BehindtheScenes.pdf.

17. Knauf, M.; Kraus, K. *Sugar Industry*, **2006**, *131*, 753–758.

18. http://www.alternative-energy-news.info/images/technical/biodiesel.jpg.

19. http://www.emerging-markets.com/PDF/Biodiesel2020Study.pdf.

9

CURRENT AND FUTURE STATE OF ENERGY PRODUCTION AND CONSUMPTION

9.1 INTRODUCTION

In 2010, China became the world's largest energy user, consuming over 2.4 million kilotons of oil equivalent, surpassing the United States for the first time (data provided by the World Bank). China and the United States also represent the world's largest contributors to CO_2 emissions, with over 7 and 5.5 million kilotons of CO_2 emissions, respectively. While the total consumption of energy between these two countries is close, there is a larger gap in the CO_2 emissions, a result of the fuel mix used for the energy supply.

The fuel mix used in the United States has changed dramatically over the years. Most of the energy originally came from wood, then by the late 1800s coal replaced wood and by the mid-1900s oil and natural gas were the primary energy sources. In the 1970s nuclear power emerged but fossil fuels still provide the bulk of our energy sources today. Figure 9.1 reveals how the historical mix has changed over time and indicates that the addition of new energy resources takes an extensive length of time to effectively propagate into the energy mix [1].

Green Chemistry and Engineering: A Pathway to Sustainability,
Anne E. Marteel-Parrish and Martin A. Abraham.
© 2014 American Institute of Chemical Engineers, Inc. Published 2014 by John Wiley & Sons, Inc.

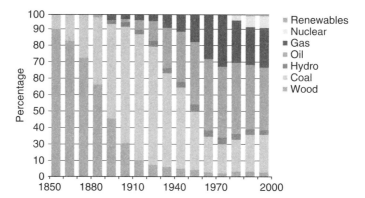

Figure 9.1. Change in the U.S. energy mix over the last 150 years [1]. (Reprinted by permission from Macmillan Publishers Ltd.)

The Energy Information Administration of the U.S. Department of Energy publishes annually a report compiling energy information for the United States, including the sources and sector analysis shown in Figure 3.9. In 2011, the United States consumed a total of 97.3 quadrillion BTU of energy, of which 36% came from petroleum, 26% was produced from natural gas, 20% came from coal, 9% was from renewable sources, and 8% was produced through nuclear power. Among the energy consumers, the transportation sector takes up the greatest share, consuming 28% of the total energy produced and 71% of the petroleum resources. The generation of electric power consumes 40% of the total energy supply, produced from a diverse resource mix, and is then distributed into final end-use sectors [2].

The current U.S. energy mix is based on a complex system of technologies for the production, distribution, and use of fuels and electricity. The Committee of America's Energy Future identified five critical characteristics of this system in 2009 [3, p. 11]:

1. Burning of carbon-based fossil fuels accounted for 85% of U.S. energy needs.
2. The impact of burning fossil fuels is seen mostly in the emission of greenhouse gases, and specifically carbon dioxide (CO_2). In 2007, 6 billion tons of CO_2 was released per year into the atmosphere by the United States alone. However, in the United States, CO_2 emissions have been declining due to the increasing use of renewables and less carbon intensive fuels including natural gas. In developing countries such as China, CO_2 emissions are still increasing due to their rapid economic development and the reliance on the cheapest fuels available: fossil fuels.

   The growth of carbon dioxide emissions is predicted to pose great challenges to international efforts to minimize the impacts of global warming. China and India together use about 19% of the world's energy resources. The U.S. Energy Information Administration expects that by 2020 China and India will consume about one-third of the world's energy resources while the United States will likely reduce its energy consumption slightly [4, p. 28].

The increase in consumption of fossil fuel energy also means a surge in the production of greenhouse gas emissions.

3. Another measure of energy use is annual gross domestic product (GDP) per unit of energy use, which could be one way to measure the energy efficiency of a country. Among OECD countries, Switzerland ranks as the most efficient, at \$11.7 per kg of oil equivalent, in 2010 (World Bank data). The United States was far behind, at \$5.9 per kg of oil equivalent.

4. Energy consumption is connected with economic prosperity. Oil consumption dropped in 2008 and 2009 due to the worldwide economic recession. In the United States, the transportation sector is almost entirely dependent on petroleum. In 2008, more than half of the petroleum consumed in the United States was imported, and in some cases, from unstable and fragile regions of the world [3, p. 14]. Competition for and access to energy plays a central role in U.S. foreign policy. Energy security remains a strong national interest. Recent advances in the use of alternative resources and new discoveries of oil and gas that can be accessed through hydraulic fracturing are beginning to change the political equation, but as with all changes in energy infrastructure, they take significant time to work through this complex system.

5. The energy system's assets are either being depleted or are getting old. There is a great debate as to the total amount of petroleum that can be recovered through existing and proposed technologies. Some experts agree that new technologies will allow increased fossil resource production, whereas others claim that the maximum level of global petroleum production has already been or will soon be reached (and that therefore international conflicts and booming oil barrel prices will arise). Economic theory indicates that as fossil resources become more scarce, prices will go up, which also means that greater cost can be expended to extract those resources. Nonetheless, alternative sources will eventually become more economical, leading to changes in the energy mix.

Infrastructure issues, particularly of transmission and distribution systems, are also of major concern. Many U.S. coal plants are 50–60 years old, beyond their normally expected lifetimes. Retrofits of these plants have kept them operational, but EPA requirements for increased environmental controls limit the economic benefits to continued improvements. Because of cost issues, power companies are beginning to switch to gas-fired power plants. These systems produce less CO_2 emissions than coal plants and have the added advantage that they can be shut down when less energy is needed. New nuclear plants are also being considered, using modular technologies that are more efficient than the 30 year old plants needing to be replaced. Replacing the current and sometimes obsolete infrastructure will be time consuming and pricey.

Since the world's energy consumption is expected to increase by 44% before 2030, there is a global recognition that reducing our reliance on fossil fuels for energy is not only necessary but is the only step toward sustainability. Transforming the way energy is produced, distributed, and consumed is the biggest challenge of the 21st century. But it is not a new challenge: since Franklin D. Roosevelt, several presidents have expressed a need to address environmental impacts and security issues arising from energy production. For example, in 1977, Jimmy Carter introduced a "National Energy Plan"; in 1997, Bill Clinton released the "Federal Energy R&D for the Challenges of the 21st Century"; and in 2001, George W. Bush's published a report on "Reliable, Affordable, and Environmentally Sound Energy for America's Future."

As the leader of one of the world's largest consumer of energy resources, Barack Obama signed Executive Order 13514 in October 2009, which "sets sustainability goals for Federal agencies and focuses on making improvements in their environmental, energy and economic performance" [5]. This Executive Order on federal sustainability focuses on federal agencies "to meet a number of energy, water, and waste reduction targets, including:

- 30% reduction in vehicle fleet petroleum use by 2020;
- 26% improvement in water efficiency by 2020;
- 50% recycling and waste diversion by 2015;
- 95% of all applicable contracts will meet sustainability requirements;
- Implementation of the 2030 net-zero-energy building requirement;
- Implementation of the storm water provisions of the Energy Independence and Security Act of 2007, Section 438; and
- Development of guidance for sustainable Federal building locations in alignment with the Livability Principles put forward by the Department of Housing and Urban Development, the Department of Transportation, and the Environmental Protection Agency" [5].

The world's reliance on fossil resources and continued economic growth pose great threats to the sustainability of the world's energy production. The Kyoto Protocol, adopted in 1997 and put in force in 2005, aimed to reduce the CO_2 emissions by 5.2% from the period 2005–2012. However, the protocol was never ratified by major emitters, such as the United States, and has therefore had limited impact. Current goals to extend the provisions of the Kyoto Protocol through 2020 have not succeeded, as individual countries evaluate the impacts of the proposed plans on their economic growth. While nations debate, the CO_2 levels in the atmosphere continue to rise, and climate variations grow more severe. The National Oceanic and Atmospheric Administration reports December 2012 atmospheric CO_2 at 394.28 ppm, an increase of approximately 1% from the year previously, and an increase of roughly 25% since 1960, as indicated in Figure 9.2 [6].

As described, conversion of a fuel is used to produce energy for the power grid and all other energy needs. Common fuels include natural gas, coal, and oil, but the same processes can be used for renewable materials such as biomass. This is also the basic process that produces energy.

The boiler works by combusting the fuel in the presence of air to produce an exhaust gas at a very high temperature. The gas exchanges heat with the water,

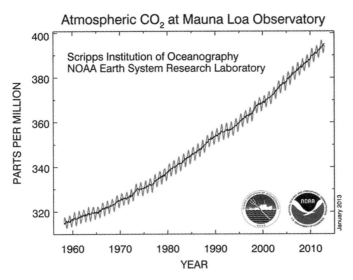

Figure 9.2. Carbon dioxide levels in the atmosphere as reported by NOAA, 1960–present [6].

converting it to steam in the process, through a series of heat exchange tubes. In addition, the radiant heat from the burning fuel can also be recovered in the burner.

The energy from the combusted fuel produces heat, which is transferred to a water stream to produce high-temperature steam, which is ultimately used to power a turbine. The full system, shown in Figure 9.3, consists of four subsystems [7]:

- The water system that transports water into the boiler.
- The fuel system, which controls the fuel flow.
- The draft system, which controls the way in which air is transported through the boiler.
- The steam system, associated with collecting the produced steam.

Most chemical reactions require an energy input for the reaction to occur in a timely fashion on an industrial scale. Given that one of the principles of green chemistry states that "energy requirements of chemical processes should be recognized for their environmental and economic impacts and should be minimized," it is important to understand energy needs of the chemical process. To do so requires one to know first how energy is produced in chemical reactions and second how it can be measured through basic thermodynamic functions.

9.2 BASIC THERMODYNAMIC FUNCTIONS AND APPLICATIONS

When a reaction occurs, there is an energy change in going from the reactants to the products; this energy change is termed the heat of reaction. To measure the total heat or energy emitted or consumed by a reaction, we use a thermodynamic function

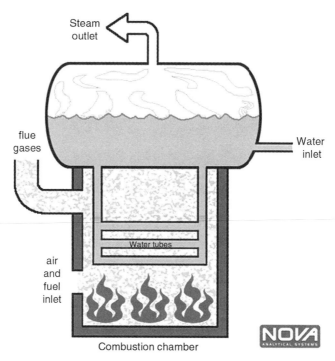

Figure 9.3. Simplified diagram of a steam boiler [7]. (Courtesy of D. Sheasby, NOVA Analytical System, Tenova Group.)

called enthalpy (H). Consider an isomerization reaction, where butylene is converted to isobutylene, for example. We can draw a diagram of this situation:

$$A \longrightarrow \boxed{\text{Reactor}} \longrightarrow B$$

If we define the reactant and the product with a common standard state, then the energy change associated with this process is the difference between the enthalpy of the product and the enthalpy of the reactant. This is defined as the heat of reaction, ΔH_r, with Δ symbolizing a change in energy between the enthalpy of the products and the enthalpy of the reactants. According to the energy balance then,

$$\Delta H_r = n(H_{out} - H_{in}) \tag{9.1}$$

with n being the number of moles of the products and reactants.

We can extend this to a situation in which the reaction is somewhat more complex, involving multiple reactants and products, each produced with different stoichiometric coefficients. In so doing, we get a summation over all of the species, which provides

$$\Delta H_r = \sum (nH)_{products} - \sum (nH)_{reactants} \qquad (9.2)$$

We can put this definition on a molar basis, by dividing by the number of moles,

$$\Delta \overline{H}_r = \sum (v_i H_i) \qquad (9.3)$$

to put this in terms of the stoichiometric coefficient, v_i. Recall that v_i is negative for a reactant and positive for a product. Then, the total heat of reaction can be expressed as

$$\Delta H_r = \frac{\Delta \overline{H}_r}{v_A} n_A \qquad (9.4)$$

All that remains is to define a common standard state, so that the evaluation of the enthalpy of the products and the reactants is consistent.

We handle the consistency problem by defining the enthalpy of an elemental species, in its normal state at $25\,^{\circ}C$, to be identically equal to zero. Thus,

$$H(C, \text{solid}, 25\,^{\circ}C) \equiv 0$$

$$H(O_2, \text{gas}, 25\,^{\circ}C) \equiv 0$$

Then, when we wish to determine the enthalpy of any compound, we write a hypothetical reaction in which the compound is formed from its component elements

$$C(s) + 2\,H_2(g) \rightarrow CH_4$$

and find that the heat of reaction is calculated as

$$\Delta H_r = H(CH_4) - 2H(H_2) - H(C)$$
$$= \Delta H_f(CH_4)$$

A reaction that leads to the formation of the compound from its elements is termed a formation reaction; the heat of formation ΔH_f is the energy generated (or consumed) through the formation of the component from its elements. These values are tabulated for many compounds and are available in the literature.

Once the values of the enthalpies for each compound are known, then we can calculate the enthalpy change using the heat of formation. For example, in the reaction

$$CH_4 + H_2O \rightarrow CO + 3\,H_2$$

the heat of reaction is given by

$$\Delta \bar{H}_r = 3\Delta H_f(H_2) + \Delta H_f(CO) - \left[\Delta H_f(CH_4) + \Delta H_f(H_2O)\right]$$

Often the heat of formation of a species is determined through a combustion reaction, so the heat of combustion is defined and tabulated specifically. The heat of combustion is simply the heat of reaction that would be obtained for the combustion of the species under consideration. Thus, for the combustion of methane

$$CH_4 + 2O_2 \rightarrow CO_2 + 2H_2O(l)$$

we can calculate

$$\Delta \bar{H}_r = \Delta \bar{H}_C = 2\Delta H_f(H_2O) + \Delta H_f(CO_2) - \left[\Delta H_f(CH_4)\right]$$

Remember that ΔH_f for O_2 is equal to zero. Note that the heat of combustion is just a specific case of heat of reaction, in which the compound reacts with the stoichiometric amount of oxygen to produce carbon dioxide, water (as a liquid), and other combustion products. Several other heats of reaction can be calculated, as shown in Highlight 9.1.

It is important to note that the heat of reaction is dependent on the phase of the species in the reacting system. Thus, it is essential to use the heat of formation for liquid water if liquid water is produced. The difference between the heat of formation of the liquid and the vapor species will be identical to the heat of vaporization of that component.

It is not possible to measure values of heat of reaction for every conceivable reaction. Fortunately, it is possible to develop equivalent chemical reactions for complex systems based on simpler reactions. Hess law states that for an equation of a reaction which is the sum of two or more equations: ΔH_r of the total reaction is equal to the sum of ΔH_r values of the underlying reactions.

Highlight 9.2 illustrates how Hess law is applied.

Highlight 9.1   Calculation of Heat of Reaction

Calculate the heat of reaction for the hydration of ethylene, given by the chemical equation

$$C_2H_4(g) + H_2O(l) \rightarrow C_2H_5OH(l)$$

Solution:

The heat of reaction is given according to the general equation

$$\Delta \bar{H}_r = \sum (v_i \Delta H_{f,i})$$

where we substitute the standard heat of formation for the enthalpy of each species. The standard heat of formation is provided for each of the species, along with the stoichiometric coefficient for each species, in the following table:

Species	v_i	ΔH_f
C_2H_4	−1	+ 52.28 kJ/mol
H_2O (l)	−1	−285.84 kJ/mol
C_2H_5OH (l)	+1	−277.63 kJ/mol

Substituting in the values provides

$$\Delta \bar{H}_r = (-1)(52.28 \text{ kJ/mol}) + (-1)(-285.84 \text{ kJ/mol}) + (+1)(-277.63 \text{ kJ/mol})$$
$$= 44.07 \text{ kJ/mol}$$

The sign of ΔH_r indicates the direction of energy transfer. In this example the heat of reaction is negative, which means that energy is transferred from the system to the surroundings and therefore this reaction is called *exothermic* (there is a production of energy). In the opposite case (positive heat of reaction), the reaction is named *endothermic* (in this case there is a transfer of energy from the surroundings to the system).

Highlight 9.2   Application of Hess Law

Using the two following thermochemical equations:

$$(1) \quad N_2O_4 \text{ (g)} \rightarrow 2 \text{ NO}_2 \text{ (g)} \qquad \Delta H_f = 57.2 \text{ kJ}$$

$$(2) \quad NO \text{ (g)} + \tfrac{1}{2}O_2 \text{ (g)} \rightarrow NO_2 \text{ (g)} \quad \Delta H_2 = -57.0 \text{ kJ}$$

find ΔH_r for the following reaction:

$$2 \text{ NO (g)} + O_2 \text{ (g)} \rightarrow N_2O_4 \text{ (g)}$$

Solution:

$2 \text{ NO (g)} + O_2 \text{ (g)} \rightarrow N_2O_4 \text{ (g)}$ is your target expression. In the target expression identify which reactants are desired in what quantities and which products are desired in what quantities.

Look at the known thermochemical expressions and decide how each needs to be changed to give reactants and products in the quantities that are in the target expression.

To have 2 NO as a reactant, all of the stoichiometric coefficients in Equation 2 need to be multiplied by 2. Proportionally the value of ΔH_2 also needs to be multiplied by 2.

$$2 \text{ NO (g)} + O_2 \text{ (g)} \rightarrow 2 \text{ NO}_2 \text{ (g)} \quad 2 \Delta H_2 = -114.0 \text{ kJ}$$

To have N_2O_4 as product, Equation 2 needs to be flipped, which also means that the sign of ΔH_1 needs to be changed to a negative one.

$$2\,NO\,(g) \rightarrow N_2O_4\,(g) \quad -\Delta H_f = 57.2\ kJ$$

When adding the two equations above, we obtained the target expression:

$$2\,NO(g) + O_2\,(g) \rightarrow N_2O_4\,(g)$$

and the overall heat of reaction for the target expression is

$$\Delta H_r = 2\,\Delta H_2 - \Delta H_1 = -114.0\ kJ - 57.2\ kJ = -171.2\ kJ$$

9.3   OTHER CHEMICAL PROCESSES FOR ENERGY TRANSFER

Traditional chemical processing relies on the transfer of energy in the form of heat or work, but it is also possible to use other processes to transfer the necessary energy to promote a chemical reaction. Because the energy transfer may be more efficient or targeted, the yield and selectivity can be improved, often eliminating the need for a catalyst, reducing the separation requirements, and eliminating waste. This section evaluates several of these less common energy transfer techniques.

9.3.1   Microwave-Assisted Reactions

Microwave frequencies are part of the electromagnetic spectrum and typically range from 110 to 140 GHz, between infrared radiation and radio waves. Energy can be transported through microwave radiation. Most kitchens now include a microwave oven, which converts electrical energy into microwaves that are transmitted to food. The absorption of the microwave energy by the food leads to an increase in the temperature of the food. Similar processes are also found in industrial processes for drying and curing products, or to generate plasma such as in the plasma enhanced chemical vapor deposition (PECVD). Because the microwave energy allows for efficient energy transfer directly into materials, it can be used to provide higher yields in shorter reaction times when compared to conventional thermal heating. Some researchers claim that there is a special microwave effect and several explanations are put forward [8]:

- Instantaneous high temperatures at the surface of the solid.
- Facilitation of contact between solid and liquid reagents.

The use of microwave chemistry is spreading rapidly wherever polar materials are present and often no solvent is required. When a polar solvent is present, the success of the chemical reaction depends on the efficient conversion of energy absorbed by the solvent. Because of the increased efficiency of energy transfer, reactions can happen in a microwave without the presence of a catalyst.

9.3.2 Sonochemistry

Ultrasound, or high intensity sound, provides another means of transferring energy through a fluid. The key element of ultrasound is the creation of acoustic cavitation, which is defined as the collapse of gas bubbles in a liquid and which generates very high local temperatures (about 5000 °C) and high pressures (over 1000 atm).

Ultrasound is fairly common for processes involving solids in a liquid or two immiscible liquids. Reactions involving ultrasound require shorter reaction times with improving rates and superior selectivity. Most of these reactions require quality mixing between species. In one example, ultrasonic baths are used to clean the surface of jewelry pieces by cavitation. While sonication has also been used widely in the field of microbiology and biochemistry (to help with digestion of cells), the use of ultrasound has been extended to the polymer field and can be scaled up to large volumes in batch or continuous flow systems. Copolymers of polyethylene and acrylamide have been built using ultrasonic cleavage of polyethylene in the presence of acrylamide [9].

Most recently, sonochemistry was used as an efficient extraction technique. Extraction of carvone and limonene from caraway seeds has been successful. Sonochemistry helped increase the yield, lower the extraction temperature, and produce a purer extract than those obtained with conventional methods [10].

9.3.3 Electrochemistry

Electrochemistry, also defined as oxidation/reduction reactions involving electron transfer between electrodes (usually metals, conductors of electricity) and ionic solutions (or electrolytes), was founded by John Daniell and Michael Faraday in the 1830s.

In previous chapters, we described the redox reaction, essentially an exchange of electrons between species. The redox reaction can be promoted through the application of an external voltage. An electrochemical process can promote a chemical reaction through the application of electricity across a cell, or one in which current is generated through the chemical reaction.

The best example of an electrochemical reaction is in the traditional battery. The earliest batteries, and those still used in automobiles, contained a liquid electrolyte, usually an acid, which would react with a solid surface, for example, lead, promoting a chemical reaction that transferred electrons through an electric circuit.

Electrochemical processes can be seen in action in nature (photosynthesis is an electrochemical process) and in commercial applications such as the coating of objects with metals through electrodeposition or through electroplating to protect metals from corrosion. It can also be used as a greener way to recover metal ions from waste streams, to remove low-molecular-weight ionic compounds through electrodeionization for the production of pure drinking water, or to regenerate an expensive metal or toxic compound in situ.

The fuel cell represents another electrochemical reaction. The fuel cell requires a constant source of fuel, usually hydrogen, in order to continue to produce electricity. The electrochemical reaction involves the oxidation of hydrogen to produce water, representing the greenest energy opportunity (except that the production of hydrogen to power the fuel cell may not be as green as desired).

9.3.4 Photochemistry and Photovoltaic Cells

Photochemical processes use photons from light sources as the source of energy. There are several advantages for using photons as they are clean with no waste production (assuming the original energy source did not produce any wastes), the reaction temperatures are lower than in traditional thermal processes, and they may provide higher selectivity as light is directly shining on the essential reagents. Photochemistry is used by nature for photosynthesis (conversion of carbon dioxide and water into glucose using sunlight) but is also used in the commercial production of vitamin A and vitamin D_3; this is one of the only industrial processes relying on photochemistry since there is no viable thermal alternative to the production of these two valuable products.

Many organic and inorganic chemical reactions are induced by light. An example of an organic photochemical reaction is the preparation of benzyl chloride ($C_6H_5CH_2Cl$) from toluene ($C_6H_5CH_3$) and chlorine (Cl_2). When the diatomic molecule of chlorine is exposed to light, breakage of the Cl–Cl bond occurs and two chlorine radicals are then formed:

$$Cl-Cl+H\nu \rightarrow 2Cl\bullet$$

The chlorine radical is then used to convert toluene to the benzyl radical:

$$Cl\bullet+C_6H_5CH_2 \rightarrow C_6H_5CH_2\bullet+HCl$$

The benzyl radical reacts with the second chlorine radical to form benzyl choride:

$$C_6H_5CH_2\bullet+Cl\bullet\rightarrow C_6H_5CH_2Cl$$

Approximately 100,000 tons of benzyl chloride is produced annually using this gas-phase photochemical reaction.

An example of a photoreactive organometallic reaction is a decarbonylation reaction (removal of a carbonyl group). When organometallic compounds (containing a metal center and organic-based ligands) are subjected to UV irradiation, their ligands are dissociated upon irradiation with UV light. For example, a solution of molybdenum hexacarbonyl ($Mo(CO)_6$) in tetrahydrofuran (THF) will become

$Mo(CO)_5(THF)$ upon UV irradiation. In this case one of the CO ligands is removed and replaced by THF. Since THF is easily dissociated, it can be replaced by another ligand in the second reaction step. This reaction is particularly useful for carbonylation chemistry since metal carbonyls resist thermal substitution and do not dissociate when heated.

Photochemically activated reactions are finding uses in modern applications such as the combustion of carbon nanotubes at high temperatures, which then break open and reorganize in the shape of horns giving the popular name "nanohorns," promising materials for chemical and bio-sensors. Their strong catalytic property is suitable for fuel cell applications and their porosity is ideal for gas storage.

Another application of converting light to electricity is photovoltaic cells. This will be discussed in Section 9.4.1.2 on solar energy applications.

9.4 RENEWABLE SOURCES OF ENERGY IN THE 21st CENTURY AND BEYOND

In order to reduce reliance on fossil fuels, new technologies that take advantage of renewable sources of energy are needed. Several clean sources of electricity, such as solar, wind, and geothermal, may be appropriate depending on the application. Hydropower represents a traditional renewable energy source. These renewable sources are described in this section.

9.4.1 Solar Energy

9.4.1.1 From Where Does It Come?

Solar energy is probably the most well-known and promising renewable source of energy. The sun generates natural nuclear energy through nuclear fusion (when the two nuclei of atoms are combined or are "fused" together to form a single heavier nucleus).

Solar energy would be sufficient to produce the entire amount of electricity needed in the world as the total amount of radiation hitting the earth is about 7000 times the total current global energy consumption. In 2009, The Union of Concerned Scientists mentioned that "all the energy stored in Earth's reserves of coal, oil, and natural gas is matched by the energy from just 20 days of sunshine" [11]. Solar energy also has the advantage of being available everywhere on earth.

However, solar energy is an intermittent energy source, as it can only be harvested during daylight hours. In addition, the intensity of the solar radiation varies significantly depending on weather conditions, location in the world, and the season. On average, the energy available varies between 3 and 9 kWh/m²/day in North America while in Northern Europe, it fluctuates between 2 and 3 kWh/m²/day. However, the variation can be substantial as the southern regions of France receive $2\frac{1}{2}$ times more solar radiation than the northern regions. Because of the intermittency, it is necessary to store solar energy, or convert it to heat or electricity and transport it where needed, even in the most remote areas of our planet.

The sun is categorized as a renewable source of energy since it has emitted radiation for billions of years, independent of how much of this energy humans have used and are currently using. There are many ways to use solar energy: either in the form of heat (solar-thermal energy) or in the form of electricity (solar-photovoltaic energy).

9.4.1.2 Where Can We Use It? Current and Future Applications

Using the sun's energy for heat is not a new process. Before early civilizations mastered the art of fire, solar energy was the only source of heat accessible. Solar thermal energy is used to maintain temperature within a greenhouse. In remote locations, solar energy can be used to heat water in a solar still and create hot water for showers in a remote location.

On average, about 14% of the total energy consumed in the United States is devoted to heating water and accounts for one-third of the average total household energy consumption [12, p. 175]. As an alternative to heating water in a conventional gas water heater, it is possible to take advantage of natural sunlight through a solar hot water system, in which the radiative energy from the sun is captured by the water. There are two types of solar-thermal collectors: concentrating and nonconcentrating. With concentrating collectors, the area exposed to the sun (collector area) is much greater than the area absorbing the radiation (absorber area); therefore, the concentration of the sun's radiation is very intense at the absorber area. Flat-plate collectors or nonconcentrating collectors usually consist of water tubes located between a transparent cover called glazing and a black absorber plate. The solar radiation passes through the transparent cover (can be glass) and the water tubes attached to the black absorber plate are used to remove the heat from the absorber. In this case, there is no difference between the area of the collector and absorber. Nonconcentrating collectors are effective in producing water at temperatures up to 95 °C, which would be sufficient to produce hot water for domestic uses or in-floor radiant heating or to heat swimming pools. Above this temperature, concentrating collectors are needed. Solar collectors are extremely popular in various countries: 90% of Israeli homes are equipped with solar water heaters; it is also the leading domestic heating technology in 10 countries of the European Union such in Germany, Greece, and France. Only $4\,m^2$ of collector is needed to supply hot water for a family of four people in France [12, p. 176].

There are also a number of ways to utilize solar energy passively, which counts on using the natural energy flows into or out of a building without the consumption of fossil fuels. This is simply done by designing the architecture of a house in a smart way. For example, one can imagine that a house made of thick stone walls could be ideal in a region where it is hot during the day and the temperatures decrease at night. The heat absorbed by the walls during the day is then released into the interior of the house at night. Most of the houses in the north of France are not equipped with air conditioning; they rely on the natural heating and cooling effect of their infrastructure.

Solar energy can also be used to cook food. Solar cookers, which do not require the use of fuel, are built and available in the shape of solar panel cookers, hotpots, and solar kettles. Since solar cooking works effectively in remote locations, it can reduce problems associated with deforestation and desertification in developing countries that rely heavily on wood and biomass as sources of energy for cooking.

A more efficient way to reach the high temperatures required for industrial applications, or produce electricity, is through the use of concentrated solar power plants in conjunction with thermal power plants. This technique is effective in regions of the world where the sun shines at least 2500 hours per year. While this type of process is not widespread, nearly 1600 MW of power was generated through concentrating solar power in 2011, in places including California and the EU. Through the use of parabolic reflectors or mirrors, the solar radiation can be directed onto an absorber tube containing a fluid such as oil or water, and this fluid can be heated to very high temperatures. Often the sunlight is concentrated into a collector on top of a "power tower," and this design is commonly termed "heliostat power plants." The steam generated is used to power an electrical generator. In California, the Luz solar power station totaling 850 parabolic trough mirrors is able to produce electricity equivalent to 380 kWh/y/m^2. The whole facility requires a ground area of 1.5 km^2 and an area of mirrors equivalent to 465,000 m^2. The solar radiation is converted to electricity with 14% efficiency. In Spain, a solar-thermal power plant near Seville contains 624 heliostatic solar mirror panels, which provides electricity for about 6000 households. It is expected that this plant will increase its capacity by 2013 and will be able to supply about 180,000 households in the near future.

Sunlight can also be converted directly into electricity using the photovoltaic effect (based on the creation of a p-n junction; p for positively doped (lack of electrons and creation of holes) and n for negatively doped (presence of extra negative charges), as illustrated in Figure 9.4.

For a semiconductor in the presence of light, an electron can be excited and moves from p-silicon (see Figure 9.4) into the n-silicon. As the electrons move, the p-silicon

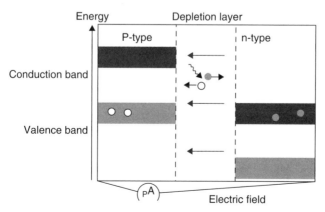

Figure 9.4. Illustration of electron–hole movement in a p-n junction [13]. (Courtesy of Marianne Breinig and James Hitchcock, University of Tennesse, Knoxville.)

becomes positively charged and the n-silicon becomes negatively charged, creating a net electric field. In the photovoltaic cell, this semiconductor device is connected to an electric circuit through which the electrons migrate back to the p-silicon, creating an electric current that can be used for power.

Photovoltaic cells are often made of silicon-based semiconductor materials, which efficiently absorb light and release free electrons, but they can also be made from more complex and less environmentally friendly materials such as cadmium and tellurium. Photovoltaic cells are typically from 1 to 10cm wide and are connected together in modules to bring the current and voltage to acceptable levels. The amount of electricity produced depends on the amount of solar radiation, which varies depending on the weather and time of the day. Usually only about 25% of the solar energy is converted into electricity under ideal conditions; the remainder is dissipated as heat.

Produced electricity can be used directly to power devices such as small calculators or road signs and lighting or communication equipment, or it may be placed directly onto the electric grid. Because solar power is intermittent, a solar energy system must include battery storage that can provide energy when it is cloudy outside. The photovoltaic cells usually produce a direct current (DC), which needs to be converted to alternating current (AC) for most households and appliances through an inverter.

9.4.1.3 How Much Does It Cost?

A typical domestic solar-thermal hot water heater system costs between $1000 and $2000. In parts of the globe where it is mostly sunny all year, a solar-photovoltaic installation can supply about 40% of all electricity needs at a cost of about $50,000. If more electricity is generated than needed, owners can sell it back to the electric utility provider and make a profit.

For commercial solar-thermal power plants, construction costs are estimated to be twice those associated with a modern coal power station and three times the costs of a gas power station. Commercial solar-photovoltaic power plants cost between $0.20/kWh and $0.60/kWh (compared to the cost of $0.035–0.060/kWh for a coal-fired power station and $0.040–0.063/kWh for a gas-fired power station) [14].

In 2011, the international cost of residential electricity averaged about $0.16/kWh. Residential solar electricity is more expensive than the average grid electricity. Improvements in performance and reduced production costs through economies of scale project that solar-photovoltaic costs will decrease to about $0.10/ kWh by 2020, placing the cost of a solar system on par with coal by 2017 and with natural gas by 2022 [15]. In addition, solar produces no greenhouse gases during the generation of electricity.

9.4.1.4 What Are the Upsides and Downsides of Solar Energy?

One of the main advantages of solar power is that sunlight is ubiquitous and despite the initially large investment, operation and maintenance are inexpensive. No waste or greenhouse gas is generated while heat or electricity is produced. Solar energy can also be used in remote locations, where connection to the electric grid is not practical.

The main drawback is that solar energy is not available during the night and its effectiveness depends very much on the location and climate conditions. While the cost of domestic solar energy cannot yet compete with electricity produced through fossil resources, it is expected that extensive development and research will lead to economically competitive photovoltaic systems. On the other hand, converting solar radiation into thermal energy can be done rather cheaply and is already used extensively in many parts of the world.

9.4.2 Wind Power

9.4.2.1 From Where Does It Come?

Wind power has been used since antiquity, mostly for sail boats and then for windmills. In the Asian world, wind was mostly used to pump water and irrigate fields. Wind is nowadays used to produce electricity via wind turbines.

Airflows on earth are created by the difference in atmospheric pressure between various locations on the surface of the earth and various altitudes. This movement of air can be tapped for its energy content and converted to electricity. A single small wind turbine can provide sufficient energy for domestic purposes, or clusters of wind turbines on land or offshore called wind farms can be used to supply electricity to large communities or for addition to the electric grid.

Commercial wind turbines are composed of several parts: on the top of a tower that can be in excess of 100 m tall, three giant blades of 30 meters or more in length rotate a horizontal shaft, turning a generator that converts mechanical energy into roughly 3 MW electricity. Inside the tower there is a cable that carries electricity to transmission lines and a computer system that controls the rotation and direction of the blades.

9.4.2.2 Where Can We Use It? Current and Future Applications

Wind power technologies have developed significantly over the past several decades, allowing for installation of wind farms especially in China, Europe, and the United States. The largest single wind farm is located in Texas, with a total capacity of 781 MW. Offshore wind energy is now also becoming more prevalent, including the largest capacity offshore wind farm located in the United Kingdom having a capacity of 300 MW. The United States ranks first in energy generated by wind, with 1.168 quadrillion BTU produced in 2011.

Larger wind turbines produce more energy, because the larger diameter blade is available to capture a greater proportion of the mechanical energy passing by the turbine. As a result, the size of turbine blades has increased dramatically since 1980. The diameter of a typical wind turbine was 15 meters in 1980 and in 2003 the diameter reached 124 meters. Research is going on to even increase the diameter to 160 meters by 2020 and therefore intensify the recovered power since the power generated is proportional to the cube of the wind velocity. Turbines are also becoming taller, to take advantage of the higher wind velocity at increased altitude, as well as allowing for longer turbine blades.

Land-based wind turbines are installed in windy areas and where the population is low. There are limited available land sites and many local populations would prefer wind turbines to not be erected in sight. Wind tends to increase along shorelines, and thus an important emerging alternative to land-based wind turbines are offshore turbines. They are similar to the ones used on land but they require reinforced foundations. The offshore wind turbines operate more regularly than the inland ones since the wind is more regular offshore and turbulence is reduced, generating increased wind speed.

The theoretical limit for efficiency of a wind turbine is 59%, but because the wind does not always blow at a constant speed, most wind farms achieve an efficiency closer to 30%. The electricity generated by wind can be harvested and delivered to the grid but because wind is intermittent, energy must be stored to be available when the wind is not blowing. This limits the contribution of wind power tremendously for electricity grid capacity.

9.4.2.3 How Much Does It Cost?

Even if wind is free, wind farms need to be built, operated, and maintained and cost about $1000 to $2000 per kW to build. Some claim that with the operating and maintenance costs, the cost of a productive wind farm is in the same range as coal, gas, and nuclear power plants [16]. Wind-based technology costs from $0.045 to $0.140/kWh. The capacity of a wind turbine is dependent on the wind speed and therefore cost can vary radically. Every little increase in wind speed (even just 5 km/h) will multiply exponentially the power generated by a wind turbine. A wind speed of 14–90 km/h is needed for wind turbines to operate productively.

Some residential wind turbines can be installed and are used to charge batteries. An interesting hybrid system that couples wind energy with solar photovoltaic provides a more continuous source of energy, since it is often windier at night and in poor weather.

9.4.2.4 What Are the Upsides and Downsides of Wind Power?

As mentioned for the sun, wind is a renewable source of energy and can be used efficiently in specific regions of the world. While it is the cheapest of all the renewable sources of energy, the production of electricity from wind power does not generate greenhouse gases (however, this is not the case in the construction and maintenance phase). The footprint of land-based wind turbines is small and therefore the land around them can be used for other purposes.

One of the main drawbacks is the initial investment associated with wind, which is still higher than that for coal, oil, or nuclear systems. The other major inconvenience is that the wind does not always blow everywhere and its speed is not consistent. Therefore, the reliability of wind power for grid electricity is low. The sites of wind turbines are often remote; storage and transportation of electricity are accumulative issues driving the cost up.

Besides the visual impact of wind turbines and their controversial aesthetic aspect, they are also blamed for their noise and for their contribution to the increased death of migratory birds.

Regarding offshore wind turbines, an additional problem lies in anchoring the turbines, for which their foundations need to resist higher winds but also waves. This also causes environmental concerns linked to the destruction of the seabed flora.

Even if wind power can only be competitive in specific situations, its price has decreased notably in the last decade and future improvements in terms of storage and transmission of wind energy are expected.

9.4.3 Geothermal Solution

9.4.3.1 From Where Does It Come?

Below the crust of our planet is a reservoir of geothermal energy coming from the original formation of the planet about 4.5 billion years ago (20%) and from radio-active decay of minerals such as uranium, thorium, and potassium (80%).

Geothermal derives from the Greek: "geo" means "earth" and "thermal" means "hot." The difference in temperature between the core of the planet and its surface implies that there is a geothermal gradient that moves the thermal energy in the form of heat from the core to the surface of the earth. This is the major factor involved in accessibility of geothermal energy; the second factor is the permeability of the rocks near the surface of the earth, which determines the rate of heat conduction to the surface.

Because of these two factors, while geothermal energy is available everywhere on earth, the intrinsic nature of the local geology and the temperature gradients are driving the use of geothermal energy. Variations in gradients can be quite large: the average geothermal temperature gradient near the surface of the earth is 3.3 °C/100 meters compared to Iceland, famous for its hot springs, where the gradients may reach 30 °C/100 meters.

9.4.3.2 Where Can We Use It? Current and Future Applications

Geothermal energy can be used either for thermal energy production or for electricity generation. In fact, geothermal energy has been used for several thousands years to heat water for homes, for cooking, and for agriculture.

Geothermal sources may either be low temperature (<149 °C) or high temperature. Low-temperature sources are most appropriate for direct application, such as the production of hot water for homes or for heat. A closed loop recycle system includes a well drilled up to 400 ft into the ground to extract high underground temperatures into a liquid working fluid, which is then distributed at the surface as heat. Slightly below the ground surface, the temperature is a fairly constant $10\,°C$. A heat pump may be employed to move heat from a building to the subsurface, acting as a natural air conditioning system.

For the production of electricity, only high- or medium-temperature sources are appropriate. For commercial applications only hydrothermal sources are suitable and, depending on their depth, several applications such as the use of dry steam at high temperature for electricity production are worth being pursued. The temperature difference between the fluid extracted from deep underground and that extracted near the surface provides the driving force for an exchange of energy that can be extracted through a turbine to generate electricity.

The United States with about 77 geothermal power plants is the leader in the production of geothermal energy, producing about 3000 MW in 2007. There are several commercial projects underway in France, Japan, and Australia. In France, the increase in geothermal energy was 30% in 2005. The average increase in the world is at a level of 4% every year [12, p. 210].

9.4.3.3 How Much Does It Cost?

The typical cost of geothermal energy is around $0.05/kWh. It is therefore a little more expensive than coal, gas, and nuclear but still cheaper than wind and solar-based energy. Prices to install a residential geothermal system vary considerably: from $7000 to $30000 for a 2000 square foot home. However, most of the cost comes from the installation and the price of the heat pump, which can be somewhere between $11,000 and $30,000. The impact on the energy bill is considerable: from $100 per month with geothermal heat compared to $600 with an oil furnace [17].

9.4.3.4 What Are the Upsides and Downsides of Geothermal Energy?

The main advantage of geothermal energy is that it is a continuous energy resource available throughout the day. There is an abundance of geothermal resources in the world but some parts of the world are particularly favored due to local geological conditions. Geothermal energy is quite flexible in that it can produce both heat and electricity and generates very low CO_2 emissions. The footprint is much smaller than for wind and solar energy and the environmental impact is very low.

However, in the majority of the cases geothermal heat pumps are necessary for large-scale production of heat and this generates some greenhouse gases. Some claim that geothermal energy is not truly renewable since hydrothermal resources for electricity production will eventually be depleted.

Even if geothermal energy will be increasingly exploited in the future and is currently the third most used renewable energy resource in the world, it only accounts for 0.4% of the worldwide energy.

9.4.4 Hydropower

9.4.4.1 From Where Does It Come?

Hydropower uses the water flows in a river or channel or the movement from the waves in a sea or ocean. It has been used for centuries to do mechanical work, such as milling grain, sawing timber, or extracting metal ore. Today, it is the most commonly used source of renewable energy for electricity generation.

To harness the energy from water, which flows or falls, water is directed into a channel or pipe and is then pushed against turbines or blades to make them rotate. There are two main types of hydropower usage, either using the current of a river or using a pumped storage or dam system. Dam systems rely on the height of the water and rate of the water flow. This is not always consistent as the availability of flowing water can vary considerably throughout the year. In a pumped water storage system, water is stored in different leveled reservoirs (one above the other). The excess electricity stored as grid electricity is then used to drive pumps that move the water back into the higher reservoir. This is then used to generate electricity during periods of peak demand.

9.4.4.2 Where Can We Use It? Current and Future Applications

Hydropower accounts for 16% of all of the electricity produced in the world. Countries that have easy access to water, such as Canada, Brazil, and Norway, use hydroelectric power stations extensively. In cases in which the water flow is insufficient or unsustained, a dam can be installed to create a lake; the water that flows through the spillway of the dam can be converted to electricity for the power grid. The largest hydropower plant in the world is Itaipu, located on the Paraná River, across the Brazil–Paraguay border. Twenty-five percent of the electricity consumed by Brazil and 90% of that consumed by Paraguay is provided by the Itaipu hydroelectric dam, which was elected one of the seven modern Wonders of the World in 1994.

Pumped storage systems are a very effective way to store energy. In a pumped storage system, two reservoirs at different elevations act together. Electricity is used during periods of low demand to pump water from one reservoir to the other, creating an elevation difference. When required, the water is allowed to flow from the higher elevation, and the mechanical energy of flow is converted to work, which is then used to drive a turbine and produce electricity.

9.4.4.3 How Much Does It Cost?

The main cost of the power station is in the construction cost, which covers 80–90% of the total lifetime cost. The capital expenditures are high but the operational costs are low. It is estimated that the lifetime cost of a new hydroelectric power station is $0.065–0.100 per kWh, which makes it more expensive than wind but cheaper than solar. When hydropower is available in a country, it is the first energy source exploited to produce electricity. China is the prime producer of electricity from hydropower, followed by Canada.

9.4.4.4 What Are the Upsides and Downsides of Hydropower?

The major downsides of hydropower are the limited number of locations where hydropower plants can be built and when these locations are appropriate, the cost is huge. However, there is no intermittency and it is an excellent way to generate peak electricity. There is no CO_2 emitted during the production of electricity but some is produced during the construction of the hydroelectric power plant.

Nevertheless, the environmental and societal impacts of hydropower are greater than for wind, solar, or geothermal energy. The construction of dams involves emissions of greenhouse gases, noise as well as sediment disturbance, and possible impact of water quality. It demands shifting rivers and extraction of millions of tons of earth and rock, therefore affecting the ecology, impacting the nearby river flows, and also relocating populations. When a dam submerges upstream land, the existing plant and fish life is impacted. Rivers become disconnected, meaning that fish can't travel the river; as a result, populations of fish, especially salmon, usually decrease as hydroelectric power plants are built.

Lastly, dam accidents have happened. Most recently in 2009, an accident occurred at RusHydro's largest plant at Sayano-Shushenskaya in eastern Siberia. Seventy-five people were killed when the turbine and engine rooms were flooded, damaging almost all turbines when the ceiling collapsed. A widespread power failure followed. More people are killed through dam failures than through nuclear reactor accidents.

9.4.5 The Case of Hydrogen Technology

The hydrogen atom is the smallest and lightest of all elements and also the most abundant element of the universe (75% of the visible mass of the universe is composed of hydrogen). It is found in a large variety of compounds such as in water, hydrocarbons such as methane, and biomass, but it is rarely found as a diatomic molecule of H_2. When this flammable gas is burned in oxygen, it forms water and also releases energy.

Hydrogen is not a primary source of energy since it does not exist as a separate entity and energy needs to be provided to remove it from other compounds. It is known as an energy carrier, a means of transmitting energy from one place or form to another. For example, hydrogen may be used to produce electricity in a fuel cell, but that hydrogen might be produced by electrolysis of water. The electrolysis consumes electricity to produce hydrogen, whereas the fuel cell converts hydrogen to water with the production of electricity.

Hydrogen is mostly produced from natural gas or coal through reforming with steam to produce syngas—a mixture of carbon monoxide (CO), hydrogen (H_2), carbon dioxide (CO_2), and water vapor (H_2O). It can also be formed by the electrolysis of water, which consists in passing an electric current through water to dissociate it into H_2 and O_2 at high temperatures or pressures. Electrolysis only accounts for 4% of the production of hydrogen while fossil fuels provide the rest [18].

Besides its use in industrial chemical processes and in the food processing industry, hydrogen was the main energy fuel for the NASA space program. Because of hydrogen's unique properties, storage and distribution limit the potential uses for hydrogen as a fuel for motor vehicles or for generating electricity. Some limited applications can be envisioned; for example, hydrogen fuel cells combined with an electric motor can replace the typical internal combustion engine. A hydrogen fuel cell functions like a battery, where the two sides are separated by a proton exchange membrane (PEM). Fuel cells make electricity from hydrogen and oxygen with the emission of only heat and water. In theory, a hydrogen fuel cell can achieve 80% efficiency, but in practice, the overall efficiency is often much less. Hydrogen has seen some success, as it was used to power buses in London and Beijing during the Olympic Games in 2012 and 2008, respectively. Air Products is the leading producer of hydrogen and has engineered about 130 fueling stations in the world. Vehicles powered by hydrogen do not generate greenhouse gas emissions, and they are quieter and more efficient than their gasoline and diesel counterparts. Hydrogen can also be made from waste and from renewable resources such as the sun or biomass.

9.4.6 Barriers to Development

Not one renewable source of energy stands out in terms of productivity, cost, or accessibility. The main issue associated with renewable electricity sources is the high cost, followed by the lack of well-established transmission and storage capacity. Integrating renewably generated electricity with grid electricity requires a significant change in the electrical grid, allowing for both inputs and outputs to the system based on availability and demand. The so-called smart grid will allow distributed energy generation but requires a substantial change in infrastructure, at a significant cost. Modernization is made difficult by regional ownership of transmission and distribution systems, and the current regulatory system that is not designed to benefit utility companies that make improvements for future innovations.

In order to reach the goal of having 20% of electricity generated by renewable sources of energy by 2030, a consistent and long-term mix of policies, regulations, public investments, and incentives must be provided to jump start future deployment of these new technologies.

9.5   CONCLUDING THOUGHTS ABOUT SOURCES OF ENERGY AND THEIR FUTURE

The current context regarding power generation is one of uncertainty. We can't predict the future. Burton Richter took the lead in ranking winners, losers, and maybes in his book, *Beyond Smoke and Mirrors: Climate Change and Energy in the*

21st Century in 2010 [18]. In his winners list, is coal (with carbon capture and storage), hydroelectric, geothermal (near-surface systems), nuclear, natural gas (as a replacement for coal), solar heat and hot water, and solar photovoltaic (for off-grid applications only). The losers are coal (without carbon capture and storage), oil for transportation, and corn ethanol and hydrogen for transportation. There are still some question marks associated with geothermal for deep mining for heat, the high cost associated with solar-thermal electric and solar-photovoltaic as well as advanced biofuels. Of course, there needs to be a spot left for new technologies not invented yet.

REFERENCES

1. Chu, S.; Majumdar, A. *Nature*, **2012**, 488, 294–303.
2. http://www.eia.gov/totalenergy/data/annual/pdf/sec2_3.pdf.
3. National Academy of Sciences, National Academy of Engineering, and National Research Council. *America's Energy Future-Technology and Transformations*, The National Academies Press, Washington, DC, 2009.
4. Miller, D.; Mann, M. *Energy Production and Alternative Energy*, Greenhaven Press, Farmington Hills, MI, 2011.
5. http://www.whitehouse.gov/administration/eop/ceq/sustainability.
6. http://www.esrl.noaa.gov/gmd/ccgg/trends/.
7. http://nova-gas.blogspot.com/2011_12_01_archive.html.
8. Matlack, A. *Introduction to Green Chemistry*, Marcel Dekker, New York, 2001, p. 465.
9. Fujiwara, H.; Tanaka, J.; Horiuchi, A. *Polym. Bull.*, **1996**, 36, 723
10. Chemat, A.; Lagha, A.; AitAmar, H.; Bartels, P. V.; Chemcat, F. *Flavor and Fragrance Journal*, **2004**, 19, 188–195.
11. Union of Concerned Scientists. "How Solar Energy Works," 2009, www.ucsusa.org.
12. Ngo, C.; Natowitz, J. B. *Our Energy Future: Resources, Alternatives, and the Environment* John Wiley & Sons, Hoboken, NJ, 2009.
13. http://web.utk.edu/~cnattras/Phys250Fall2012/modules/module%204/conduction_in_solids.htm.
14. http://www.iea.org/textbase/nppdf/free/2010/projected_costs.pdf.
15. http://solarcellcentral.com/cost_page.html.
16. Nicholsen, M. *Energy in a Changing Climate*, Rosenberg Publishing Pty Ltd, Australia, 2009, p. 61.
17. http://www.qualitysmith.com/request/articles/articles-hvac/geothermal-heat-pump-cost/.
18. Richter, B. *Beyond Smoke and Mirrors: Climate Change and Energy in the 21st Century*; Cambridge University Press, New York, 2010, p. 74.

10

THE ECONOMICS OF GREEN AND SUSTAINABLE CHEMISTRY

David E. Meyer and Michael A. Gonzalez

US Environmental Protection Agency Office of Research and Development
National Risk Management Research Laboratory
Sustainable Technology Division Cincinnati, Ohio

10.1 INTRODUCTION

In the previous chapters, the reader has been introduced to a brief history of chemistry and the transition currently occurring in the approaches and thinking that are needed to address the 21st century environmental and sustainability challenges that society globally is now facing. The reader has been introduced and versed in the 12 principles of green chemistry [1] and the 12 principles of green engineering [2] and the contributing roles they play in society and the science and engineering disciplines. Novel concepts have also been introduced to the current and next generation of research and researchers that includes the interaction of chemistry with the environment, the development and application of sustainability indicators and metrics and life cycle assessment (LCA) methodologies for assessing the current and potential states of sustainability for a system (e.g., a process, reaction, or supply chain), and

Green Chemistry and Engineering: A Pathway to Sustainability,
Anne E. Marteel-Parrish and Martin A. Abraham.
© 2014 American Institute of Chemical Engineers, Inc. Published 2014 by John Wiley & Sons, Inc.

the use of renewable materials as feedstocks for sustainable materials management. Additionally, for readers who are chemists, they have been introduced to chemical engineering concepts such as reactor design and kinetics, reactor and process engineering, thermodynamics, separations, and energy and heat transfer. These are concepts chemists now need to consider and integrate into their research during the chemical synthesis design phase. What is being developed, up to this point, is the demonstrated need for research to no longer be focused on one stage of the process or be constrained to a closed system, but the need for research and researchers to think holistically about the challenge they are solving, or the new technology they are developing. This chapter introduces and demonstrates the economic and, in turn, the correlated societal and environmental benefits that are gained when the principles of green chemistry and green engineering are introduced into a technology.

To place this chapter, and in fact the entire book, into perspective, the reader must understand the roles that green chemistry and green engineering play in the concept of sustainability. While there are many definitions for sustainability and these can also be modified to meet one's needs or desires, the definition the authors employ is a combination of the Bruntland Commission definition from 1992 [3] integrated with the mission of the U.S. Environmental Protection Agency (EPA) [4]. This combination arrives at a definition of "protecting human health and safeguarding the environment to meet the needs of current and future generations." The authors feel this definition captures the three pillars of sustainability—environmental, societal, and economics— as well as emphasizing the role of environmental protection and its contribution to increasing the sustainability of a system.

While the contributions of green chemistry and green engineering over the past 20 years have been significant, it must be recognized they are tools, which contribute to achieving an increased level of sustainability for a system as a whole. And as tools, they comprise a larger methodology, and that methodology is *sustainable chemistry*. In this context, sustainable chemistry applies a life cycle perspective to the social, economic, and environmental impact of a good, service, chemical, or product across its entire life cycle.

As evidenced in literature and practice, the goal of sustainability is now being applied to the chemical sector to reduce the negative effects on the environment and its health. To achieve sustainability, for the lifecycle of a chemical, researchers must have the ability not only to minimize or eliminate this risk across the lifecycle but also to be able to access and quantify any remaining risk and ensure the research direction taken is in a more sustainable direction. As the lifecycle of a chemical is mapped out, it is evident there are many opportunities that exist for improvement to current technologies as well as research areas for development of novel and innovative processes. Sustainability of a chemical synthesis not only occurs at the synthesis stage but also can be manipulated at any stage of the process lifecycle with direct and indirect benefits and consequences.

The chemical industry has made drastic improvements in the quality of life for society, but at the same time has not fully considered the effect on the health of the environment. For many decades, dilution was the solution to pollution. As a result, we are now seeing the effect of this approach of years of hazardous materials entering the environment from a number of human, animal, and environmental causes. Natural resources once seen as abundant and meeting every need of the human population have reached a point where it is obvious that our consumption exceeds the current supply. With concerns rising for pollution generation and natural resource consumption, sustainability has moved beyond the status of a buzzword to the forefront of industrial management. The concept of sustainability necessitates a shift to renewable resources, nonhazardous materials, and a decrease in the amount of waste being produced and released into the environment. Identifying the best means to achieve these goals is still a difficult task for research in any discipline. Even if we develop new and innovative processes or products with their own increased sustainability, sustainability does not have an endpoint and there is always more that can be done to achieve global sustainability. It is routinely acknowledged and demonstrated that the current methods for many industrial chemical production routes are inefficient. After scale up, many of these processes must be reexamined at a smaller scale to increase efficiency and production volume, decrease waste production and cost, and minimize energy consumption. This results in wasted time, personnel-hours, and increased costs to redesign the existing process to be more sustainable. While this process is transpiring, the continued large production of the inefficient process is still consuming copious amounts of materials, energy, and capital.

The premise of employing green chemistry and green engineering approaches is to contribute to the development of sustainable manufacturing processes for chemicals and products that simplify the reaction strategy and minimize resource and energy consumption, process time, and environmental impacts throughout the product or chemicals life cycle. In the absence of green chemistry and green engineering advancements, a major factor of any technology is the economic impact that can be expected. Oftentimes, the perceived bottom line is the final determinant companies contemplate when investigating a new process or synthesis option. Therefore, the economics of green chemistry are vital to its introduction into the marketplace and its eventual success. This chapter introduces the concepts, economic benefits, and needed thinking in order to increase the viability and introduction of technologies that employ green chemistry and green engineering into practice and the marketplace.

10.2   CHEMICAL MANUFACTURING AND ECONOMIC THEORY

To understand the economics of green chemistry and green engineering, it is first necessary to understand basic economic theory as it applies to chemical manufacturing. Traditionally, chemists at the bench scale develop viable product pathways to meet scientific and corporate demands while engineers design optimized processes to implement these pathways using production costs as the primary design criterion. This approach to product design is limited because it treats chemical manufacturing as an isolated action and neglects the multiple levels of economic influence that can impact a chemical product. In actuality, economic theory can be applied with increasing complexity to three levels of scale: (1) microscale or plant scale, (2) corporate scale, and (3) macroscale or economy scale. Each level is now briefly defined only as a means to provide a context for understanding the economic influence of green principles.

10.2.1   Plant (Microscale) Scale Economics

Economic theory at the plant scale is an integral part of product design and development. The profit derived from a given product is largely impacted by the cost to produce it. Although typical chemists and engineers might be aware of key cost factors, there is a much larger set of factors that govern production costs, many of which might not be considered during product design. The complete set of factors can be divided into two types—capital expenditures and operational expenditures.

Capital expenditures, or CAPEX, include all necessary costs to build an operational process or plant and are comprised of land, equipment, construction, and miscellaneous administrative costs [5]. Companies must first purchase land for development, with prices determined by local real estate prices. Equipment costs account for the purchase of reaction vessels, separators, furnaces, boilers, heat exchangers, pumps, piping, control systems, heat tracing, insulation, spare parts, and such. These costs will depend on manufacturers' prices for the specified equipment quantities and sizes. Typical construction costs involve land development, infrastructure (buildings, roads, sewer systems, etc.), labor (assembly, welding, steel and concrete fabrication, etc.), equipment (cranes, mixers, etc.), contractors' fees and incentives, utilities, auxiliaries, insurance, engineering services, and project management. Once a plant or process has been constructed, there are additional administrative costs that must be incurred prior to operation. These include training, testing, inspection and permitting, and start-up. Companies will often put up only a portion of the capital costs with cash on hand (equity) and finance the rest through low-interest loans (debt). A process or product will only have net profitability once the total CAPEX has been recovered. The timetable to achieve this profitability is set by the period of time to attain a return on investment or ROI.

The operational expenditures, or OPEX, account for all fixed and variable costs associated with product manufacturing [5]. Fixed costs cover all yearly costs that are not dependent on the production levels (running time). Examples of fixed costs are capital expenses (CAPEX including equipment depreciation), maintenance, basic payroll and benefits, insurance premiums, safety, property taxes, and utilities for workspace facilities. Conversely, variable costs are costs directly related to production time. Examples of variable costs are raw materials, process utilities, payroll overtime, waste disposal, and product storage. Once a company makes the decision to move forward with implementing a new process or product, it will specify a desired rate of return on investment and set product prices accordingly.

10.2.2   Corporate Economics

At the corporate level, the economics associated with chemical manufacturing are less related to actual manufacturing and instead focus on product management including research and development, logistics, and product image. Successful companies must first be willing to invest in research and development (R&D) to identify novel and/or improved products. This investment can be quite costly in terms of personnel and resources. However, a thorough R&D effort can help ensure the scale-up and transition from concept to production is more easily achieved in a shorter amount of time. Without R&D or at least a strong R&D effort, a company's market position can grow stagnant and become detrimental to the well-being of the company as consumers' needs evolve. Highly profitable companies have the ability to quickly reinvest profits in R&D to maintain continued growth and strengthen their market position. So how do companies operating with negligible profits compete? For smaller companies, it may become necessary to solicit funds from venture capitalists, government grants, and other private investors to help raise the necessary capital for R&D investment. Larger companies can raise this capital by selling a stake in the company as stock on public stock exchanges or issuing corporate bonds to purchasers guaranteeing future repayment of debt. Regardless of the approach, the use of external capital will reduce the potential profit of a product because these investments must be paid back and are always made with the expectation of a profitable return to the lender. Lenders exposed to higher risk will demand a larger ROI.

A second key aspect of product management is the logistics of product storage and distribution. At the plant level, the primary objective is to manufacture the product with minimal production costs. At the corporate scale, companies must develop networks to efficiently deliver the product to customers. For example, after gasoline is refined from crude oil, it is transported to regional storage terminals for further delivery to customers (regional distributors and/or gas stations). The proper location of these storage terminals within the distribution network relative to both the manufacturing plant and their customers is critical because the cost of transporting the gasoline can be significant given current fuel costs. For gasoline, product storage itself will also impact the associated cost and profit because it can be an indicator of supply relative to demand. When gasoline stocks become significant, product demand is considered low and the price customers are willing to pay will be lowered accordingly. To counteract this effect, oil companies will control the fraction of raw materials (crude oil) converted to gasoline to avoid excess production levels.

Product image is vital to the success of any product. The three main aspects a company can use to control its product image are through marketing, industry trade, and regulatory guidance [6, 7]. Marketing costs are a necessary part of business to help sway customer preference, as well as maintain customers. Traditional costs involve ad campaigns and distribution of product literature. The instant availability of information that has accompanied the technological revolution spurred by the growth of the Internet has forced marketing tactics to evolve and added to these marketing costs. Companies must pay to design and maintain Internet domains for product marketing either using contractors or hiring information technologists. In addition, companies must be proactive against the threat of negative product press presented online in customer and product reviews as these can shape customer product perceptions and affect sales. For many industries, companies with similar products find it advantageous to form trade associations to help develop product standards and distribute the costs associated with building a common product image. For example, the American Wood Protection Association develops and implements standards for wood treatment solutions that can be used to sell the benefits of the various products [8]. Trade associations can also help companies with shaping regulatory policy for product markets. Traditional economic theory predicts environmental regulation negatively impacts company profits [7]. Therefore, companies typically incur costs associated with engaging policy makers on regulatory issues. Individually, a company can hire lobbyists to represent its needs. However, a collective industrial voice will carry more weight in decisions. Trade associations can carry out product safety testing and lobby on behalf of their constituents to positively influence regulatory decisions for the industry. These services will only increase the resulting product costs for companies.

10.2.3 Macroeconomics

The chemical industry as a whole functions within the larger context of regional, national, and global economies, which are subject to societal and environmental stressors. At the regional level, product demand dictates price and can fluctuate in response to local stressors. For example, pesticide manufacturers will see a decline in profits when customers are experiencing drought conditions. If the pesticides are manufactured using bioprocesses, the company could experience a further decline in profits because raw material prices would escalate in response to drought-impacted crop production. At the national level, societal concerns and government policy will impact product markets and profitability. In some cases, the impact will be negative like the soaring cost of cigarettes in response to antismoking health campaigns and tobacco taxes. Recent government-mandated changes to product labeling are designed to further minimize profits for tobacco companies by eliminating consumer demand. At the same time, various national governments are providing incentives through tax credits and purchase rebates to increase demand (and profits) for alternative energy products in response to evolving societal preferences for sustainability. In some cases, national governments are passing legislation to prioritize sustainable development within industries. In 2005, China passed initiatives to promote a circular economy through the promotion of green chemistry and green engineering [9]. Globally, product economics are largely influenced by the politics of cultural differences and the distribution of raw materials. As societies have advanced technologically, so has the demand for raw materials. The raw materials needed to produce fuels and chemicals, mainly crude oil, are unevenly distributed throughout the world with a select number of oil-producing countries controlling the market, most of which are still developing socially. As wars and civil unrest destabilize these regions, oil-dependent countries are facing steep increases in raw material costs that permeate throughout regional and national economies. Likewise, electronic manufacturers are facing ever increasing costs for specialty materials such as rare earth elements because approximately 97% of the global supply is controlled by one country [10]. Another major influence on product economics at the global scale is the transition from local product markets to global supply chains. Business transactions at this level will be subject to trade policies and taxation that will influence pricing and profitability.

The combination of economic factors affecting chemical manufacturing when considering all levels of scale should convey the complex economic system companies must navigate when making decisions. These various factors make it possible to provide an understanding for how the principles of green chemistry and green engineering can be related to economic theory. The hope is this knowledge can then be applied to develop better "green solutions" with more attractive incentives to persuade companies to implement them.

10.3 ECONOMIC IMPACT OF GREEN CHEMISTRY

Although the principles of green chemistry and green engineering have been established for well over a decade, only recently have companies begun large-scale implementation and use of these concepts for industrial applications, often with emphasis on developing renewable feedstocks for chemical processes. The slow growth in the application of the green principles can be attributed to their misguided worth within the business model. Green principles are most often associated with "environmentally friendly" products. Thus, the only perceived benefits for the company are the environmental outcomes of the manufacturing processes or products themselves. If consumers can be swayed to use these products based on the "ecofriendly" label, then a company will likely incorporate green principles into its strategy to increase market share. However, this business view of green principles is naive and excludes many of the fundamental economic benefits that are offered from their use. To better understand this point, Table 10.1 demonstrates how each of the green chemistry and green engineering principles are related to the economic factors discussed in Section 10.2 and the respective relevance from a lifecycle perspective.

At the microscale, each of the 12 green chemistry principles can be related to a number of the economic factors discussed in Section 10.1. In all, 10 of the 12 principles can impact safety and insurance costs, 9 principles can impact waste disposal costs, 7 principles can impact material costs, 6 principles can impact equipment costs, 4 principles can impact utility costs, and 4 principles can impact land use costs. At the corporate scale, all 12 principles will impact R&D costs, 11 of the 12 principles can impact product image costs, and 3 principles can impact logistics costs. At the macroscale, all 12 principles can impact global supply chain costs while 9 principles can impact government policy costs. More important than the numbers are how and why the principles will impact these costs.

For the sake of this discussion, first consider the simple one-step reaction sequence:

$$A + B \rightarrow C$$

where A and B are reactants and C is the desired product, a high-demand ultrapure precursor for specialty chemicals. The stoichiometric coefficient for all chemicals in this reaction are one (1). This reaction is carried out at room temperature with a 100% conversion and 100% yield and involves no solvents or downstream processing

TABLE 10.1. Relating the 12 Principles of Green Chemistry and Green Engineering to Economic Factors and Relevance to a Life Cycle Perspective

Green Chemistry and Engineering Principles	Upstream Life Cycle	Process Level	Downstream Life Cycle	Economic Factors		
				Microscale	Corporate	Macroscale
Prevention (overall). It is better to prevent waste than to treat or clean up waste after it has been created. **Prevention instead of treatment.** It is better to prevent waste than to treat or clean up waste after it is formed.	Less feedstock load; therefore, reduce resource depletion.	Elimination of waste treatment units, decreasing capital and manufacturing	Impact minimization; waste prevention helps in the minimization or elimination of LCI/A data needs for the impact assessment of EHS risks (mid and end points), toxicological tests, and degradation studies	Raw materials, waste treatment and disposal	R&D, product and corporate image, regulatory guidance	Global supply chain, government policy
Real-time analysis for pollution prevention. Analytical methodologies need to be further developed to allow for real-time, in-process monitoring and control prior to the formation of hazardous substances.	reduce energy consumption, increase capital utilization	no energy load for waste treatment units; reduce recycling		Raw materials, utilities, waste treatment and disposal, equipment, safety, insurance, land use	R&D, product and corporate image, regulatory guidance	Global supply chain, government policy

Proactive

Synthesis							
Atom economy. Synthetic methods should be designed to maximize the incorporation of all materials used in the process into the final product.	Reduce resource depletion, feedstock processing, and the need for extra raw materials used for intermediate steps or stoichiometric synthesis; reduce number of feed components; reduces data needs for LCI/A.	Equipment reduction such as purification units; reaction equipment size reduction and/or residence time; reduction of intermediate steps; process simplicity; reduce energy consumption; decrease pollution control systems	Impact minimization, minimization of by-products reduces data needs for LCI/A; decrease waste treatment; reduce separation wastes	Raw materials, utilities, waste treatment and disposal, equipment, safety, land use	R&D, product and corporate image, regulatory guidance	Global supply chain
Design for separation. Separation and purification operations should be designed to minimize energy consumption and materials use.	Increased capital utilization; decrease need for separation agents; decrease need for upstream energy related inputs (processes)					
Conserve complexity. Embedded entropy and complexity must be viewed as an investment when making design choices on recycle, reuse, or beneficial disposition.						

(continued)

TABLE 10.1. (*Continued*)

	Green Chemistry and Engineering Principles	Upstream Life Cycle	Process Level	Downstream Life Cycle	Economic Factors		
					Microscale	Corporate	Macroscale
Synthesis	***Reduce derivatives*.** Unnecessary derivatization should be minimized or avoided if possible, because such steps require additional reagents and can generate waste. ***Design for separation*.**				Raw materials, utilities, waste treatment and disposal, equipment, safety, insurance, land use	R&D, product and corporate image, regulatory guidance	Global supply chain
	***Catalysis*.** Catalytic reagents are superior to stoichiometric reagents. ***Maximize mass, energy, space, and time efficiency*.** Products, processes, and systems should be designed to maximize mass, energy, space, and time efficiency.				Raw materials, utilities, waste treatment and disposal, equipment, safety, insurance, land use	R&D, product and corporate image, regulatory guidance	Global supply chain

Toxicity	*Less hazardous chemical syntheses.* Wherever practicable, synthetic methods should be designed to use and generate substances that possess little or no toxicity to human health and the environment. *Inherent rather than circumstantial.* Designers need to strive to ensure that all materials and energy inputs and outputs are as inherently nonhazardous as possible.	Minimization of risks in the manufacturing and handling of feedstocks	Minimization of EHS risks in the workplace, and risks in case of failure; increased worker safety, reduced capital costs	Reduction of life cycle inventory data for the impact assessment of hazard risks (mid and end points); fewer hazard tests	Raw materials, waste treatment and disposal	R&D, product and corporate image, regulatory guidance	Global supply chain, government policy
Design	*Designing safer chemicals.* Chemical products should be designed to effect their desired function while minimizing their toxicity. *Inherent rather than circumstantial.*				Insurance, safety, waste treatment and disposal	R&D, product and corporate image, regulatory guidance	Global supply chain, government policy

(*continued*)

TABLE 10.1. *(Continued)*

| | Green Chemistry and Engineering Principles | Upstream Life Cycle | Process Level | Downstream Life Cycle | Economic Factors | | |
					Microscale	Corporate	Macroscale
Design	**Design for degradation.** Chemical products should be designed so that at the end of their function they break down into innocuous degradation products and do not persist in the environment. **Durability rather than immortality.** Targeted durability, not immortality, should be a design goal. **Design for commercial "Afterlife."** Products, processes, and systems should be designed for performance in a commercial "afterlife." **Minimize material diversity.** Material diversity in multicomponent products should be minimized to promote disassembly and value retention.	Safer products and degradation products may use less hazardous raw materials; immortality requires excessive use of resources; limit waste in unused inventory (i.e., resources)	Minimization of *EHS* risks in the workplace, and risks in case of failure; immortality requires much processing; processing is simpler, with less equipment and potentially less energy use	Minimization of product life cycle impacts and toxicity tests, potential biodegradable products; eliminates downstream impacts	Insurance, safety, waste treatment and disposal	R&D, product and corporate image, regulatory guidance	Global supply chain, government policy

298

Safety						
Safety	*Output-pulled versus input-pushed*. Products, processes, and systems should be "output pulled" rather than "input pushed" through the use of energy and materials. *Meet need, minimize excess*. Design for unnecessary capacity or capability (e.g., "one size fits all") solutions should be considered a design flaw.					
	Safer solvents and auxiliaries. The use of auxiliary substances should be made unnecessary wherever possible and innocuous when used.	Minimize data needs; eliminated solvents and auxiliaries can be excluded from LCI/A studies; reduce resource use, reduce energy consumption				
	Design for separation.	Equipment reduction such as purification units, reaction equipment size and/or residence time reduction, reduce energy consumption, process simplicity	Minimize data needs, safer/no solvents and auxiliaries can be excluded for LCI/A studies			
	Inherently safer chemistry for accident prevention. Substances and the form of a substance used in a chemical process should be chosen to minimize the potential for chemical accidents, including releases, explosions, and fires.			Safety, insurance premiums	R&D, logistics, product and corporate image, marketing, regulatory guidance	Global supply chain, government policy

(continued)

TABLE 10.1. (Continued)

Green Chemistry and Engineering Principles	Upstream Life Cycle	Process Level	Downstream Life Cycle	Economic Factors		
				Microscale	Corporate	Macroscale
Energy — ***Design for energy efficiency.*** Energy requirements of chemical processes should be recognized for their environmental and economic impacts and should be minimized. If possible, synthetic methods should be conducted at ambient temperature and pressure. ***Maximize mass, energy, space, and time efficiency.*** ***Integrate material and energy flows.*** Design of products, processes, and systems must include integration and interconnectivity with available energy and materials flows.	Reduce energy demand; reduce environmental and ecological costs for fulfilling the energy needs	Minimization of *EHS* risks in the workplace, and risks in case of failure (fire, explosion, pressurized equipment)	Minimize high-temperature releases, reduce GHG emissions	Utilities, safety, insurance, equipment	R&D, product and corporate image, regulatory guidance	Global supply chain, government policy

Renewable						
Use of renewable feedstocks. A raw material or feedstock should be renewable rather than depleting whenever technically and economically practicable. ***Renewable rather than depleting***. Material and energy inputs should be renewable rather than depleting.	Reduce or eliminate nonrenewable resource use	Simplify separation / purification systems	Minimization of the effects of releases, biodegrad-ability, reduce GHG emissions	Raw materials, utilities, waste treatment and disposal, equipment, safety, land use	R&D, product and corporate image, regulatory guidance	Global supply chain

Source: Adapted from Ruiz-Mercado et al. [11].

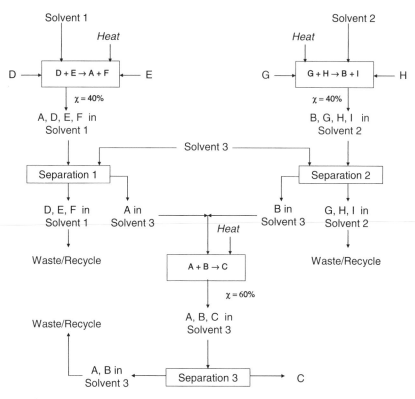

Figure 10.1. A theoretical traditional process for the production of a chemical product illustrating the complex and costly nature of processes designed without incorporating green chemistry.

for product recovery. Thus, it is an ideal reaction from a chemist and chemical engineer viewpoint. This ideal chemical process represents the minimum cost scenario because it (1) utilizes maximum reaction efficiency, atom economy, and yield; (2) uses a minimal quantity of raw materials (A and B) to produce a given quantity of product (C); (3) requires a minimal energy input; and (4) eliminates the need for waste disposal.

Now suppose we compare this ideal chemical process to a more complex process that is currently in operation by a company to obtain product C, as shown in Figure 10.1. Instead of the ideal case above, there are now four high-cost raw materials (D, E, G, H), two nonvaluable waste products (F, I), three organic solvents, three reaction steps, three separation steps, and a need for supplied heat to make the desired product C. The individual reaction steps themselves are inefficient with conversions much less than 100%. The microscale costs associated with such a process can be substantial. Based on the conversions listed, only 24% of the initial reactants actually end up as viable product. Thus, a large quantity of raw materials (reactants and solvents) would be required to produce a significant quantity of product, resulting in a larger cost of materials. If the addition of recycle feed loops was to be considered, further complexity and cost to this scenario would be the result. In addition to the cost of acquiring materials, a company must also be mindful of the cost of disposal for generated waste materials. Even if all of the solvents and reactants are recovered and recycled as in Figure 10.1, the company must dispose of the waste products F and I. Depending on the hazardous nature and volume of these materials, disposal cost can quickly drive up the operating cost of the process. Another consideration that will impact production cost is product monitoring. Because the product, C, must be ultrapure, the presence of so many other chemicals increases the chances of contamination and makes it necessary to continuously analyze process outputs, as well as adding to the cost of purification and decreasing process throughput.

Use of this process will also require energy to supply the required heat for reaction and/or separation, power the process equipment (pumps, blowers, etc.), and provide process control. The large number of process steps will require more personnel to run the equipment. This translates into the company being required to pay more for insurance premiums due to a number of factors including a larger number of potential victims and the use of hazardous chemicals. The additional costs associated with security, training, certification and permitting, and transportation cannot be overlooked. Likewise, the large number of processes will require a larger plant footprint in terms of land use. At the corporate scale, the complexity of the process will require a substantial R&D investment to obtain the needed product purity. The combination of low yield and high raw material costs will make it challenging to establish a cost-effective distribution network. Regional locations for product storage sites will be limited, which could possibly drive up land costs based on the greater number of constraints narrowing real estate choices. If the price for the product is on the high side of the industry average, the company will have to spend more on marketing strategies to convince downstream customers why their product is the best choice and worth the added cost. The added cost of manufacturing could then impede the ability of the company to participate in trade associations to aid with oversight and regulation of the product market. At the macroscale, governments are increasingly encouraging policy development with sustainability in mind. Given the typical long-scale return on investment for most new processes, it is possible that investing in such a wasteful process now could lead to unforeseen penalties and cost in the future if the process must be modified to remain within newly imposed regulatory guidelines. In the end, all of these factors will translate to a more expensive product and cost of operation with a higher risk for achieving the desired return on investment.

How will the principles of green chemistry impact this scenario? The answer to this question will illustrate the true economic value of green chemistry from a business perspective. For the example above, now assume that a second company, which also manufactures chemical product C, uses a newly designed modular process developed through the application of the principles of green chemistry. This novel process is detailed in Figure 10.2. When compared to the original process, this process requires less raw materials and energy, generates no wastes,

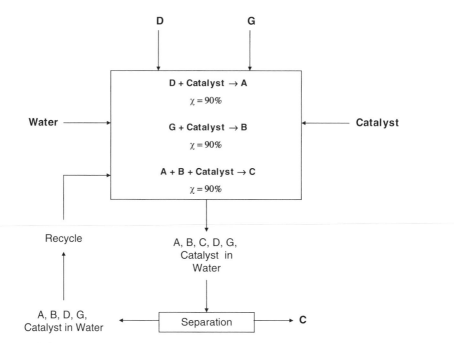

Figure 10.2. An alternative modular process for the production of a chemical product demonstrating the advantages of incorporating green chemistry into process design.

utilizes fewer processing steps (one reaction and separation), and is overall more efficient. Most of these improvements can be attributed to the beneficial influence of encouraged catalysis use (principle 9) for more efficient reactions. In addition to material benefits, the catalytic pathways can require milder operating conditions (temperatures, pressures, and solvents), which drive process energy requirements. For this example, a common catalyst is used to perform a "one-pot" reaction where the two reactants (D, G) can be transformed into the necessary intermediates (A, B) without additional chemicals and then directly reacted to form the product (C). In addition, the reaction can be carried out in water at room temperature, eliminating the use of energy and organic solvents. The modular nature of the process allows production volumes to be continuously scaled to reflect product demands and maintain optimal profit margins. More detailed analysis of process intensification and modularization will be saved for subsequent discussions on green marketing and business strategies in the next section.

At the microscale, this process is more beneficial in terms of both capital and operating costs when compared to the original process. In general, the capital costs will be less for this process due to the application of green chemistry principles 1, 2, 5, 8, and 9 leading to fewer processing steps requiring less equipment to achieve desired production quantities. Additional savings in capital cost are possible given the assumption that smaller processes will require less land and construction time. The associated operating costs will also be smaller when compared to the original process. Although the use of catalyst will add cost to the process, it has eliminated the need for reactants E and H (principles 1 and 2) and now allows for water to be used in place of an organic solvent(s) (principle 5). The reaction efficiency at each step has been increased to 90%, which means 81% of the primary reactants (D, G) are converted to viable product C (principle 2), versus 24% in the traditional process. This increase translates into more efficient use of feedstocks and material conservation. The elimination of wastes and the use of a common solvent provides for easy recycle loops for the unreacted starting materials. When all of this is combined with the lower operating temperature (principle 6), the raw material and energy costs for this process will be much less than the costs for the traditional process. The fewer processing steps will also mean a reduced size operating staff, which should contribute to a reduction in payroll and benefits. The benign nature of the process could also reduce the associated insurance and permitting costs. In the end, all of these cost reductions can result in a lower price for product C in order to obtain the same desired return on investment, while offering higher profitability.

At the corporate scale, the development of this improved process may require more R&D investment to identify the necessary catalytic pathways. This increase should be offset by the reduced need for the development of separation processes after eliminating the use of organic solvents and nonvaluable waste products. The modular and scalable nature of the process when combined with a smaller plant footprint should remove a number of constraints from the design of the product distribution network and contribute to a reduced cost of logistics for the company. These reduced costs should make the company highly competitive in the product market based on price alone. This inherent competitive edge from pricing should require smaller marketing investments and further increase net revenue. This additional revenue can be applied toward active participation in trade groups and regulatory product development for the product market. The benign nature of the process will eliminate the burden (and therefore costs) of studying health effects during production to maintain regulatory compliance. The gains in profitability provide the company with capital to reinvest into the company and offer additional greener processes to other product lines. This should lead to a strengthening of market position and recognition as a leader in sustainability.

At the macroscale, the reduction in raw material and energy requirements will help protect the profitability of the process from externalities such as price fluctuations in the upstream global supply chain that have become a dominant effect of escalating sociopolitical tensions throughout the world. The greener nature of the process will increase the demand for the resulting product with downstream chemical manufacturers responding to shifting societal demands for sustainable development. The improved environmental performance will guarantee that the long-term returns on the process will be realized even as national environmental policies evolve with society. Ultimately, implementation of green chemistry and green engineering can help maximize long-term profitability while minimizing the perceived investment risks.

This example, although ideal, represents the benefits that should frame the mindset of implementing green chemistry within industry. Although the environmental benefits of green chemistry were not discussed in the example above, they should not be dismissed. However, these benefits alone will not be enough to compel industries to adopt the principles of green chemistry (see Section 10.4). The ability to alter the bottom-line will ultimately govern business decisions. Providing a clear understanding to decision makers of how green chemistry can influence not only the microscale economics of a process but also the larger scale economic factors for corporate profitability is essential to the successful incorporation of the principles of green chemistry into process/product development and design.

10.4 BUSINESS STRATEGIES REGARDING APPLICATION OF GREEN CHEMISTRY

According to Pike Research, the current value of the global chemical industry is roughly US$3 trillion [12]. At the same time, the green chemical industry only accounts for 0.1% or US$3 billion with a projected value of US$100 billion by 2020. These numbers are significantly smaller than one would expect based on the potential economic benefits outlined in Section 10.2. From that discussion, companies have significantly resisted the implementation of green chemistry because of a misguided belief by investors and business managers that the development of green processes is motivated by environmental consciousness. The key challenge facing chemists and engineers who realize the broader benefits of green process design is identifying strategies to help grow support for deploying green concepts on a wider scale within the chemical industry. Effective implementation will require stakeholders throughout the entire life cycle of a chemical product (i.e., investors, manufacturers, consumers, governments) to be aware of the inherent benefits that can be realized when incorporating green chemistry (and engineering). This concept of sustainability marketing has been discussed by Iles and is based on the Porter hypothesis, which states that companies can improve commercial success through an efficient use of resources resulting from strict environmental regulations [13]. However, the Porter hypothesis itself poses a challenge to the acceptance of green chemistry because its reasoning is counterintuitive to traditional economic theory, which holds "environmental requirements are unavoidably increasing the private costs of the economy, resulting in the decreased competitiveness of the government and companies" [7]. Therefore, it is crucial to understand the challenges facing the application of green chemistry as a result of these competing economic views and how these challenges might be resolved. This knowledge, when combined with the numerous demonstrated and potential economic benefits of green chemistry, will provide a solid business strategy for the application of green chemistry.

Application of green chemistry begins with R&D. Not coincidentally, R&D investment poses one of the largest challenges to green chemistry because it is often viewed as an environmental investment [6, 7, 14]. As discussed in Section 10.2, companies at the corporate level require sufficient R&D investment to develop new processes and products, which are vital to maintaining a strong market share. Larger companies that typically have the necessary capital on hand or can easily obtain external funding are reluctant to invest in projects with unproven technologies, especially technologies viewed as environmental, because the resulting processes carry a much higher risk as a result of uncertainties surrounding scale-up and reliability [14]. Murovec and co-workers note that studies have shown successful companies experiencing growth are more likely to gamble on innovative, environmentally friendly technologies, provided the company possesses a more proactive attitude toward environmental stewardship and sees it as a means to gain market shares [7]. For smaller companies, profits are typically sufficiently reduced and limit the ability to internally fund R&D projects. These companies undergo much greater scrutiny when trying to secure external funding in response to their perceived economic weakness [7]. To complicate matters, external funding sources (banks, venture capitalist, etc.) may be even less inclined to back the development of "green processes" because the implied environmental gains would suggest a negative return on investment when placed in the context of traditional economic theory.

The negative perception regarding investment in green chemistry is indicative of a larger aversion to sustainable development and has been described by Pacheco and co-workers as a "prisoner's dilemma problem" [6]. In traditional game theory, the prisoner's dilemma is typically used to demonstrate how interrelated individuals who reject a group mentality and act in their own best interest will do worse than had they embraced the collective good. The idea is that the best outcome for an individual occurs when he/she alone rejects the group, while the worst outcome occurs when an individual chooses the group and everyone else rejects the group. In fact, it is actually better for an individual if everyone rejects the group versus everyone but him/her rejecting the group. All participants of the game are aware of these outcomes and fear they will be the only one to choose the group. This fear drives everyone to reject the group, knowing that they will at least perform better than the worst possible outcome. When applying this theory to green process development, companies and investors are the "players" in the game and industrial sectors are the "groups." The underlying fear keeping each player in the "green prison" is that he/she will be the only one to incur the substantial R&D costs needed to pursue green technologies. This will result in a power shift within the group whereby competitors can sell cheaper products and gain a competitive edge. According to market estimates, a US$40 billion savings in process costs for the global chemical industry is possible through waste minimization alone [12]. This value could be much higher if the other benefits provided by green chemistry are included. As opposed to groups realizing the mutual benefits of green design, companies are willing to settle for the option of maintaining a level playing field and everyone rejecting the need to invest in green chemistry because this is at least a better outcome than the worst case scenario inspiring the fear.

Perhaps the underlying cause contributing to the reluctance to develop and implement green chemistry and green engineering is society's lack of a clear voice when it comes to the question of environmental preservation and sustainability. In the

United States, environmental protection based on national regulations has recently (as of 2013) been sensationalized in the media as a "job killer," supporting classic economic theory about the impact on companies and economies by pursuing environmental benefits. This message is driven by the agendas of powerful industrial lobbies responding to what had been a growing trend in consumer preferences for sustainable products and services such as alternative energy and green buildings. With unemployment levels hovering at all time highs, the "job killing" message has created a societal "prisoner's dilemma." While individuals may understand the ultimate good of society is best served in the long run by pursuing sustainability, the fear of near-term financial suffering can potentially mute the willingness of consumers to absorb the extra costs associated with this endeavor. If companies perceive a negative shift in consumer preference, this too will serve as a deterrent when contemplating future investments in green chemistry and green engineering. A shift in public attitude can also impact the ability of governments to enact environmental regulatory policies. For example, recent efforts by the U.S. government to regulate greenhouse gas emissions and address climate change have been blocked by political forces responding to the efforts of lobbyists acting on behalf of industries and citizens fearing the economic ramifications of positive environmental actions. For government decision makers, this was both discouraging and confusing given the regulations were developed to meet the demands set forth during global environmental summits. Without enforceable environmental standards to incentivize against economic risk, companies will continually default to the prisoner's mentality when considering emerging environmental technologies and pass on the opportunity to realize the full benefits of sustainability and green product development.

Although these challenges may seem daunting, they are not insurmountable. When studying entrepreneurship and investment in green technologies, Pacheco and co-workers propose a solution to escape from the "green prison" by changing the rules of the game [6]. This can be accomplished in one of three ways. The first way to change the rules is by instituting industry norms. Like society, norms in industry will instill a uniformity of behavior. If the companies in an industrial sector can agree upon a self-imposed code of conduct to implement green principles, then the group option of the prisoner's dilemma will become the most attractive outcome for the game. The second way to change the rules is through the use of property rights. For example, industries that establish an emissions trading policy do so with the expectation that only members who are actively pursuing improvements to their own environmental performance will be allowed to trade. The apparent benefits of the trade system will force companies to comply or risk giving competitors an advantage.

The final way Pacheco and co-workers propose to change the rules of the game is through government legislation. The use of government legislation is the most effective way to address the prisoner's dilemma because the promise of rewards (tax breaks, grants, etc.) or penalties (fines, legal charges, etc.) will reinforce the need to act for the best interest of the collective and discourage rejection of the group. A key example of the use of government policy and legislation to encourage implementation of green chemistry is the efforts of China to create a national circular economy [9, 15]. For many years during the latter parts of the 20th century, the primary goal of the Chinese government was to become a leading world economy. The demand for growth was pursued with little regard for the potential environmental impacts it might bring. As environmental problems mounted and societies worldwide became increasingly interested in sustainable development, China recognized its own need to better manage its environmental resources and began implementing policies to this effect. In 2001, the Institute for Process Engineering was established as a center for research on green chemical processes [9]. By 2003, China adopted a national regulatory program to manage a variety of chemicals. In 2005, the National Program for Experimental Units of Circular Economy was launched, the foundation of which has significant overlap with the 12 principles of green chemistry. The goal of this program is to create a closed loop of material flows (a circular economy) within the Chinese economy [15]. A recent study by Matus has examined the impact of these programs in response to policy changes and identified the current barriers companies are facing. While companies want to participate in the cultural shift, many are ill-equipped to handle the development of green technologies in-house. Instead, R&D is contracted out to academic centers, oftentimes using guidance from the Chinese government to help form the right industry–academia relationships. A drawback to this approach is the control it has given industry to define the goal and scope of green chemistry research to be more applied to the extent of eliminating basic science-oriented research. The advantage is the Chinese government is willing to help financially support these collaborations. This is extremely beneficial for smaller companies operating with a much tighter profit margin. Now that the demand for green technologies is escalating, other factors such as employee training must also be considered. Despite the remaining barriers, the actions of the Chinese government have helped the chemical industry escape the prisoner's dilemma problem.

Governments aren't the only entities with the ability to enact policies to promote green technologies. In 2009, Walmart, the largest retailer in the world, helped found The Sustainability Consortium (TSC), whose mission is to implement standards for communicating the sustainability of consumer products [16, 17]. The TSC is made up of 83 major retailers working with 9 NGOs and government agencies [17]. The product labels being developed should be considered norms because they offer no chance for rewards or threat of penalty, but instead will provide uniformity in retail. This can be a powerful tool to help break the prisoner's dilemma because the desire of Walmart and other retailers to promote sustainable products will force manufacturers to decide either to go with the group and implement green technologies throughout the product supply chain (including upstream vendors) or to continue with their current manufacturing processes and face potential revenue losses as TSC retailers begin to focus their advertising campaigns on compliant products. The power of large-scale policy is necessary for encouraging the use of green chemistry and green engineering. As TSC demonstrates, this power can be evenly distributed between governments and industry.

The purpose of this discussion was to provide a business strategy to improve the ability to incorporate green chemistry and green engineering in chemical manufacturing. The key to a successful strategy is that it must be based on economic theory that can accept economic growth as a positive effect of environmental technology.

The strategy should first address the stereotypical belief that the business value of environmental technologies is primarily the ability to market sustainability. This will shift the cost analysis from a society base to an industry base and allow for incorporation of a greater number of the economic factors discussed in Section 10.2. The inclusion of a broader set of factors will provide the potential for a better return on investment to offset the normal risks associated with new process development. Next, the strategy should encourage investment in R&D by providing investors with a means to escape the "green prison." To this end, companies should make a stronger effort to work more collaboratively to develop industry norms that make green technologies more favorable. Such collaborations will help provide a strong industry voice to influence formal policy development at a larger level and help lock industry players into the "group" mentality. Finally, a company should work with partners upstream and downstream in the supply chain to develop more sustainable products in a manner that could more evenly distribute the associated costs.

10.5   INCORPORATION OF GREEN CHEMISTRY IN PROCESS DESIGN FOR SUSTAINABILITY

The availability of literature explicitly detailing the economic benefits of green chemistry is limited [18]. Instead, the benefits of green chemistry in process design fall under the larger topic of sustainable process design. This is because the true value of green chemistry technically transcends economic savings and can encompass environmental and societal benefits too. The goal of sustainable design is to consider environmental and societal issues early in the design process as opposed to the end-of-pipe mentality that has dominated traditional process design. This type of thinking encourages chemists and engineers to work together from the bench-top to the final scaled-up process to develop processes and products that satisfy the criteria for sustainability by using the principles of green chemistry and green engineering. However, the broad system boundaries that must be considered for sustainable design pose a challenge to the design process because of the difficulty in estimating the true benefits and impacts of green chemistry across the supply chain, especially the economic benefits [18]. The development of tools to aid in the design of sustainable processes is a growing field. Over time, these tools have included decision-support methodologies and metrics, design algorithms, and computer software. These tools can be all-encompassing, addressing an entire process, or focus on specific aspects of the process such as solvent selection.

The starting point when developing design tools is establishing suitable metrics for evaluation of green and/or sustainable processes. A set of sustainability metrics typically includes environmental, social, and economic factors [18, 19, 20, 21]. In addition, green chemistry metrics have been developed to characterize reaction pathways. Chemical factors can include the effective mass yield, E-factor (environmental impact factor), atom economy, and mass intensity [22]. These factors are used to assess how efficient a chemical process is when converting starting materials to product. Differences among the four factors are related to how nonreactants and nonproducts (waste, catalyst, and solvents) are treated. Evaluation of the eco foot-print associated with chemicals as they go from raw materials to final disposal is often made using the *life cycle assessment* (LCA) methodology as defined in ISO 14044 (ISO 2006). Typical impact categories include global warming/climate change, stratospheric ozone depletion, human toxicity, ecotoxicity, photo-oxidant formation, acidification, eutrophication, land use, and resource depletion [23, 24]. Economic or cost factors can include any of the following: capital cost (material, equipment, and labor), net present value (NPV), product revenue, waste treatment, training, insurance, and health and safety compliance. Societal metrics include provision of employment, the health and safety of workers and area residents, odor, noise, and public acceptability of products and/or processes [19].

When comparing potential processing or product alternatives, the various metrics can be calculated and weighted to determine which option is the best. However, this can prove to be a difficult task given the lack of standard weighting methods and has led to numerous publications offering the aforementioned tools for green and/or sustainable design. These methodologies will now be discussed, beginning with the simple environmental tools and following the incorporation of such tools into larger sustainability frameworks that encompass economic and societal concerns.

A major factor in the application of green chemistry is the choice of solvents for reaction media. An ideal solvent will provide maximum product yield and efficiency while minimizing the environmental, health, and safety (EHS) impacts of the process. Unfortunately, many solvents that produce favorable reaction conditions do so at the expense of EHS impacts. Therefore, it is necessary to identify solvents that opti-mize the trade-offs between the various performance criteria. Gani and co-workers have proposed a framework to guide the selection of a green solvent for an organic reaction [25]. The selection process involves two levels of evaluation. During initial evaluation, all possible solvents are listed and ranked based on basic user-defined reaction constraints using computer-aided molecular design and preexisting reaction data. The best candidates from this ranking are further evaluated at the second level using detailed calculations and experimentation that determine the specific performance and EHS impact of the solvent candidates. These results are then used to identify the optimal solvent. Folic and co-workers have extended the application of this methodology to multistage reaction systems [26].

Capello and co-workers have also proposed a comprehensive framework for the environmental assessment of solvents (both single and mixture) [27]. The tool com-bines EHS analysis of potential solvent hazards with LCA results for environmental impact in a simple three-step procedure. First, a solvent is scored using the EHS method for nine effect categories. These include the potential for release, acute toxicity, chronic toxicity, fire/explosion and reaction/decomposition, persistency, and air and water hazard. The second step involves application of LCA, as described above using the software tool Ecosolvent to calculate the impact scores for the solvent. Finally, the two assessment scores are combined and used to rate the solvents.

Andraos has devised an algorithm to evaluate processes using the green chemistry metrics reaction yield, atom economy, reaction mass efficiency, and E-factor [28]. In this algorithm, a process is first analyzed at the kernel (process step) level. These results are then combined at the global level to rate the performance of the entire process. This decomposition technique accounts for the potential use of by-products from one reaction in subsequent reactions within the process. When applied to existing processes, the kernel level scores can be used to identify potential areas of improvement within a process. For comparison of alternative processes, the global score can be used to select the "greenest" option.

Halim and Srinivasan have proposed an intelligent simulation-optimization framework for waste (E-factor) minimization of batch processes [29]. The procedure involves five steps: material flow representation, decomposition analysis for identifying design alternatives, detailed simulation of design alternatives, synthesis of a recycle network, and integrated simulation-optimization analysis for multi-objective solutions. This approach combines both qualitative and quantitative design tools through its use of heuristic design, process simulation, network synthesis, and optimization. This framework has resulted in the creation of a software tool, BATCH-ENVOPExpert, for batch process design.

Charpentier proposes a different view of green process design [30, 31]. Instead of stepwise frameworks, Charpentier discusses the need for multiscale and multidisciplinary design practices through the use of a triplet molecular processes–product–process engineering (3PE) approach. His premise is that emerging technologies such as nanotechnology coupled with growing trends in cost control and labor management will require a more integrated approach to process design and intensification to obtain more sustainable (or greener) manufacturing practices and products that are economically viable. Essentially, successful process design must utilize tools that enable multiscale modeling accounting for system performance at all length and time scales from the atomic to the macro level. These tools can encompass traditional engineering theory (reactor design, separations, thermodynamics, and material transport), molecular simulation, computational fluid dynamics, various forms of microscopy, environmental and health analysis (LCA, EHS, risk assessment), and waste minimization. The goal is to first design processes at the atomic scale using property-based simulations that can be used as a basis to predict scaled-up plant performance. At each stage of the design, the incorporation of environmental and health impact assessments can be used to guide the choices for material and processing alternatives to produce more sustainable systems

For all of the tools discussed, the basic goal of the design process is to minimize material usage and waste generation while alleviating the environmental impacts that often accompany chemical processing. Even when considering process options from this environmental perspective only, the task of identifying a "best alternative" can be challenging given the multi-objective nature of the problem. This process becomes even more daunting when economical and societal factors are also to be considered due to the large number of unknowns that must be evaluated and optimized. In addition, the selection, evaluation, and weighting of suitable metrics can be subjective in nature and produce varying results depending on the stakeholder and the decision criteria. This can be especially difficult when considering novel processing options because of the lack of knowledge and data concerning actual scaled-up performance. For these reasons, numerous decision-support tools for sustainable process design have been created. Some focus on how to incorporate sustainability analysis in general into the design process while others offer detailed accounts of how to obtain the optimized solution to the multi-objective design formulation.

GREENSCOPE (Gauging Reaction Effectiveness for Environmental Sustainability of Chemistries with a multi-Objective Process Evaluator) is a systematic methodology and software tool that can assist researchers from industry, academia, and government agencies in developing more sustainable processes [21, 32]. The sustainability of a process is measured in terms of environmental, efficiency, energy, and economic indicators (the four E's), with each indicator being mathematically defined. The indicators (140) express diverse aspects of performance in a format that is easily understood, supporting realistic usage. The indicators enable and demonstrate the effectiveness of the application of green chemistry and green engineering principles in the sustainability context.

To evaluate the environmental aspects of alternative chemistries or technologies, GREENSCOPE employs the Waste Reduction (WAR) algorithm [33]. The WAR algorithm determines the potential environmental impacts of releases from a process in eight impact categories: human toxicity by ingestion, human toxicity by dermal/inhalation routes, aquatic toxicity, terrestrial toxicity, acidification, photochemical oxidation, global warming, and ozone depletion. While these potential impacts are defined as midpoint indicators (as opposed to endpoint indicators), the measures for the categories are well defined, which is a substantial improvement over arbitrary environmental or mass-based scores.

Efficiencies for chemical reactions are reflected in values such as conversion and selectivity, which track yields, product distributions, and recycle flows needed to make a desired amount of product. Another measure of how green a reaction is can be obtained from the atom economy (i.e., how many atoms from the feeds are in the product). These measures, which are well known in green chemistry, are related to environmental impacts since the product distribution defines what chemicals (and amounts) may leave a process. These efficiencies represent a bridge between the lab-scale experiments of a chemist and further engineering calculations.

Energy is a basic component of chemical processes. Its use depletes resources and creates potential environmental impacts. Connecting to yet another sustainability indicator, a less efficient process can be expected to use more energy.

Without a positive economic performance, no industrial process is sustainable. The economics of processes are measured according to their costs. For economists, this is an oversimplified view of markets, but for engineering calculations, the annualized costs are significant measures. The costs are tied into the process through efficiencies, energy, and environmental impacts.

A novel aspect of the GREENSCOPE methodology and tool is that each indicator is placed on a sustainability scale enclosed by scenarios representing the best target (100% of sustainability) and the worst case (0% of sustainability). This sustainability scale allows the transformation of any indicator score to a dimensionless form using the worst and best scenarios. A process that is better in environmental, efficiency, energy, and economic terms will most likely be sustainable, although one can expect that trade-offs will need to be made.

Azapagic and co-workers offer a "systems approach" method for sustainable design that is based on four basic stages: project initiation, preliminary design, detailed design, and final design [19]. As opposed to traditional process design with boundaries that isolate the desired process as a unique system, the system in sustainable design is expanded to include the product and process life cycles. Project initiation involves identification of stakeholders and high-priority design criteria. The various process alternatives are listed and evaluated using the selected metrics. Preliminary design begins with selection of the best sustainable process based on the initial criteria. A working flowsheet for the process is constructed including material and energy flows, equipment design, process control, and identification of potential safety considerations. This data can then be used to conduct preliminary cost analysis consisting of microeconomic (capital and operating costs and profitability) and macroeconomic (added-value, potential environmental liability) indicators. The list of sustainability metrics is expanded to include all desired constraints and an environmental analysis of the system is performed using LCA. These analyses can then be coupled with societal considerations (health and safety, public acceptance, odor, noise, visual impact) to evaluate the sustainability of the system. The detailed design stage involves an iterative optimization of the preliminary design calculations to maximize the sustainability of the process. The final design stage occurs once a process has been designed that satisfies the defined criteria and involves the preparation of drawings and plans for construction.

Simlarly, Diwekar uses a systems analysis perspective to generate a framework for sustainable process design [34]. Unlike Azapagic and co-workers, Diwekar's methods couple this perspective with multi-objective problem solving to address the stochastic nature of sustainability. With multi-objective problem solving, a mathematical function is developed for each of the desired metrics based on quantifiable process variables and is assigned a weighting factor. The choice of metrics and weighting factors is subjective and left to the discretion of the stakeholder. Metrics for all three aspects of sustainability (environment, economics, society) can be incorporated. Nonlinear computational theory is then applied to the system to arrive at an optimized solution. While this may not sound difficult, the actual theory can involve highly sophisticated algorithms that use artificial intelligence to account for both the subjective weighting scheme and process variable uncertainty. The end result is a ranking of alternatives for decision makers based on trade-offs between the metrics. Diwekar provides several examples to aid in the formulation of metric functions.

The use of multi-objective optimization is a growing trend in sustainable design. Hugo and co-workers offer a process design methodology that focuses only on the minimization of environmental impacts while considering process cost and excluding societal impacts [35]. The Eco-Indicator 99 assessment model is incorporated into traditional economical-based process design to generate a set of objective functions. Unlike previous methodologies, the solution algorithm minimizes subjectivity in the calculations by analyzing all possible weighting schemes. The output is a set of

Pareto optimal solutions, or best alternatives. The subjective decision making is then carried out based on these alternatives. Li and co-workers have adopted this same approach, but using their own environmental impact model in place of the Eco-Indicator 99 methodology [36, 37]. Khan and co-workers have developed the software GreenPro-I to aid with multicriteria decision making during sustainable design based on LCA [38, 39]. This is another computational tool that is capable of performing multi-objective optimization of sustainability criteria to identify the best compromise solution for a specified design objective.

A prevailing thought in sustainable design is that green thinking must be incorporated into the design process as early as possible. Kralisch and co-workers illustrate this approach when developing the ECO (ecological and economical optimization) method to identify the most sustainable solution to a design problem during research and development (R&D) [40]. The methodology uses a simplified version of LCA combined with rudimentary economic analysis similar to life cycle costing (LCC) to overcome the data limitations often encountered during R&D. LCC offers an added benefit over traditional cost analysis because it accounts for the costs associated with a chemical throughout its life cycle, including equipment and energy costs for manufacture and disposal. The multicriteria ECO problem formulation involves three objective functions describing energy consumption (EF), health and environmental risk (EHF), and cost (CF) that can be applied to each step in the proposed process. The Pareto optimal solution can be identified using decision theory, resulting in the ability to optimize each process step to arrive at the best process. Even if an optimal solution cannot be identified, the ranking of alternatives can be used to eliminate obvious adverse options and focus research efforts to reduce the cost of R&D.

Yu and co-workers recommend the use of an analytic hierarchy process (AHP) approach when considering the environmental and economic trade-offs during sustainable product design [41]. LCA and LCC are both used to assess product performance. AHP is a multicriteria decision making method that can be used to organize the complexity of a design problem into four manageable levels: goal, criteria, subcriteria, and alternatives. For sustainable design using this method, the goal is an optimized life cycle assessment of a product based on criteria of total life cycle cost and total environmental impact. The subcriteria are the individual impact categories (LCA and LCC categories) calculated for each of the alternative products to be considered. AHP is subjective in nature because it requires user input to establish a weighting matrix for ranking criteria during calculation of total cost and impact.

Yan and co-workers have devised a sustainable product conceptualization system (SPCS) as a framework for sustainable design [42]. The system first uses a design knowledge hierarchy to break a potential product down by category, then component, and finally part and part option. This hierarchy is then used in conjunction with the initial design criteria (e.g., materials, manufacturing feasibility, marketability, distribution) as specified by the product developers to generate a Hopefield network that can be solved to identify the best design options based on limited and preliminary data that may contain varying degrees of uncertainty. The best alternatives are then subjected to a more rigorous evaluation of sustainability using environmental and cost objectives specified by higher-level decision makers. The sustainability results are then used by decision makers to select the most feasible product design. This approach is interesting because it attempts to divide the decision making responsibilities among the various stakeholders and allow "experts" to make the necessary decisions at each level. So chemists and engineers can handle the selection of materials and processes while business managers can evaluate the acceptable trade-offs between product performance and environmental and economic impact.

Lapkin and co-workers address sustainable design through the use of a hierarchy ranking scheme to organize selected indicators [18]. The first indicator level is for products and processes and includes green chemistry metrics and energy usage as defined by process chemists and engineers. The next level describes company criteria set by business managers. It can include added process/product value based on gains in utility usage, environmental impact, and product safety. The third level examines infrastructure indicators based on environmental and energy requirements set by governments. The final society level indicators examine the total sustainability of a product or process based on the needs of end users and the general public. This approach allows the boundaries for analysis to be successively widened at each level and provides a way to integrate the technical, economical, and societal concerns of the various stakeholders in an efficient manner. The solution of the resulting multi-optimization problem can be accomplished using any of the techniques outlined above.

Not all process design will involve new processes. GREENSCOPE and Sustain-Pro by Carvalho and co-workers were created to evaluate process retrofits. Sustain-Pro, an Excel-based software, can be used to retrofit existing processes for a more sustainable operation [20]. The methodology consists of six steps: (1) steady-state data collection; (2) flowsheet decomposition to understand material and energy interactions among the various units within the system; (3) calculation of safety, environmental, and cost indicators; (4) sensitivity analysis to optimize indicators; (5) identification of design variables most affecting indicators; and (6) creation of alternative process steps that satisfy indicator targets. Much like GREENSCOPE, the benefit of this tool is its ability to identify the hot spots for the various indicators within a process, thereby focusing the redesign efforts. This approach captures the essence of green chemistry and green engineering principles by forcing the reevaluation of traditional process design practices to develop cheaper, more efficient processing strategies. They differ from other methodologies because the system boundaries are drawn around the process and its interaction with the environment, excluding the upstream and downstream implications of the process during its life cycle.

The various methodologies all attempt to provide the user a systematic approach to sustainable design. The difficulty of applying these tools will increase as the number of design parameters (metrics) increases. As with any decision making process, the ranking of criteria (weighting) will depend on the stakeholders and their decision criteria. The economic impact of green decisions can involve more than mere processing costs and will require a global assessment including potential upstream and downstream benefits to fully realize their influence.

10.6 CASE STUDIES DEMONSTRATING THE ECONOMIC BENEFITS OF GREEN CHEMISTRY AND DESIGN

Although many benefits can arise from the use of green chemistry and green engineering and design, the economic impact of such technologies is probably the most important from an industrial perspective. For this reason, a number of comprehensive technology assessments have been published which include detailed economic analyses. Some of these case studies were developed as examples of how to apply sustainable design methodologies described previously, while others have been performed using the traditional principles of process economics. Applications ranging from biofuels refining to traditional solvent processing have been addressed. Selected case studies will now be presented to offer the reader a better understanding of the potential economic impacts by applying green design.

The production of vinyl chloride monomer (VCM) from ethylene and chlorine is a common case study used when discussing sustainable design because of the large-scale use of VCM, mainly for the production of polyvinyl chloride (PVC).The key synthesis steps involve the formation of ethylene dichloride (EDC) followed by cracking to form VCM. Multiple undesirable by-products (carbon dioxide, trichloromethane, chloroethane, trichloroethane, and tetrachloroethane) are formed during the EDC synthesis process, while by-products hydrochloric acid and trichloroethane result from the hydrocracking process. Azapagic and co-workers apply a simplified case study to this example to demonstrate the application of their PDfS methodology [19]. This case study considers potential changes to traditional VCM production processes, including alternative feedstocks and changes to a key reaction step, and examines how to identify the most sustainable alternative early in the design process. In the end, the PDfS methodology found that economic sustainability is attainable using present technologies while environmental sustainability will require creation of new processing alternatives.

Carvalho and co-workers also attempted to redesign a typical VCM plant using their Sustain-Pro flowsheet to improve the sustainability [20]. The VCM process is first divided into the corresponding five sections and then analyzed using sustainability metrics to identify the areas of potential improvement. Once a hot spot is located, in this case the formation of EDC, alternatives are evaluated based on available data and used to find the most sustainable solution. These alternatives include inserting a recycle purge, improving existing separation units, inserting new separation processes, and improving reaction conversion. Of these, only the use of a new separation technology, membrane pervaporation, is found to increase the environmental sustainability of the process while maintaining the economic constraints. However, improving the reaction conversion is not considered because of a lack of feasible technologies.

Khan and co-workers applied their multicriteria GreenPro methodology and software to design a VCM plant [38]. Again, conventional processing alternatives are considered, including the use of air and an improved wastewater purification stripping process. A multicriteria model consisting of 125 environmental and economic constraints is solved for each alternative and compared to determine which is the most sustainable. Applying this approach, the use of air is both environmentally and economically favored. However, the optimal solution for all constraints is a trade-off between the optimal cost and environmental scenarios.

These examples do not draw heavily on green chemistry, but they still illustrate how sustainable design will lead to the need for applying the principles of green chemistry to achieve economically viable environmental solutions. After applying the three different methodologies, only marginal gains in sustainability for the VCM process can be achieved using conventional technologies. None of the alternatives put forth in these case studies looked at developing new benign catalytic pathways that increase the atom efficiencies or alternative synthesis pathways. Such an alternative could reduce the material and energy cost and consumption and result in a truly greener process. A better case for this point can be seen when considering the role of green design and solvents.

Ionic liquids (ILs) are an interesting development in the pursuit of green solvent technology. On the one hand, they offer a low vapor pressure, are nonflammable, and can dissolve a number of organic, inorganic, and polymeric materials. These attributes can lead to reduced emissions, improved reaction kinetics, and cheaper solvent recycling for chemical processes, all which are attributes of green chemistry. On the other hand, typical IL synthesis involves alkylation of a suitable organic compound followed by anion exchange. Thus, the synthesis of these materials is costly because of a need for large quantities of organic solvents and energy-intensive processes to recover them. In addition to the adverse environmental impacts arising from their synthesis, many ILs have been shown to possess some level of toxicity, making them a questionable choice for sustainable design. For these reasons, Kralisch and co-workers have applied their ECO methodology to the synthesis of ILs to try to identify greener alternatives for the alkylation step based on varying the temperature, organic solvent, reactant concentration and molar ratio, and reaction time [40]. The optimized solution involved a moderate-temperature, solvent-free synthesis, which reduced the EF by 78%, EHF by 98%, and CF by 87%. These improvements can be associated to the proposed green processes involving ILs to help offset the potential toxic risks the ILs themselves may present. In a more fundamental sense, these improvements provide a clear example of how the principles of green chemistry can lead to enhanced economic benefits. Interestingly, this example also suggests that there will always be room for improvement in green processes until true sustainability is achieved.

ILs are just one type of solvent system which have shown promise for green processing. Dunn and Savage examine the use of high-temperature water (HTW) during the synthesis of terephthalic acid, the precursor for polyethylene terephthalate (PET) used in injection-molded plastics [43]. Commercially, terephthalic acid is produced via the catalytic reaction of p-xylene with oxygen, and a bromide initiator in an acetic acid solvent in the presence of a Co-based catalyst. The reaction is carried out at high temperature and pressure, uses large amounts of solvent, and involves the formation of several undesirable by-products, including methyl bromide. Further complicating the process, one of the by-products, 4-carboxybenzaldehyde, prevents polymerization to PET and requires intensive downstream processing steps to separate. In addition, the azeotropic separation of acetic acid from the by-product water requires large quantities of energy. HTW is a promising solvent alternative because it is has minimal environmental impact, offers tunability based on the adjustability of its properties, and is relatively cheap. Its use for the proposed synthesis will provide a completely recyclable solvent requiring no azeotropic separation and eliminate the formation of hazardous by-products such as methyl bromide. However, these environmental benefits will only be a viable option if the process economics are favorable. Dunn and Savage have simulated four alternative HTW processes and assessed their economic and environmental impact based on capital cost and the sustainability metrics for energy intensity and pollutant intensity. Environmentally, the results show that the use of HTW reduces the associated environmental footprint, provided the water can be recycled with little or no makeup volume needed. Economically, the HTW process has roughly the same capital cost as the current acetic acid process, making it appear to be a viable alternative. An evaluation of operating costs is needed to fully understand the economic impact. Dunn and Savage concede that the operating conditions necessary for the HTW process (300–380°C, 150–250 bar) could result in significant operating costs because of larger energy demands. However, they neglect to consider the potential cost savings arising from factors such as reduced waste management or reduced worker protection costs and insurance when making this point. Also, no comparison of productivity (conversion/yield and processing time) for the HTW and conventional processes is given, which will greatly impact the profitability of the processing plant. These factors are important when trying to understand and assess the economic impact of green chemistry and design.

Perhaps the most recognized application of green chemistry and design is the biorefining industry. As such, extensive work has been published examining the benefits associated with the conversion of renewable feedstocks into fuels and chemicals. For example, Zhang and co-workers have evaluated the acid-catalyzed transesterification of waste cooking oil into biodiesel fuel [44]. As opposed to petroleum-based diesel, biodiesel is renewable, has a low-impact emission profile, and is biodegradable. It is typically produced via the transesterification of triacylglycerols in virgin vegetable oils and animal fats using alkali catalyst in the presence of methanol. This alkali-catalyzed process has a much faster reaction rate than the acid-catalyzed mechanism, but is sensitive to the presence of free fatty acids in the feedstock and requires pretreatment of the feed if low-grade feedstocks (i.e., waste oil) are used.Thus, biodiesel production using alkali-based catalysis is limited commercially because the high cost associated with obtaining the raw feedstock (70–95% of the total production cost) translates to noncompetitive product pricing and with biodiesel costing up to one and a half (1.5) times that of petroleum-based diesel. The ability to use cheaper feedstocks such as waste cooking oil could greatly increase the economic viability of biodiesel (concept of materials and waste management) due to the price of waste cooking oil being two to three times cheaper than virgin materials. Also, additional cost savings could be realized by avoiding costly pretreatment of the feed to control the fatty acid content. With this in mind, Zhang and co-workers conducted their assessment of utilizing the acid-catalyzed process. The economic analysis included fixed capital cost, total manufacturing cost, after-tax rate of return, and the break-even price of biodiesel. On the basis of capital cost, alkali-catalyzed plants are cheaper. However, the high price of feed materials greatly offsets this advantage through elevated manufacturing costs. Ultimately, the break-even biodiesel price is reduced from US$857/tonne for the alkali-catalyzed plant to US$644/tonne for the acid-catalyzed alterative. The reduced break-even price is still more than the US$500–600/tonne of petroleum diesel, but close enough that the potential environmental benefits could offset this minor economic disadvantage. This is another example of how green chemistry and design can lead to environmental gains without disrupting the economic flow of society.

Biodiesel represents only one area of interest in a much larger market of biofuels and biochemicals. Henrich and co-workers have studied the economics for converting biomass into liquid synthetic fuels in a manner analogous to coal-to-liquid (CTL) and gas-to-liquid (GTL) technologies [45]. Such technologies first convert a carbon source into syngas (CO/H_2 mixture), which can then be reformed into a number of desired hydrocarbon-based products (fuels, solvents, plastics). For biomass-to-liquid (BTL) conversion, this technology offers a renewable, clean alternative to petroleum. Potential drawbacks include energy-intensive processing supply issues for the feedstock based on the rather large quantity of solid biomass that would need to be transported to supply daily biorefinery demands. The latter issue can be resolved if lignocellulosic biomass (wood, straw, etc.) is used. This is because these materials can be pretreated locally using pyrolysis and converted to a suitable liquid form that is much easier to transport in bulk quantities. However, the issue of energy usage makes this technology economically viable only for plant sizes where product throughput is high enough to overcome operating costs. This is a good example for illustrating how traditional economic factors may not capture the full benefits of green chemistry and design. For example, a technology that is not reliant upon petroleum has the potential for tremendous savings in the future given the uncertainty and volatility of the crude market. In addition, it is possible that tax incentives and other financial benefits will be made available to companies offering "cleaner" fuels and services as societal concerns for environmental issues escalate. This really illustrates the greater point that economic assessment of sustainable technologies needs to be expanded to the national or global level to gain a better understanding.

High operational costs are a prevailing theme with biochemical production, prompting researchers to look for ways to increase both the profitability and sustainability of biorefineries. Eckert and co-workers undertake this problem by looking at the recovery of additional high-value products using a novel solvent extraction process [46]. During bioethanol fermentation, as much as 40% of the biomass is unconverted and typically burned as fuel for its heat value. However, the chemicals that make up this fraction are considered high value and could provide additional profit for a biorefinery. Some of the more prominent chemicals include vanillin and syringol, which are used in food and fragrances and syringaldehyde used in dyes and cancer pharmaceuticals. These materials can sell for anywhere from $5 to $25 per pound. Eckert and co-workers have demonstrated the ability to recover these compounds in high purity using methanol containing dissolved CO_2, a gas-expanded liquid. Gas-expanded solvents are relatively green and offer tunability based on the reaction conditions, content, and properties of the gas. For vanillin and syringaldehyde, this simple extraction can replace the multistep syntheses, often involving toxic chemicals, used in the past. Syringol has not been synthesized commercially. This added option to recover these high-value chemicals contributes to the economic and technologic viability of biorefining. Discussions about the use of process intensification and green solvent technology to enhance the economic benefits of biorefining are readily available [47, 48].

A growing application for green chemistry that can provide insight about its benefits is the manufacture of pharmaceuticals. For example, Steffens and co-workers have applied a multicriteria design methodology to the production of penicillin to find a more sustainable alternative [49]. Classic penicillin production involves fermentation, filtration, solvent extraction, crystallization, and drying. There are known issues with this process, including significant water usage, generation of large quantities of solid waste, and discharge of an extraction solvent, butyl acetate, in the process wastewater stream. The applied design methodology combines the total annualized plant cost, LCA results, and sustainable process index (SPI) to optimize and improve the classic penicillin process. The resulting best set of optimized solutions require only 50 kg of water per year represented as critical water mass index (CTWM), a yield and SPI between 2 and $3.5 \times 10^7 \, m^2/y$, and an annualized cost savings between US\$1.5 and US\$1.8 M, at the time the study was completed. The reduced environmental impact observed is the result of a process substitution that uses filtration in place of the organic solvent-based extraction steps. No improvement to the solid waste generation is possible unless new methods for penicillin fermentation are developed. In the end, this example required no radical changes to realize the benefits of green chemistry and design. Instead, it illustrates the benefit of using sustainable design practices early in the design process.

A final application worth mentioning is the use of modular process intensification. Lier and Grünewald have performed a net present value (NPV) analysis to compare the performance of a modular chemical plant with a conventional large-scale plant of the same capacity [50]. The NPV calculations found the modular plant was able to achieve profitability quicker than the large-scale plant because it can be set up and made operational in a much shorter period of time. The other advantages of modular plants are reduced R&D time and cost, configurability to control output (avoid overproduction), staggered module construction, faster installation, reduced operating costs, and a shorter time to reach profitability.

10.7 SUMMARY

The focus of this chapter has been to provide a better understanding of how green chemistry and green engineering should be construed within the economic structure of chemical manufacturing. The benefits of green technologies are only fully realized by considering economic factors at the microscale, corporate scale, and macroscale. This knowledge will help corporate decision makers embrace green chemistry and green engineering during the product or process development phase by transcending the misguided conventional belief that investment in environmental technologies cannot support economic growth and profitability. The key challenge facing deployment of green technologies is the risk of investment in R&D given the changing societal preferences regarding sustainability. Given the holistic view that must be adopted to best understand the benefits of green chemistry and green engineering in achieving sustainability, chemical assessment must transition from traditional cost–benefit and NPV calculations to larger sustainability assessments. Several methodologies and subsequent software packages have been developed with the goal of guiding decision makers through the process of multicriteria decision support. The successful evaluation of these tools using real-world examples illustrates the value of green chemistry and green engineering when pursuing sustainable development.

REFERENCES

1. Anastas, P. T.; Warner, J. C. *Green Chemistry: Theory and Practice*, Oxford University Press, Oxford, 1998.
2. Anastas, P. T.; Zimmerman, J. B. *Environ. Sci. Technol.*, **2003**, *37*, 94A–101A.
3. U.N. GAOR, Agenda Item 21. In *46th Session*, **1992**; p UN Doc A/Conf.151/26.
4. U.S. EPA. EPA Mission Statement. Available at http://www.epa.gov/aboutepa/whatwedo.html, last (accessed September 10, 2012).
5. Huisman, G. H.; Van Rens, G. L. M. A.; De Lathouder, H.; Cornelissen, R. L. *Biomass and Bioenergy*, **2011**, *35*, S155–S166.
6. Pacheco, D. F.; Dean, T. J.; Payne, D. S. *J. Business Venturing*, **2010**, *25*, 464–480.
7. Murovec, N.; Erker, R. S.; Prodan, I. *J. Clean Production*, **2012**, *37*, 265–277.
8. American Wood Protection Association (AWPA). Available at www.awpa.com (accessed september 10, 2012).
9. Matus, K. J. M.; Xiao, X.; Zimmerman, J. B. *J. Cleaner Production*, **2012**, *32*, 193–203.
10. BCC Research. *Market Research Report — Rare Earths: Worldwide Markets, Applications, Technologies*, **2009**.
11. Ruiz-Mercado, G. J.; Gonzalez, M. A.; Smith, R. L. Expanding GREENSCOPE beyond the gate: a green chemistry and life-cycle perspective, *Clean Technology & Environmental Policy*, **2014**, DOI 10.1007/s10098-012-0533-y.
12. Pike Research. Green chemistry: biobased chemicals, renewable feedstocks, green polymers, less-toxic alternative chemical formulations, and the foundations of a sustainable chemical industry (Executive Summary Reprint), *Ind. Biotechnol.*, **2011**, *7*, 431–433.
13. Iles, A. *Bus. Strat.Environ.*, **2008**, *17*, 524–535.
14. Harmsen, J. *Chem. Eng. Process.*, **2010**, *49*, 70–73.
15. Geng, Y.; Fu, J.; Sarkis, J.; Xue, B. *J. Cleaner Production*, **2012**, *23*, 216–224.
16. Walmart. Available at www.walmart.com, (accessed September 10, 2012).
17. The Sustainability Consortium (TSC). Available at www.sustainabilityconsortium.org (accessed September 10, 2012).
18. Lapkin, A; Joyce, L.; Crittenden, B. *Environ. Sci. Technol.*, **2004**, *38*, 5815–5823.
19. Azapagic, A.; Millington, A.; Collett, A. *Chem. Eng. Res. and Des.*, **2006**, *84*, 439–452.

20. Carvalho, A.; Gani, R.; Matos, H. *Process Safety and Environmental Protection*, **2008**, *86*, 328–346.

21. Ruiz-Mercado, G. J.; Smith, R. L. Gonzalez, M. A. *Ind. Eng. Chem. Res.*, **2012**, *51*, 2309–2328.

22. Constable, D. J. C.; Curzons, A. D.; Cunningham, V. L. *Green Chemistry*, **2002**, *4*, 521–527.

23. Bare, J. C.; Norris, G. A.; Pennington, D. W.; McKone, T. *J. Ind. Ecol.*, **2002**, *6*, 49–78.

24. Baumann, H.; Tillman, A. *The Hitch Hiker's Guide to LCA: An Orientation in Life Cycle Assessment Methodology and Application*, Studentlitteratur, Sweden, **2004**.

25. Gani, R.; Jiménez-González, C.; Constable, D. J. C. *Comp. Chem. Eng.*, **2005**, *29*, 1661–1676.

26. Folic, M.; Gani, R.; Jimenez-Gonzalez, C.; Constable, D. J. C. *Chinese J. Chem. Eng.*, **2008**, *16*, 376–383.

27. Capello, C.; Fischer, U.; Hungerbuhler, K. *Green Chem.*, **2007**, *9*, 927–934.

28. Andraos, J. *Org. Proc. Res. Dev.*, **2009**, *13*, 161–185.

29. Halim, I.; Srinivasan, R. *Chem. Eng. Res. Des.*, **2008**, *86*, 809–822.

30. Charpentier, J. *Comput. Chem. Eng.*, **2009**, *33*, 936–946.

31. Charpentier, J. *Chem. Eng. Res. Des.*, **2010**, *88*, 248–254.

32. Ruiz-Mercado, G. J.; Smith, R. L.; Gonzalez, M. A. *Ind. Eng. Chem. Res.*, **2012**, *51*, 2329–2353.

33. Young, D. M.; Cabezas, H. *Comput. Chem. Eng.*, **1999**, *23*, 1477–1491.

34. Diwekar, U. *Resources, Conservation and Recycling*, **2005**, *44*, 215–235.

35. Hugo, A.; Ciumei, C.; Buxton, A.; Pistikopoulos, E. N. *Green Chemistry*, **2004**, *6*, 407–417.

36. Li, C.; Zhang, X.; Zhang, S. *Chem. Eng., Res. Des.*, **2006**, *84*, 1–8.

37. Li, C.; Zhang, X.; Zhang, S.; Suzuki, K. *Chem. Eng. Res. Des.*, **2009**, *87*, 233–243.

38. Khan, F. I.; Natrajan, B. R.; Revathi, P. *Journal of Loss Prevention in the Process Industries*, **2001**, *14*, 307–328.

39. Khan, F. I.; Sadiq, R.; Husain, T. *Environmental Modeling & Software*, **2002**, *17*, 669–692.

40. Kralisch, D.; Reinhardt, D.; Kreisel, G. *Green Chemistry*, **2007**, *9*, 1308–1018.

41. Yu, Q.; Zhixian, H.; Zhiguo, Y. A. N. *Chinese J. Chem. Eng.*, **2007**, *15*, 81–87.

42. Yan, W.; Chen, C.; Chang, W. *Comput. Ind. Eng.*, **2009**, *56*, 1617–1626.

43. Dunn, J. B.; Savage, P. E. *Green Chemistry*, **2003**, *5*, 649–655.

44. Zhang, Y.; Dubé, M. A.; McLean, D. D.; Kates, M. *Bioresource Technol.*, **2003**, *90*, 229–240.

45. Henrich, E.; Dahmen, N.; Dinjus, E. *Biofuels, Bioproducts and Biorefining*, **2009**, *3*, 28–41.

46. Eckert, C.; Liotta, C.; Ragauskas, A.; Hallett, J.; Kitchens, C.; Hill, E.; Draucker, L. *Green Chemistry*, **2007**, *9*, 545–548.

47. Cardona, C. A.; Sánchez, Ó. J. *Bioresource Technol.*, **2007**, *98*, 2415–2457.

48. Gutiérrez, L. F.; Sánchez, Ó. J.; Cardona, C. A. *Bioresource Technol.*, **2009**, *100*, 1227–1237.

49. Steffens, M. A.; Fraga, E. S.; Bogle, I. D. L. *Comput. Chem. Eng.*, **1999**, *23*, 1455–1467.

50. Lier, S.; Grünewald, M. *Chem. Eng. Technol.*, **2011**, *34*, 809–816.

11

GREEN CHEMISTRY AND TOXICOLOGY

Dale E. Johnson

Department of Nutritional Sciences & Toxicology
University of California, Berkeley;
Department of Environmental Health Sciences, University of Michigan;
Emiliem, Inc.

Grace L. Anderson

Department of Nutritional Sciences & Toxicology
University of California, Berkeley

11.1 INTRODUCTION

The principles of green chemistry have closely been related to the field of toxicology since the concepts were first articulated in the early 1990s. At least three of the 12 principles of green chemistry involve either human health issues or environmental toxicity and risk identification [1]. These are:

Principle 3. "Whenever practicable, synthetic methodologies should be designed to use and generate substances that possess little or no toxicity to human health and the environment." An assumption is made that hazards of specific chemicals can be defined a priori.

Green Chemistry and Engineering: A Pathway to Sustainability,
Anne E. Marteel-Parrish and Martin A. Abraham.
© 2014 American Institute of Chemical Engineers, Inc. Published 2014 by John Wiley & Sons, Inc.

Principle 4. "Chemical products should be designed to preserve efficacy of function while reducing toxicity." An assumption is made that levels of exposures of specific chemicals can be correlated to developing health issues or environmental problems. Additionally, it is assumed that chemicals of concern with known or suspected hazard traits can be identified and the levels in products then related to potential health or environmental issues. This is highlighted in a case study on safer consumer products (see Section 11.6.2).

Principle 11. "Analytical methodologies need to be developed further to allow for real-time, in process, and control prior to the formation of hazardous substances." This principle is especially relevant as interpreted today for toxicology because it can be viewed as defining emerging technologies that are changing the approaches used in toxicological hazard identification and risk assessment [2].

Chemicals that exist in or that are entering the environment fall into two toxicological categories: data rich and data poor, with 80–90% of all chemicals falling into the data-poor category. Over the last several years, new technologies have emerged that attempt to resolve the data-poor category to allow clear hazard-based decisions and regulatory actions to take place. These technologies have included multiple genomics technologies, high-throughput screening against known toxicity related targets, high-content screening where targets, time frames, and chemicals are evaluated in multiplexed formats, and more relevant cell-based assays that presumably replicate human systems [3]. More recently, the fields of systems biology and computational toxicology have evolved into primary tools used to identify data gaps and suggest prioritized screening programs.

In many respects, these newer approaches have changed the focus of toxicological inquiry from standard testing schemes in animals to a more integrated approach that identifies critical events leading to toxicity and indicators of hazards in a variety of nonanimal systems. The majority of publicly available open access databases of toxicological findings were derived from older, standardized testing procedures in which newer concepts were not incorporated. The newer concepts include attempts to define mode of action (MoA) [4], adverse outcome pathways (AOPs) [5], and the threshold of toxicological concern (TTC) [6].

This chapter highlights both older and newer methodologies and discusses the challenges of using animal-based, cellular, or subcellular assays to predict human toxicological outcomes. A critical question will be raised in Section 11.7: How much toxicological science must a chemist know and understand to function in a high-level green chemistry mode?

11.2 FUNDAMENTAL PRINCIPLES OF TOXICOLOGY

11.2.1 Basic Concepts

Several textbooks and articles have been written on various aspects of the field of toxicology, most notably *Casarett & Doull's Toxicology: The Basic Science of Poisons* [7], which offers clear, concise descriptions of key concepts of toxicology.

However, a useful starting point for a nontoxicologist is an open access toxicology tutorial at `http://sis.nlm.nih.gov/enviro/toxtutor/Tox1`. The following sections summarize key points in the online tutorial and can also be studied in more detail in the Casarette & Doull's textbook.

11.2.1.1 Definitions

The discipline of toxicology draws upon a combination of chemistry and biology. More specifically, toxicology is the study of the adverse effects of chemicals and physical agents on living organisms. A toxicologist investigates the inherent properties of toxicants, their interactions with biomolecules, and the subsequent effects on exposed organisms and systems.

Many types of chemicals may cause detrimental effects in living individuals. A *toxic agent* is simply anything that produces an adverse biological effect, including physical agents such as radiation or temperature. More narrowly, a *toxicant* is a substance that produces an adverse effect. This definition excludes physical and biological (living) agents. A *toxin* is a specific harmful protein produced by a living organism. **Poison** is a general term for a toxicant that causes immediate illness or death in relatively small concentrations. Note that biological agents such as bacteria and viruses that invade and multiply are not considered toxicants, though they possess the potential to harm their hosts. Toxicology typically addresses nonliving agents such as elements and molecules. Toxicology, therefore, is intricately linked with the understanding and advances of chemistry.

11.2.1.2 Effects

Many contributing factors determine the effect of a given toxicant in a particular situation. The age, species, and sex of an exposed organism each influences the toxicant's action. Additionally, the chemical form, dosage, and route of exposure (dermal, gastrointestinal tract, etc.) of the toxicant are critical factors. Together, these variables govern the amount of the substance that enters the body and thereby its ultimate effect. Toxicokinetics will be discussed more fully in subsequent sections.

Adverse effects can also be classified as chronic, subchronic, subacute, or acute. Chronic toxicity refers to cumulative damage after months or years of exposure to a toxicant. Subchronic usually describes an incidence of exposure that lasts several weeks or months. Subacute indicates an exposure event that is limited, but repeated more than once. Acute toxicity is the term for an immediate and often severe effect that is apparent after a single dose. A single compound may exert different effects at different exposure levels. For example, one acute effect of benzene is central nervous system depression, while chronic benzene exposure may cause bone marrow toxicity.

Toxic effects may be categorized further by the tissue, organ, or system that is targeted. Some examples to demonstrate the variety of toxic pathways are (1) reproductive toxins that harm the gonads, reproductive organs, or fetus, (2) hepatotoxins that specifically damage liver cells, or (3) genotoxins that alter DNA or chromosome structure or number. In this way, toxicants may act on whole organisms, specific cell types, or a single biological molecule. The diversity of toxicants is reflected in the myriad effects they produce in living systems.

11.2.1.3   Traditional Testing Methods

The experimental process is essential to the discipline of toxicology. Scientific studies allow toxicologists to identify toxicants and understand their mechanisms. This information is imperative to medical and regulatory professionals who work to preserve human health and environmental sustainability. Toxicological findings and expression of risk will appear differently based on whether the compound in question is an environmental chemical or a therapeutic agent. With therapeutics, there are always risk/benefit evaluations, whereas with environmental chemicals there are typically no benefits, simply risk. For chemicals in products, such as consumer products including intentional food additives, one must assume that the chemical has a desired function that is qualified by potential toxicity risk. Toxicological research, however, can be challenging to conduct as it necessarily investigates adverse effects in living systems, in many instances at much higher doses than those expected to reach humans. Toxicologists generally rely on three sources of scientific data to potentially minimize human injury: (1) accidental or routine exposure cases in human populations, or clinical trials with therapeutics in humans, (2) animal studies (in vivo), and (3) cellular studies (in vitro). Accidental or routine exposure scenarios afford toxicologists the rare opportunity to collect human data on environmental chemicals. This is invaluable in understanding human metabolism and susceptibility. Animal studies have traditionally been used to mimic and predict human reactions to toxicants in various levels and circumstances. Unlike the human exposure cases, animal studies allow the researcher to control more variables and produce statistically stronger data. The task of extrapolating from animals to humans, however, remains a potential challenge of the method. Cellular studies have grown in popularity and utility in recent decades. They offer the advantages of being highly controllable and economical and they do not use animal subjects. Cellular studies, however, are generally not as comprehensive as other test methods, and extrapolation to humans is even more challenging from cell lines than animal models. Currently, scientific advances and ethical concerns are pushing the field of toxicology toward alternative testing methods that spare human, animal, and environmental distress.

11.2.1.4   Dose and Dose Response

A *dose* is the amount of a substance administered at one time. As simple as this term is, it is the foundation of one of toxicology's founding precepts: "dose determines response." Based on this premise, there is a safe and dangerous level for every substance, whether it is typically considered benign or toxic. A dose–response graph depicts the relationship between exposure and result or toxicological endpoint. A typical dose–response graph uses experimental data to plot a variety of doses on the

x-axis against the frequency of some measurable effect on the y-axis. Most often the graph resembles an "S," with low doses eliciting no response and high doses eliciting the maximum response. Between these areas, the response frequency increases with the dose.

Key data points on a dose–response curve and which appear in the literature and public databases are the *no observed adverse effect level* (NOAEL) and *lowest observed adverse effect level* (LOAEL). These are actual data points on the graph that reflect the highest level in which no adverse effect is seen, and lowest level at which an adverse effect is seen. In some cases, these levels are the subject of interpretation by a toxicologist as to what is considered adverse in a study. If the level reflected any effect, adverse or not, it would be called the *no observed effect level* or NOEL. Other important pieces of information are the *effective dose* (ED), in therapeutic terms, *toxic dose* (TD), and *lethal dose* (LD). These doses can be extrapolated from the dose–response graph for any desired portion of the population. For example, the LD_{50} is the dose that is lethal in 50% of the population being tested. One measure of safety used by toxicologists studying therapeutic entities is the *therapeutic index* (TI). Usually this is calculated as the ratio of the TD_{50} over the ED_{50}, or the dose that is toxic in 50% of the population tested over the dose at which 50% of the population is expected to experience the relevant effect. These 50% levels are usually extrapolations from the dose–response curve. The TI is sometimes misleading, however, as it does not consider the unique shape of the dose–response curve for each determination; that is, the shape and slope may be different for toxicity than it is for efficacy.

The shape of a dose–response curve is highly informative. A compound is generally considered safer the further its effective (beneficial) dose is from its toxic dose. The slope of the dose–response line between the TD_0 and TD_{100} visually represents this relationship. A larger or steeper slope describes a potentially dangerous compound for which the toxic dose is close to the effective dose. A smaller or less steep slope implies a safer compound with a greater range between the effective and toxic doses. The *margin of safety* (MOS) uses this information to categorize and compare the safety of different compounds. The MOS is computed as the ratio of the LD_1 over the ED_{99}. The larger the MOS, the safer the compound is generally considered. Again, this is a calculation generally used for therapeutic compounds. For environmental chemicals, other calculations are used, some of which use an estimate of acceptable daily intake versus levels of intentional or unintentional exposure.

11.2.1.5 Risk Assessment

A few definitions are needed to understand risk assessment. *Risk* is the likelihood that a hazard will occur in a given set of circumstances. *Hazard* refers to the capability of a substance to cause an adverse effect. Furthermore, *risk assessment* is the practice of identifying hazards, exposures, and risks, and *risk management* is the regulatory action based on scientific, social, and economic factors.

The first publication of standardized risk assessment concepts and terminology was written by the National Academy of Sciences in 1983 [8]. This paper outlines four basic steps of risk management: (1) hazard identification, (2) dose–response assessment, (3) exposure assessment, and (4) risk characterization. In the first step, a scientist attempts to identify the harm a compound could cause.

This can be determined by animal, cellular, and epidemiology studies or predicted by the chemical structure using a structure–activity relationship (SAR) model. In the dose–response assessment step, available data are mathematically manipulated to address human metabolism and exposure parameters. This usually entails extrapolating a relevant low dose for humans based on high-dose animal studies. A useful value to calculate is the *acceptable daily intake* (ADI), which is calculated by multiplying the NOAEL by a safety factor, which accounts for uncertainty in cross-species extrapolations. The ADI reflects the amount an individual can be exposed to daily without experiencing adverse effects. The exposure assessment portion identifies the route, frequency, severity, and duration of an exposure event in a given population. Finally, the risk characterization step estimates a substance's effects in given environments and populations. The characterization must consider the intensity and nature of an effect and the consequences of exposure to multiple substances. Furthermore, this step should evaluate the previous three steps and note areas of uncertainty.

11.2.1.6 Regulation

Regulation is the intersection of scientific data and law. Regulation is based in risk assessment and describes the action taken by a government to protect the public from dangerous toxic exposures. Governmental bodies such as the Environmental Protection Agency, Food and Drug Administration, and Occupational Safety and Health Administration are responsible for enacting these protective measures. Regulatory practices include setting exposure limits, mandating protective occupational procedures, and requiring sufficient warning labels. The challenge of regulation is finding an acceptable, realistic exposure level. Calculations must include a variety of factors that may affect exposure such as water intake, air intake, time spent indoors/outdoors, and compounding exposures. With so many untested chemicals and variable exposure factors, regulation is one of the most challenging and important areas of toxicology.

11.2.2 Toxicokinetics

11.2.2.1 Introduction

Toxicokinetics describes the journey of a toxicant within a living system. The process includes four fundamental steps: absorption, distribution, metabolism, and excretion. Simplistically, *toxicokinetics* may be thought of as "what the body does to a chemical." This is contrasted by the term *toxicodynamics*, which may be thought of as "what the chemical does to the body." The amount of toxicant in the blood over certain time periods becomes the *internal dose* and is the most relevant measurement in determining potential effects in humans [9].

11.2.2.2   Absorption

For a toxicant to exert a biological effect, it must first gain access to a living individual. The route by which a chemical enters the body can dramatically influence its action. The absorption path is determined by a combination of the toxicant and organism's inherent chemical properties. In general, lipophilic, nonpolar compounds pass most easily through the body's lipid cellular membranes. Specialized receptors, however, may bind and uptake selected substances that could not independently enter the cell. Additionally, a process called endocytosis in which a cell engulfs surrounding molecules may enable cellular absorption of polar, hydrophilic molecules. Some of the most common routes of absorption are listed below.

GASTROINTESTINAL TRACT. The gastrointestinal (GI) tract is a complex system comprised of several diverse environments including the acidic stomach and alkaline intestines. Substances that are taken into the body through the mouth follow this absorption path.

The digestive solvents of the stomach provide a favorable environment for absorption of acidic compounds. Within the stomach, acidic molecules are non-ionized due to the excess of protons H^+. As nonionized compounds, they pass through the stomach's cellular membranes more easily into general circulation. Most compounds, however, continue through the GI tract to the small and large intestines after the stomach, where different pH gradients exist. Compounds absorbed via the gastrointestinal tract will be distributed to the liver first and subject to *first-pass metabolism* prior to entering the systemic circulation. This is discussed later in more detail (see Section 11.2.2.4).

RESPIRATORY TRACT. Volatile substances that are inhaled enter the body through the respiratory tract. This includes the nasopharyngeal region, tracheal area, and lungs. The lungs are the exchange site of vital gaseous compounds present in the air and volatile waste compounds from the body. The exchange relies on a large surface area within the lungs composed of a single layer of cells. While efficient for gas exchange, the slim barrier makes the lungs especially vulnerable to volatile toxicants.

DERMAL ROUTE. Substances that are able to penetrate the skin may follow the dermal absorption path. The skin, or epidermis, is a relatively difficult barrier to penetrate. It has many layers of cells, is comparatively dry, and provides only limited access to blood vessels. The outer layer of the skin, the stratum corneum, is packed with the protein keratin that makes it especially difficult to penetrate.

OTHER ROUTES. The GI tract, respiratory tract, and epidermis are the most important absorption paths of most toxicants, but other routes do exist, especially in a medical context. These include intradermal, subcutaneous, intramuscular, intraperitoneal, and intravenous injections.

11.2.2.3   Distribution

Distribution is the step of toxicokinetics in which a toxicant moves from the site of absorption throughout the body. The toxicant may pass cell barriers, enter the circulation system, or enter the lymphatic system. The distribution mechanism can have profound effects on the action of a toxicant. A substance absorbed through the GI tract will be transported through the portal system to the liver. Compounds absorbed through the lungs, skin, or intravenous injection, however, will immediately enter systemic circulation. These two routes are metabolically distinct, as demonstrated in the next section.

Distribution often follows blood flow. Organs that receive the most blood, therefore, are at heightened risk of toxicant exposure. Some toxicants may have special affinity for certain tissues regardless of blood circulation levels. The body also possesses several barriers such as the blood–brain barrier or placental barrier that specifically limit distribution. Distribution is not a uniformly beneficial or detrimental process. A toxicant may be distributed to its site of toxic action or it may be transported to an inert storage tissue. For example, lead is a potent neurotoxin when it reaches the brain. It is chemically similar to calcium, however, and may safely be taken up and stored by bone.

Several distribution models have been developed to describe a toxicant's movement through the body. These models help estimate the concentration and duration of a toxicant in a living individual. The simplest distribution models treat the body as one compartment. This assumes that the chemical is spread evenly through the body and eliminated at a steady rate proportional to the amount left in the system. In reality, few toxicants follow a one-compartment model. A two-compartment model is slightly more complex and usually more realistic. In this estimation, the first compartment clears the toxin from the body steadily with time, similarly to the one-compartment model. The toxicant concentration in a second compartment, however, rises as the first declines. This represents the toxicant's movement from one area of the body to a second compartment, for example, from the blood to adipose tissue (body fat). The concentration in the second compartment also declines with time, but at a different rate. This concept can be expanded to multicompartment models that consider additional regions of the body.

11.2.2.4   Metabolism

The body uses metabolic processes to transform toxic chemicals to more hydrophilic, thus more easily excreted, compounds. In some cases, however, metabolism "bioactivates" a molecule, increasing its toxicity. Metabolism can be broken into two main phases. In phase I, a functional chemical group is added to the chemical compound in preparation for phase II. Phase I enzymes can oxidize, reduce, or hydrolyze a toxicant. The most numerous phase I enzymes are called cytochrome P450s, or CYPs. In phase II, enzymes that recognize the phase I chemical motifs or functional groups add large polar groups to the toxicant. This prevents the compound from passing through cellular membranes and being absorbed and increases the elimination of the conjugate (substance in its hydrophilic state). Two important phase II reactions are glucuronidation which adds glucuronic acid (carboxylic acid $C_6H_{10}O_7$) or sulfation which adds sulfate (SO_4^{2-}) to the chemical structure. Each phase I and II enzyme has a range of accepted substrates, which are well known and in several instances are predictable.

The liver possesses the most metabolic enzymes. This is why it is important for toxicants to pass through the liver and undergo detoxifying metabolic reactions before entering systemic circulation. This protective process is called first-pass metabolism and, as mentioned earlier, is the main pathway for oral absorption. Substances absorbed dermally, intravenously, or inhaled bypass first-pass metabolism and the portion delivered to the liver is proportional to the systemic blood supply to the liver. The kidneys and lungs are lesser sources of detoxification and have approximately 10–30% of the metabolic capacity of the liver.

11.2.2.5 *Excretion*

Excretion is the process by which a compound exits the body. Urinary excretion, fecal excretion, and exhaled air are the most common pathways. Compounds removed from the blood by the kidney are excreted in urine. Within the kidney, functional units called nephrons filter the blood. Some bases and weak acids are actively transported out of the blood. Ions are especially subject to the kidney's filtration. Large molecules are primarily excreted in fecal matter because of the size exclusion in the glomerulus of the kidney. In addition, large molecules may be secreted in the bile and therefore excreted in the feces. Gases with low solubility and liquids with high volatility are efficiently excreted in exhaled air. Other less prominent excretion pathways include breast milk, sweat, and tears. Certain heavy metals such as Cr, Cd, Co, W, and U will accumulate in hair follicles and be detectable in hair.

11.2.3 Cellular Toxicity

The body is an adaptive system that constantly adjusts in response to its changing surroundings. *Homeostasis* refers to the body's ability to retain a relatively stable state in variable conditions. In the presence of a disruptive toxicant, cells undergo various response pathways to cope with the imbalance. This may involve an increase or decrease in cellular activity or an alteration to the cell's morphology and function. At times, the repair system is detrimental to the organism, and this is referred to as *pathological adaptation*. A relatively new approach in toxicology research deals with the identification of targets or biological molecules that are perturbed and that create a signal of potential outcomes. These signals are called *biomarkers* and will be discussed later (see Section 11.3.2).

11.2.3.1 Cellular Responses

Atrophy is the cellular response of size reduction. This response lessens the cell's oxygen, organelle (specialized subunit within a cell with a specific function), and nutrient needs. This may be an effective reaction to diminished resources. *Hypertrophy* is an increase in cell size. This can be beneficial when increased capacity is demanded of a cell that does not normally divide, such as cardiac or skeletal muscle cells. *Hyperplasia* refers to an increase in cellular number by division. This is only possible in cells capable of mitosis. During *metaplasia* one mature cell type transforms to another mature cell type. This process is used when scar tissue, for example, replaces normal functioning tissue in response to chronic irritation or inflammation.

11.2.3.2 Cellular Death

Some toxins induce irreversible cell damage. When cells are unable to effectively adjust to an imbalance, they undergo apoptosis or necrosis. *Apoptosis* is a normal cellular process in which a cell undergoes programmed death. The cell shrinks and fragments into bodies that are naturally phagocytosed or cleared from the area. This does not initiate an inflammation response or effect surrounding cells. *Necrosis*, conversely, is a disorderly and hazardous cellular death pathway. When a toxicant induces cellular necrosis, the cell may lose membrane integrity and leak its internal contents. This initiates an inflammatory response that is potentially dangerous to surrounding cells. These attributes of cells are used to determine the effects of compounds in specific cell types in cellular assays. Such assays are used to develop in vitro dose–response curves in the studied cells.

11.2.3.3 Neurotoxicity

The central and peripheral nervous systems are composed of specialized neural cells. Toxicants may affect neurons and supporting cells in three main fashions. The first is by inducing cell death. The second is by disrupting electrical transmission. Neurons function by passing electrical signals down the length of their cell bodies to adjoining cells. When this process is disrupted, sensory, motor, and thought functions are diminished or lost. Some signals and commands are transmitted between neurons by neurotransmitters, or chemicals that are released from one neuron and received by another. Toxicants that interfere with this process may increase or decrease neurotansmitter activity and alter normal signaling.

11.2.3.4 Cancer

Cancer is a specialized response pathway that is a health concern of high importance to toxicologists. In cancer, a cell frequently experiences a mutation that confers unregulated growth. In a promotion event, the cell proliferates and forms a mass, or tumor. During progression, individual cells dislocate from the tumor. This is the process of metastasis. Carcinogens are agents that induce mutagenesis, promotion, or progression events. The formation of cancer in animal studies is the endpoint that causes the most widespread concern both for environmental chemicals and therapeutic agents. This will be discussed in more detail later in this chapter (see Section 11.6).

11.3 IDENTIFYING CHEMICALS OF CONCERN

Newer approaches to screening for potential toxic effects in humans rely heavily on the precedent of novel, useful screens that have been incorporated in the drug discovery process of the biopharmaceutical industry [10]. For the most part, procedures that have been most successful are those that are focused on under-standing chemically induced toxicity, where a correlation between chemical structure and potential toxicity is assessed [11, 12]. These include both in vitro and in vivo studies designed to uncover chemical reactivity or metabolic instability, unintended off-target effects possibly related to unanticipated interaction with targets or receptors other than those targeted for therapeutic reasons, physiochemical properties of compounds and/or metabolites which affect the absorption, distribu-tion, and elimination of these molecules, and induced or altered biological inter-actions that could lead to unintended effects. Compounds or classes of compounds that are known to confer toxicity to a specific tissue will typically lead to the development of an in vitro assay with results correlated with potential toxicity endpoints in either animal or human studies. These can become defined, quantifiable endpoints that can lead to a series of predictive toxicological practices. To facilitate the discovery process of therapeutics, chemical libraries are maintained that group chemicals by structural similarities [13]; therefore, detailed quantitative structure–activity relationship (QSAR) modeling can be used as a predictive process. It should be pointed out that these screens are used primarily to rank order compounds in analog series [14]. The challenge is greater for chemical compounds where distinct analog series do not exist, such as industrial chemicals or environmental pollutants. Years of work in this field have led to some important conclusions. First, screens must be predictive, and the predictive endpoint must be human toxicity. If the goal of any screening effort is to predict toxicity in rodents because rodents were used in pivotal toxicology studies, then the toxicity induced in rodents must be relevant and predictive of the same effect in humans. Typically, the most relevant cellular in vitro assays are done in human cell-based systems, which maintain both genotype and phenotype [15]. Second, assays developed retrospectively based on experience or a previously observed effect tend to validate the previous finding but may not be predictive in previously untested compounds. Third, it must be clear whether the screening process is predicting potential toxicity or a plausible mechanism that could potentially lead to toxicity in humans [14]. This lesson was learned in large-scale toxicogenomics screening programs where a gene expression endpoint was a potential mechanism marker but not necessarily predictive of toxicity [16].

Because of these challenges, and particularly the advantages and disadvantages of relying on results from single endpoint screens, the field has been turning to the integrated approach of incorporating emerging technologies into chemical safety assessment [17]. Three important aspects of this approach are highlighted below.

11.3.1 Mode of Action Approaches

The mode of action (MoA) approach seeks to gain an understanding of the key events along a causal pathway that lead to a toxicological endpoint. Extensive reviews of MoA examples exist in the literature [18, 19, 20]. Additionally, The International Programme on Chemical Safety (IPCS) has published Human Relevance Frameworks [21, 22], a process that incorporates a weight of evidence approach relying heavily on robust mechanistic and experimental data. The evaluation sequence is listed below as a series of questions.

1. Is there sufficient evidence to conclude that a MoA has been established in animals?
2. Are there fundamental differences between the experimental systems and humans that would make it unlikely that the toxicity would occur in humans?
3. Assuming that the key events are plausible in humans, are there quantitative differences in either kinetics or dynamics that would indicate a differential sensitivity in humans?
4. To what extent do any quantitative differences in the key events impact selection of dose–response approaches and uncertainty factors used to estimate margins of safety? This is a key question in designing new methods and screening studies to supplement retrospective data or fill data gaps.

A good example of some of the uncertainty and differences in response between species is the evaluation of tumors in rodent studies and the corresponding relevance in humans [23]. Rodent carcinogenicity studies are evaluated and peer reviewed but only in reference to the chemical being tested and the validity of the findings (usually statistical significance) in the study. Studies and results appear in public access databases, and these data become the evidence from which to draw human relevance. Using examples of rat thyroid and pancreatic tumors, relevance conclusions may contain the following type of information.

The rat thyroid is more susceptible to secondary (nongenotoxic) carcinogenesis than the human thyroid because rats lack the high-affinity thyroxine-binding globulin present in humans. Instead, rats utilize a low-affinity albumin (103-fold lower affinity). Consequently, the thyroid hormone half-life in rats is 10 times shorter than in humans and turnover is more rapid, requiring higher "work" to maintain homeostasis. Furthermore, a greater amount of thyroid hormone is present in the follicular colloid of humans than rats. Thus, greater demand for the thyroid hormone in humans is addressed by the ready reserve available from thyroid binding globulin and colloid. The rat, however, requires synthesis of more hormone and is associated with a greater thyroid follicular proliferative response. Accordingly, rats are more susceptible to hyperplasia and neoplasia with the disruption of the synthesis, secretion, or metabolism of thyroid hormones. In rats, proliferative changes are primarily due to a prolonged stimulation by TSH released by the pituitary (endocrine gland connected to the hypothalamus) in response to decreases in circulating T3 and T4 levels (T3 and T4 are tyrosine-based hormones produced by the thyroid gland and responsible for metabolism regulation). Alterations in the normal feedback mechanisms usually occur from:

Interference with iodine uptake and thyroid hormone synthesis or secretion
Interference with peripheral metabolism of T3 or T4
Increased metabolism and excretion of thyroid hormone

Rats in an altered metabolic condition, particularly with an increased metabolic load and a stimulated pituitary gland, would be likely candidates for thyroid follicular tumors [19, 23, 24, 25, 26, 27]. Furthermore, unlike rats, hypothyroidism in humans, associated with increased TSH, is not related to an increased risk of thyroid cancer. The only recognized human thyroid cancer-inducing agent is radiation.

Pancreatic cancers of the acinar cell type (acinar refers to a cluster of cells resembling berries) can be experimentally induced in rats through a sequence of changes beginning with hyperplasia, progressing to adenomas and ultimately carcinomas. In contrast, humans characteristically develop ductal carcinomas and their pathogenesis does not involve acinar cell hyperplasia or adenomas. High unsaturated fat diets, corn or safflower oil by gavage, trypsin inhibitors (e.g., raw soy flour), and gastrointestinal surgical procedures have all been shown to induce pancreatic acinar cell cancer in rats. The gastrointestinal hormone cholecystokinin (CCK) is a trophic factor for the normal rat pancreas leading to increased acinar cell proliferation and may enhance pancreatic tumors in rats. There is evidence that stimulation of endogenous CCK levels by different xenobiotics will lead to rat pancreatic hypertrophy and hyperplasia [28, 29, 30, 31]. CCK does not act as a trophic factor in mice, hamsters, dogs, nonhuman primates, or humans; CCK does not increase acinar cell proliferations in these species and does not appear to be involved in carcinoma induction in these species.

These two examples show the extent of evaluation needed to establish a MoA approach and judge the relevance of toxicological findings from animals in humans.

11.3.2 Adverse Outcome Pathways

The adverse outcome pathway (AOP) approach is characterized by identifying a sequence of events based on chemical–biological interactions at the molecular level [5]. It describes in vivo chemical perturbations that trigger subcellular, cellular, tissue, organ, organism, and ecotoxicological population effects from exposure to the toxicant. Since several of the event endpoints or perturbations are speculated or predicted, this creates the need for robust predictive algorithms that can substitute for experimental data. Endpoints in these toxicological systems are frequently referred to as biomarkers. Biomarkers are characteristics objectively measured to become indicators of normal biologic processes, pathogenic processes, or, in the case of therapeutics, pharmacologic response(s) to therapeutic intervention. Biomarkers are frequently used to validate an in vitro system and establish that it characterizes toxicity in a way similar to in vivo studies. Biomarkers are an integral part of targeted therapeutic approaches and personalized medicine because they serve as an objective indicator that a molecule is reaching its intended target and eliciting a desired effect. In environmental research and risk assessment, biomarkers are frequently referred to as indicators of human or environmental hazards [32]. The key to discovering or predicting biomarkers through computational means involves the prediction or identification of the molecular targets of toxicants and the association of these targets with perturbed biological pathways.

11.3.3 Threshold of Toxicological Concern

The threshold of toxicological concern (TTC) describes a level of toxicant exposure that represents negligible risk to human health or the environment. In some situations, this is also referred to as a *de minimis* level. Data-poor chemicals often are classified using TTC as a surrogate for definitive toxicity data [6]. Decision trees are often constructed using the original classifications by Cramer and co-workers [33], which are as follows:

- *Class I.* Simple-structure chemicals that are efficiently metabolized and have a low potential for toxicity
- *Class II.* Chemicals of intermediate concern that are less innocuous than Class I substances but lack the positive indicators of toxicity that are characteristic of Class III chemicals.
- *Class III.* Chemicals with structures that suggest significant toxicity or for which it is not possible to presume safety.

Over time the TTC principles have been vetted against a number of diverse datasets and differing opinions have been raised about the robustness of the approach.

11.3.4 Chemistry-Linked-to-Toxicity: Structural Alerts and Mechanistic Domains

Structural alerts or chemical motifs known to be associated with toxicity through either the parent compound or reactive metabolites have also been used to predict potential toxicity from chemical structures. These expert algorithms appear in commercial software programs and in online open access sites. For instance, there are seven chemical domains that are used to define and predict the covalent interaction between a chemical and a macromolecule (biological target) that leads to an initiating event at the beginning of an AOP [34]. These include Michael addition, acylation, Schiff base formation, aromatic nucleophilic substitution, unimolecular aliphatic substitution, bimolecular aliphatic nucleophilic substitution, and reactions involving free radicals. The structural alerts from ToxTree define 57 unique structural features that are categorized into various toxicological domains [34]. These are relevant from a chemical structure standpoint but it must be remembered that they lack biological context. These alerts have been relevant and particularly useful in predicting in vitro genotoxic potential where results can be validated back through higher throughput assays. The website for OpenTox, `www.opentox.org`, is a source for both the TTC and structural alerts and also for characterizing compounds through a "read across" approach, where groups of chemicals are categorized based on physicochemical and toxicological properties. Using the assumption that "like structure" correlates with "like effects," these approaches are used to fill data gaps on data-poor compounds [35].

11.4 TOXICOLOGY DATA

11.4.1 Authoritative Sources of Information

Chemicals of concern are typically included in lists maintained by authoritative bodies internationally. An excellent open access source of the lists is contained in the PLuM (Public Library of Materials) database maintained by the Berkeley Center for Green Chemistry (`http://bcgc.berkeley.edu`). The lists included in PLuM along with the number of compounds represented on each list (last updated October 2011) are as follows:

- Canada DSL   22,017
- SIN List 1.1   480
- EC PBT Info   127
- REACH Restricted Substances   1012
- REAC SVHC Candidate List   55
- IARC Monographs   182
- Stockholm Convention POPs   37
- California Prop 65   187
- Washington PBTs   86
- Neurotoxic (Grandjean & Landigan)   215
- EC Endocrine Disruptors   421
- Asthmagens on the AOEC Exposure Code List   303
- US NIOSH Occupational Carcinogen List   146

Several databases are available for detailed toxicological information on chemicals. Judson [36] and Voutchkova et al. [37] have compiled database lists along with web addresses for several of the most useful resources. The most comprehensive source is from the National Library of Medicine and The Division of Specialized Services (SIS), which produces the Toxicological and Environmental Health Information Program (TEHIP). The TOXNET component at `http://toxnet.nlm.nih.gov/` is a compilation of several databases on toxicology, hazardous chemicals, environmental health, and toxic releases. Some of the key databases are listed below.

ChemIDPlus. Chemical information on over 370,000 chemicals.
HSDB. Hazardous Substances Data Base; over 4700 chemicals with peer-reviewed toxicology data.
TOXLINE. References for toxicology literature.
CCRIS. Chemical Carcinogenicity Research Information System; carcinogenicity and mutagenicity data on over 8000 compounds.
DART. References to developmental and reproductive toxicology literature.
GENETOX. Peer-reviewed genetic toxicology test data on over 3000 compounds.
IRIS. Integrated Risk Information Systems provides hazard identification and dose–response for over 500 chemicals.
ITER. International toxicity estimates for risk for over 600 compounds.
LACTOMED. Drugs and other chemicals to which breastfeeding mothers may be exposed.

11.4.2 Data Gaps: The Challenge and the Opportunity Arising from New Technologies

As mentioned previously, between 80% and 90% of all chemicals reaching the environment, present in the workplace, and/or contained in consumer products lack sufficient toxicological data to develop clear and unambiguous risk assessments for human health. Several new technologies now being used to address these issues include high-throughput screening (HTS), high-content screening (HCS), systems biology and pathway mapping, stem cell screening, and virtual tissues. With HTS, there is a long history of cellular and target screening in the pharmaceutical industry where millions of compounds have been tested against a multitude of targets with potential therapeutic indications and for off-target toxicity. However, for the most part, these screens have involved single concentrations to set molar threshold levels for target interaction. The goal has been to obtain potency information, rank order series of compounds, and create analogs through chemical structure manipulations and test these analogs for improved potency and specificity of target interactions. Compounds that pass certain screening criteria, or "filters," progress through several series of tests including animal assays and eventually human trials. In this way, the relevance of the initial screens can be assessed. For environmental chemicals lacking relevant human data except epidemiological studies, the HTS has evolved into a quantitative approach where chemicals are typically tested at seven or more concentrations in systems that can process over 100,000 compounds per day. Presumed relevance comes from the use of human cells or cell lines and human proteins. The use of assay cells derived from stem cells continues to be researched in terms of applicability to in vitro screening because the cells are capable of self-renewal and exhibit pluripotency. The goal of high experimental throughput is to quickly identify bioactivity signatures that potentially relate to the induction or exacerbation of adverse biological effects [3]. The balance between experimental throughput and human relevance will continue to be debated and hopefully revealed in the near future. Questions that will need to be addressed for toxic responses are the necessity for tissue or organismal context and the need for exposure and/or metabolic context in isolated cell preparations. While it is safe to say that "we aren't there yet," the major effort that is crossing between environmental and food and drug agencies suggests there will be a finish line in sight in the future.

Another challenge arising from new technologies is the status of legacy data both in government agencies and in industrial laboratories. Most current data exist in electronic form; however, a majority of older studies exist in paper format. In addition, certain software applications used by both industry and government agencies utilize arbitrary scoring systems that are specific to certain commercial software applications. As we move into an integrated system utilizing all information and data, these issues must be addressed and resolved.

11.5 COMPUTATIONAL TOXICOLOGY AND GREEN CHEMISTRY

Computational toxicology involves the application of computer technology and mathematical/computational models to analyze, model, and/or predict potential toxicological effects from (1) chemical structure (parent compound or metabolites), (2) similar compounds, (3) exposure, bioaccumulation, and persistence, (4) differential indicators or patterns related to exposure, and (5) networks of biological pathways affected by the chemical [38]. In addition, it allows the further understanding of mechanisms of toxicity, whether they be organism specific, organ specific, and/or disease specific [3, 39, 40, 41]. Also, it is used to attempt to explain why certain individuals, ethnic groups, or populations are more susceptible to chemical exposures and to draw associations between chemical exposure and increased risk for certain diseases [42]. Key data components are relational databases that allow cross-referencing existing data to create informed predictions about data-poor compounds. Since predictions are only as good as the data models they are built on, a crucial step in this field is innovative approaches to generate and utilize newer data sources. Below we highlight some data sources and tools, primarily open access, which are widely used by toxicologists.

11.5.1 Tools for Predictions and Modeling

11.5.1.1 EPA Resources

ACToR (WWW.ACTOR.EPA.GOV). ACToR stands for Aggregated Computational Toxicology Resource and is a data warehouse maintained by the EPA. ACToR incorporates many specialized databases within the EPA including ToxRefDB (animal toxicity data), ToxCastDB (high-throughput screening data, discussed further below), ExpoCastDB (exposure data for chemical prioritization), and DSSTox (chemical structure data). Additionally, ACToR compiles data from over 1000 public sources. In total, ACToR provides toxicity data information for over 500,000 environmental chemicals, searchable by chemical name, identification numbers, or structure. The data warehouse contains a wealth of information including physicochemical properties and in vitro and in vivo toxicological data. ACToR is a comprehensive, easily searchable center for a broad range of chemical and toxicological queries.

ToxCast (HTTP://WWW.EPA.GOV/NCCT/TOXCAST). ToxCast is a system maintained by the EPA to rapidly screen and prioritize chemicals based on potential risk to human health and the environment. ToxCast utilizes advances in high-throughput (HTP) technology to economically screen large numbers of chemicals. Launched in 2007, the program has already screened more than 300 chemicals. For comparison, the EPA estimates that it took 30 years and $2 billion to screen an equivalent number of compounds with traditional animal toxicity tests. These 300 compounds were part of Phase I, the "proof of concept" portion of the process, in which well-studied chemicals were analyzed with the HTP screening methods. Phase I was completed in 2009 and positively demonstrated the HTP screens' ability to deliver results similar to the previously performed animal toxicity tests. Phase II is currently in process. In this phase the EPA is investigating 2000 diverse chemicals from various industries. Data produced thus far is available through the ToxCast database on the EPA website. In addition to being an economical and fast option, ToxCast helps reduce animal testing by using and validating alternative methods. ToxCast has helped illustrate the utility and accuracy of emerging alternative assessment tools.

VIRTUAL ORGANS (HTTP://WWW.EPA.GOV/NCCT/VIRTUAL_TISSUES). The Virtual Tissues Research Project undertaken by the EPA aims to create reliable in silico models (performed via computer simulations) to predict chemical responses and disease progression in important human tissues and organs. The tissue models are constructed from data gleaned in both in vitro and in vivo models. Animal in vivo assays typically provide the high-dose–response data for the model. Alone, this information is difficult to apply to low-dose exposure scenarios in humans. To construct the virtual organ, engineers combine this in vivo data with low dose, human cell-based in vitro assays to complete the dose–response curve. Together, a more complete and predictive in silico model is possible.

Due to its central role in toxicant metabolism, the liver is one of the first organs being constructed in the Virtual Tissue Research Project. Physiologically based pharmacokinetic modeling, cellular systems, and molecular networks are integrated to mimic the multitude of activities performed by the liver. Once completed, this innovative project will be an invaluable resource for accessible, accurate, and responsible prediction of liver toxicity.

Another virtual tissue model being developed is the v-embryo. This simulation investigates teratogenesis, or the production of birth defects, resulting from chemical exposures in a pregnant woman. The model is being constructed largely from zebrafish and stem cell research, two areas that are especially informative for developmental processes. The developing eye has been chosen as the prototype organ for the proof of concept phase of the virtual embryo. The eye is an excellent starting point because it is well studied previously, is susceptible to both genetic and environmental interferences, exhibits a range of phenotypes, and utilizes a multitude of signaling pathways. The next processes to be incorporated into the model will be vascular, limb, and embryonic stem cell development. The virtual embryo is an ambitious and important project that could contribute to improved human health from the moment of conception.

EPI SUITE (HTTP://WWW.EPA.GOV/OPPT/EXPOSURE/PUBS/EPISUITE.HTM). The Estimation Program Interface (EPI) Suite is a collection of predictive tools also maintained by the EPA. The suite is comprised of approximately 17 individual models that estimate various physiochemical properties. Such properties include log octanol–water partition coefficient, aerobic and anaerobic biodegradability of organic chemicals, and melting point, boiling point, and vapor pressure of organic chemicals. One model of special importance to toxicologists and green chemists is the Ecological Structure Activity Relationships (ECOSAR) program. This model predicts acute and chronic aquatic toxicity in fish, aquatic invertebrates, green algae, and some saltwater and terrestrial species. EPI Suite is useful in estimating the important physiochemical parameters that must be considered when designing a compound or product.

11.5.1.2　QSAR

Structure–activity relationships (SARs) are models founded on the concept that the activity of a compound is a direct function of its chemical structure. These models may be based on qualitative (noncontinuous) or quantitative (continuous) data and together are referred to as QSARs. QSARs are constructed by grouping chemicals by common structural characteristics or descriptors such as a functional group, hydrocarbon chain length, or polarity. A collection of well-studied compounds within the group of interest serves as a training set. The training set compounds are graphed according to structural characteristic and a given endpoint of interest. For example, hydrophilicity may be graphed against acute aquatic toxicity. The models range in complexity from two to many input parameters. Additionally, the graph may be built with a simple linear regression model, or may incorporate complex algorithms depending on the depth of information available, complexity of the relationship investigated, and purposes of the model. The model is evaluated for accuracy by a replacement process in which each compound is removed from the set then replaced according to the model's predictions. A reliable model will accurately reinsert the test compound. At this point, less researched compounds may be evaluated by the model for the endpoint of interest. This is an extremely efficient way to utilize existing data to fill data gaps without additional animal tests and at far less expense. The most reliable models are constructed from structurally related compounds and attempt to predict values associated with compounds that fit into the chemical space represented by the training set. When a newly evaluated compound falls outside the chemical space of the model, the predictions carry a high level of uncertainty. It is generally recognized that QSAR models should be structured to provide, (1) a defined endpoint, (2) an unambiguous algorithm, (3) a defined domain of applicability, (4) appropriate measures of goodness-of-fit, robustness, and measures of predictability, (5) and a mechanistic correlation. Ideally, the selection of chemical descriptors that are used in a model take into consideration the mechanism of action and the rate-limiting step of the biological process being modeled [43].

OECD QSAR TOOLBOX (HTTP://WWW.QSARTOOLBOX.ORG). The Organisation for Economic Co-operation and Development (OECD) was founded in 1960 and has 34 member countries including the United States. Their free QSAR Toolbox is especially helpful during the difficult task of grouping compounds for QSAR modeling. The toolbox will evaluate a compound for previously recognized structural motifs and indicate if it is currently part of any regulatory inventories or chemical categories. The toolbox will provide data previously generated for the compound of interest and similar compounds. Additionally, the toolbox can group compounds by common metabolites or mechanisms of action. The OECD QSAR Toolbox provides accessibility to existing QSAR data and assists in the building of new QSAR models by accurately categorizing compounds of interest.

11.5.1.3 OpenTox

OpenTox (www.opentox.org) is designed to assist in the creation of predictive computational toxicology tools. The program supports the formation and validation of models for both publicly available and private datasets. OpenTox includes two subprograms—ToxPredict and ToxCreate. Both modules allow those with limited QSAR training to take advantage of the predictive capabilities of QSAR technology. ToxPredict uses public datasets to predict a compound's toxicity from its structure. ToxCreate aids people in the generation of their own predictive tools, especially for private datasets.

11.5.1.4 Systems Biology Tools

GENEGO (WWW.GENEGO.COM). Genego is a commercial data mining and analytic tool that uses a systems biology approach. Genego consolidates information from a huge library of gene expression studies, SNPs, metabolic profiles, and high-content screening (HCS) assays into interactive maps and pathways. These networks can be filtered, combined, and organized according to the user's preferences. Additionally, the program maintains several canonical pathways of key importance to human toxicity. Private data or other compounds and networks of interest may be inserted into these canonical pathways to measure their potential influence on that pathway. Genego is a tool to discover and illustrate the complex relationships between compounds and physiological molecules. It is an excellent way to place a compound in the context of greater biological systems.

STITCH (www.stitch.embl.de). STITCH is a free online tool that takes information from experiments, literature, and other databases to map connections between chemicals and proteins. The user may choose how many intermediary compounds are allowable between the input components. The networks indicate by color what type of relationship has been identified between two compounds (e.g., inhibitory, cofactor, activation). Conveniently, these networks may be exported. Each relationship and compound may be clicked on to reveal the source of information. In this way, the user may have access to and evaluate the strength of the supporting data. STITCH is a simple way to find and illustrate chemical and physiological interactions.

THE COMPARATIVE TOXICOGENOMIC DATABASE (www.ctdbase.org). The Comparative Toxicogenomic Database (CTD) [44] is a public research tool primarily intended to help investigate the role of environmental chemicals on human health. The online database compiles research from a variety of reputable sources, such as the U.S. National Library of Medicine, so that users may perform broad or specific searches regarding the relationships of chemicals, genes, and diseases. One may search the database from a variety of starting points: chemical, disease, gene, gene ontology, organism, pathway, or reference. Initial input is returned with known interactions from all the other categories. For example, if "toluene" is typed into the chemical search box, the CTD returns information such as genes, diseases, and pathways associated with toluene in humans and other organisms. Additionally, specific information such as chemical codes, official names, basic chemical properties, and citations are provided. These lists are interactive, so any entry may be clicked to find further information. Each search also returns links to external citations, websites, and databases that may be relevant (e.g., GENETOX, household product database).

More experienced users can take advantage of the CTD's sophisticated analysis and organization tools. Under the "analyze" tab a variety of additional searches are available. One such function is the "batch search" that investigates several input points simultaneously. The search delivers a compact list of basic identifiers and known interactions within a variety of organisms. Other special analysis tools include the Venn diagram based inquiries. This function takes a few entries from the user then maps which genes, chemicals, or diseases are common between all or some of the different queries. The CTD is extremely useful for many problems related to environmental toxicology. For example, if one is concerned about a certain chemical being emitted into the environment, the CTD is an appropriate place to begin research. Upon looking up the chemical, genes that may be affected by the compound and diseases already known to be associated with the chemical will be returned. If the mechanism of action in the human body is poorly understood, the genes, pathways, related chemicals, and references to previous research may help piece together the mechanism. In the case of laboratory work, the CTD can provide invaluable preliminary information regarding which genes and processes to specially monitor in assessing the chemical's effects. Using the analysis tools, one may also perform these searches in tandem with another chemical of concern or gene of interest to yield more specific data. The Comparative Toxicogenomic Database is an excellent resource to begin, support, or expand research on the molecular relationships between chemicals and human health.

11.5.2   Interoperability of Models for Decision Making and the Case for Metadata

There are both opportunities and challenges in using computational methods to fill data gaps for risk assessment decisions on data-poor chemicals. Opportunities are detailed elsewhere in this chapter, but the major challenge will be the transparency of data and sources used. All computational models carry uncertainty factors, some of which result from the compounds used to construct the model and the structural similarities of the new compounds under evaluation. Detailed analyses of compounds will involve the movement of results of one model into another to finally reach a relevant decision point. This will test the ability of different information technology systems and software applications to communicate, exchange data accurately, and use the information that has been exchanged. Computationally generated data must eventually exist in a metadata format and contain tags to identify (1) the models used (and when), (2) uncertainty factors and how these change when the data travel from one model to the next, and (3) the validity of previously generated data in the context of new information on the chemical from emerging technologies.

Two recent events highlight these challenges in expert meetings and scientific workshops. The Council of Canadian Academies convened an Expert Panel on the Integrated Testing of Pesticides and published a document in 2012 entitled *Integrating Emerging Technologies into Chemical Safety Assessment* [17]. This expansive document discusses topics highlighted in this chapter and it will become a major relevant source of information in this field for years to come. The Society of Toxicology convened a SOT CCT Meeting—Building for Better Decisions: Multiscale Integration of Human Health and Environmental Data—in May 2012 at the U.S. EPA site in Research Triangle, North Carolina. Topics in this meeting included the interoperability of models used for decision making on chemicals in the environment. Opinions and recommendations from the meeting will be published in future issue(s) of *Environmental Health Perspectives*.

Large-scale integration issues present both software and data challenges requiring software engineering as well as toxicology data generation solutions.

11.6   APPLICATIONS OF TOXICOLOGY INTO GREEN CHEMISTRY INITIATIVES

11.6.1   REACH

REACH is the European Commission's regulatory system for chemicals produced and used in industry. REACH stands for Registration, Evaluation, Authorization, and Restriction of Chemicals and was enacted into law June 1, 2007. This relatively new system aims to increase the collective knowledge of industrially important chemicals, protect human health, and promote environmental sustainability. REACH is committed to obtaining these goals while conserving resources and respecting animal life to the highest possible degree. This is achievable with collaborative research and reliable alternative testing methods [45].

REACH is currently in the process of implementation. Part of the REACH strategy includes shifting a substantial amount of the chemical safety responsibilities from public authorities to manufactures. Registration deadlines are determined by the amount of a compound a company manufactures per year. For example, the deadline for substances produced at greater than or equal to 100 tons per year is June 2012, while the deadline for those produced at 1 ton or more per year is June 2018.

Safety information is also required of manufacturers according to the amount produced per year. High quantity producers must submit a Chemical Safety Report (CSR) that includes key information such as intended uses of the compound, exposure and risk assessments based on these uses, and risk management strategies to lessen the risks. This information is used to create accurate safety data sheets (SDSs) and labels. Downstream users must alert manufacturers of their intended uses so that they may be included in the initial assessments, or they must perform their own assessments. Additionally, downstream users and smaller manufacturers must incorporate the safety information and risk reduction strategies identified by upstream manufacturers.

Compounds classified as mutagens, carcinogens, or reproductive toxins of high concern must gain special authorization from The Commission before commercial use. This authorization is also required for compounds shown to be especially persistent in the environment, bioaccumulative, or toxic (PBT compounds). Authorization is awarded if proper containment protocols are demonstrated that significantly lessen the risk associated with these compounds of concern or if external circumstances make replacement of the compound unfeasible.

The Commission uses compound registration documents to distill and distribute safety information as well as identify gaps in knowledge. Additionally, this is an effective method to reduce animal testing, one of REACH's central goals. REACH mandates that data generated from vertebrate studies be shared among researchers, in part through the registration process. New vertebrate studies must be specially approved to ensure that results will be nonrepetitive and of high scientific quality and relevance. Researchers investigating similar compounds gather in a "Substance Information Exchange Forum" to share vertebrate data and define future needed studies. These strategies help preserve the natural world, protect animal life, unite scientists, consolidate data, and reduce research costs [46].

Dedication to alternative testing methods makes REACH regulation unique. Assuming proper validation, REACH accepts QSAR and read-across data in place of traditional *in vivo* assays. New animal studies are only permissible when the desired information is necessary and may not be reliably obtained with alternative methods. Additionally, the European Commission operates the European Center for the Validation of Alternative Methods (ECVAM). This agency investigates novel testing strategies and promotes sound alternative methods. The assays approved by ECVAM encompass a range of endpoints, from genotoxicity, to skin sensitization, to acute aquatic toxicity. REACH and ECVAM are leaders in the global shift toward responsible, alternative scientific methods.

11.6.2   State of California Green Chemistry Initiatives

In 2008, the State of California signed into law two green chemistry bills (AB 1879 and SB 509) to establish a broad policy to be coordinated through the Department of Toxic Substances Control (DTSC). This groundbreaking legislation utilizes toxicological information of chemicals to improve the safety of consumer products. In AB 1879, the DTSC was authorized to establish a process for product life cycle (cradle to grave) evaluation, identify and prioritize chemicals of concern (CoC), evaluate the presence of CoC in consumer products sold in California, and assess alternatives of CoC in products with the aim to create safer alternatives to "products of concern" in California. The bill also established a Green Ribbon Science Panel (DEJ is a member) to advise the DTSC. Bill SB 509 required the DTSC to establish and maintain a Toxics Information Clearinghouse for detailed information on specific chemical hazard traits and environmental and toxicological endpoints.

The regulations, referred to as "Safer Consumer Products," apply to all consumer products that contain a CoC and are sold, offered for sale, supplied, distributed, or manufactured in California. There are several exemptions, such as dangerous prescription drugs and devices and their associated packaging, dental restorative materials, medical devices, food, pesticides, and products used solely to manufacture a product exempted by law. Also exempted are products to be used solely out-of-state and those that are regulated by other regulatory bodies with the same ultimate purpose of safeguarding public health.

There are four steps in the regulatory process:

1. The DTSC must establish the CoC list, which initially will contain approximately 1200 compounds as of July 2012.
2. The DTSC must create a Priority Products list of prioritized products that contain CoC. The DTSC will give priority to products meeting the following criteria:
    a. CoC in the product pose a significant potential to cause adverse public health and environmental impact.
    b. The product is widely distributed and used by consumers.
    c. There exists the potential for the type and extent of exposures that can adversely impact public health or the environment.
    d. The products may result in potential adverse exposures to CoC through inhalation or dermal contact.
    e. For formulated products, the intended use is application directly to the body, dispersion as an aerosol or vapor, or as applied to hard surfaces there is likelihood of runoff or volatilization.
3. Manufacturers, importers, and retailers must notify the DTSC when their product is listed as a priority product, and they must perform an alternative assessment for the product and the CoC it contains. These assessments are designed to limit potential human exposures or the level of potential adverse public health and environmental impacts that the presence of the CoC in the product imposes.
4. The DTSC must regulate and impose regulatory responses on the products and alternatives, particularly to avoid regrettable substitutions.

These Safer Consumer Product regulations are far reaching and represent one of the first (if not first) legal standards for requiring alternative assessments of the scope envisioned. They also require and demand relevant, validated, and curated toxicological information on CoC that can be viewed and accepted by all parties. With this is mind, the selection of CoC takes advantage of several lists of compounds by authoritative bodies internationally and does not require the DTSC to create new information. All of these processes will take place in the public forum through the DTSC website. The more difficult process will be the alternative analyses and substitution of CoC with less toxic compounds. This will necessitate the ability to assess compounds with less data than the original CoC and to create a process of data gap identification and intuitive data generation so the alternative can be judged to be safe, or at least safer, than the CoC that it replaces. Predicting human toxicity is challenging, as can be seen in therapeutic research where one of the major stumbling blocks to successful drug development is the less than optimal prediction of human toxicity. With new medications, the factors involved in predicting adverse effects are complex and include the chemical (drug) itself, the individual with or without unique susceptibilities and concomitant medications, errors, particularly in compliance, and the combined effects of multiple medications and dietary supplements. All of these factors vary based on underlying conditions or diseases in the affected individual. Relying strictly on the chemical structure—including reactive chemical motifs and the generation of reactive metabolites—has not provided all the answers, particularly in controlled clinical trial situations [16]. This becomes even more critical with environmental chemicals because there are no controlled clinical trials and, in several cases, no relevant human data.

These regulations will require the use and implementation of newer technologies and predictive algorithms and an approach to integrate emerging technologies into chemical safety assessment [17].

11.7   FUTURE PERSPECTIVES

The field of toxicology has always been linked directly to the field of chemistry, simply because toxicology describes the adverse interactions between chemicals and biological systems. Years of research on how specific structural features of chemicals influence biological responses when exposure is relevant either from route or measured dose have emerged and can now be accessed via open access online sources as discussed earlier. These so-called structural alerts have been developed through the examination of large sets of data such as in TOXNET and ACToR and in the millions of compounds screened in the biopharmaceutical industry. This has led to several proposals of using similar directed synthesis programs to reduce potential toxicity of newly synthesized chemical compounds. Voutchkova et al. [37] present an excellent review of these proposals. As mentioned earlier, most of these structural modification processes have been the result of years of practical application in the pharmaceutical industry. It is important to remember, however, that synthesis to modify potential biological effects of potential therapeutics is always balanced by the counter screening for therapeutic potency, and new synthetic schemes have to remain within a novel intellectual property chemical space. Frequently, this also involves the synthesis of specific stereoisomers where both potency and toxicity can be significantly different [47]. The most important aspect of green chemistry in the biopharmaceutical industry is in the large-scale process chemistry production of the active pharmaceutical ingredient (API), and the impacts on worker safety and the environment.

With the rapid changes occurring in the field of toxicology there remains the question: "How much toxicology does a chemist have to know?" The majority of toxicologists practicing in the field today, including in academia, industry, and the government, do not have a complete up-to-date understanding of all the new technologies and approaches that are changing the field into an information science. The toxicology field will soon be divided into data generation and "tox-bioinformatics." There is a need for a specific green chemistry bioinformatics section within the field of toxicology where chemists and other nontoxicologists can access relevant information and models to solve specific green chemistry problems. This is different than simply incorporating toxicology course work into chemistry curricula where the chemist must then access a series of databases just to get relevant information. As mentioned earlier, the interoperability of models and data becomes an issue at the individual level.

A green chemistry tox-bioinformatics system currently does not exist but should be one of the priorities for the application of toxicology into green chemistry in the future.

REFERENCES

1. Anastas, P. T.; Kirchhoff, M. M. Origins, current status, and future challenges of green chemistry, *Acc. Chem. Res.*, **2002**, *35*, 686–694.

2. Richard, A. M.; Yang, C.; Judson, R. S. Toxicity data informatics: supporting a new paradigm for toxicity prediction, *Toxicology Mechanisms and Methods*, **2008**, *18*, 103–118.

3. Kavlock, R. J.; Dix, D. Compuational toxicology as implemented by the U.S. EPA: providing high throughput decision support tools for screening and assessing chemical exposure, hazard, and risk, *J. Toxicol. Environ. Health Part B: Crit. Rev.* **2010**, *13*, 197–217.

4. Dellarco, V. L.; Baetcke, K. A risk assessment perspective: application of mode of action and human relevance frameworks to the analysis of rodent tumor data, *Toxicol. Sci.* **2005**, *86*, 1–3.

5. Schultz, T. W. Adverse outcome pathways: a way of linking chemical structure to in vivo toxicological hazards: In: *In Silico Toxicology: Principle and Applications*, Cronin, M.; Madden, J. (Eds.), The Royal Society of Chemistry, London, 2010, pp. 351–376.

6. Munro, I. C.; Renwick, A. G.; Danielewska-Nikiel, B. The threshold of toxicological concern (TCC) in risk assessment, *Toxicol. Lett.* **2008**, *180*, 151–156.

7. Klaassen, C. D. (Ed.). *Casarett & Doull's Toxicology: The Basic Science of Poisons*, McGraw-Hill, New York, 2008.

8. Committee on the Institutional Means for Assessment of Risks to Public Health, National Research Council, *Risk Assessment in the Federal Government: Managing the Process*, The National Academies Press, Washington, DC, 1983.

9. Johnson, D. E.; Wolfgang, G. H. I.; Giedlin, M. A.; Braeckman, R. Toxicokinetics and toxicodynamics. In: *Comprehensive Toxicology*, Williams, P. D.; Hottendorf, G. H., (Eds.). Elsevier Science, Ltd., Amsterdam 1997, Vol. 2, pp. 169–181.

10. Johnson, D. E.; Wolfgang, G. H. I. Predicting human safety: screening and computational approaches, *Drug Discovery Today*, **2000**, *5*, 445–454.

11. Green, N.; Naven, R. Early toxicity screening strategies. *Cur. Opin. Drug Discov. Dev.*, **2009**, *12*, 90–97.

12. Johnson, D. E.; Wolfgang, G. H. I. Assessing the potential toxicity of new pharmaceuticals, *Curr. Top. Med. Chem.*, **2001**, *1*, 233–245.

13. Johnson, D. E.; Blower, P. E. Jr.; Myatt, G. J.; Wolfgang, G. H. I. Chem-Tox informatics: data mining using a medicinal chemistry building block approach, *Curr. Opin. Drug Discov. Dev.*, **2001**, *4*, 92–101.

14. Johnson, D. E.; Sudarsanam, S. Molecular challenges in frontloading toxicity testing of anti-cancer drugs in drug discovery. In: *Encyclopedia of Drug Metabolism and Interactions*, Lyubimov, A. (Ed.), John Wiley & Sons, Hoboken, NJ, 2012, pp. 1–20.

15. MacDonald, J. S.; Robertson, R. T. Toxicity testing in the 21st century: a view from the pharmaceutical industry, *Toxicol. Sci.*, **2009**, *110*, 40–46.

16. Johnson, D. E. Predicting drug safety: next generation solutions, *J. Drug Metab. Toxicol.*, **2012**, *3*, 1–4.

17. Canadian Report. *Integrating Emerging Technologies Into Chemical Safety* Assessment, 2012.

18. Klaunig, J. E.; Babich, M. A.; Baetcke, K. P.; Cook, J. C.; Corton, J. C.; David, R. M. PPARalpha agonist-induced rodent tumors: modes of action and human relevance, *Crit. Rev. Toxicol.*, **2003**, *33*, 666–780.

19. Meek, M. E.; Butcher, J. R.; Cohen, S. M. A framework for human relevance analysis of information on carcinogenic modes of action, *Crit. Rev. Toxicol.*, **2003**, *33*, 591–653.

20. Seed, J.; Carney, E. W.; Corley, R. A.; Crofton, K. M.; DeSesso, J. M.; Foster, P. M. Overview: using mode of action and life stage information to evaluate the human relevance of animal toxicity data, *Crit. Rev. Toxicol.*, **2005**, *35*, 664–672.

21. Boobis, A. R.; Cohen, S. M.; Dellarco, V. L.; McGregor, D.; Meek, M. E.; Vickers, C. IPCS framwork for analyzing the relevance of a cancer mode of action for humans, *Crit. Rev. in Toxicol.*, **2006**, *36*, 781–792.

22. Boobis, A. R.; Doe, J. E.; Heinrich-Hirsch, B.; Meek, M. E.; Munn, S. IPCS framework for analyzing the relevance of a noncancer mode of action for humans, *Crit. Rev. Toxicol.*, **2008**, *38*, 87–96.

23. Ward, J. M. The two-year rodent carcinogenesis bioassay—will it survive? *J. Toxicol. Pathol.*, **2007**, *20*, 13–19.

24. McClain, R. M. Mechanistic consideration for the relevance of animal data on thyroid neoplasia in human risk assessment., *Mutation Res.*, **1995**, *333*, 131–142.

25. Capen, C. C.; Martin, S. L. The effects of xenobiotics on the structure and function of thyroid follicular and C-Cells, *Toxicol. Pathol.*, **1989**, *17*, 266–293.

26. Alison, R. H., Neoplastic lesions of questionable significance to humans, *Toxicol. Pathol.*, **1994**, *22*, 179–186.

27. Dellarco, V. L.; McGregor, D.; Berry, C. Thiazopyr and thyroid disruption: case study within the context of the 2006 IPCS human relevance framework for analysis of a cancer mode of action, *Crit. Rev. Toxicol.*, **2006**, *36*.

28. Dethloff, L. Gabapentin-induced mitogenic activity in rat pancreatic acinar cells, *Toxicol. Sci.*, **2000**, *55*, 52–59.

29. Longnecker, D. Experimental pancreatic cancer: role of species, sex, and diet, *Bull. Cancer*, **1990**, *77*, 27–37.

30. Watanapa, P.; Williamson, R. C. N. Experimental pancreatic hyperplasia and neoplasia: effects of dietary and surgical manipulation, *Br. J. Cancer*, **1993**, *67*, 877–884.

31. Rao, K. N.; Takahashi, S.; Shinozuka, H. Acinar cell carcinoma of the rat pancreas grown in cell culture and in nude mice, *Cancer Res.*, **1980**, *40*, 592–597.

32. Larson, H.; Chan, E.; Sudarsanam, S.; Johnson, D. E. Biomarkers. In: *Computational Toxicology Vol II.* Reisfeld B and Mayeno A (Eds.). Humana Press, Springer Science, New York, 2012. Chapter 11, pp. 253–273. DOI 10.1007/978-1-62703-059-5_11.

33. Cramer, G. M.; Ford, R. A.; Hall, R. L. Estimation of toxic hazard: a decision tree approach, *Food and Cosmetics Toxicology*, **1978**, *16*, 255–276.

34. Enoch, S. J.; Cronin, M. T. D. A review of the electrophilic reaction chemistry involved in covalent DNA binding. *Crit. Rev. Toxicol.*, **2010**, *40*, 728–748.

35. Benigni, R.; Tcheremenskaia, O.; Jeliazkov, V.; Hardy, B.; Affentranger, R. *Initial Ontologies for Toxicity Data: OpenTox D3.1 Report on Initial Ontologies for Toxicology Data*; OpenTox, 2009.

36. Judson, R. S. Public databases supporting computational toxicology. *J. Toxicol. Environ. Health Part B: Crit. Rev.*, **2010**, *13*, 218–231.

37. Voutchkova, A. M.; Osimitz, T. G.; Anastas, P. T. Toward a comprehensive design framework for reduced hazard, *Chem. Rev.*, **2010**, *110*, 5845–5882.

38. Raunio, H. In silico toxicology—non-testing methods, *Frontiers Pharmacol.*, **2011**, *2*, 33.

39. Johnson, D. E.; Rodgers, A. D.; Sudarsanam, S. Future of computational toxicology: broad application into human disease and therapeutics. In: *Computational Toxicology: Risk Assessment for Pharmaceutical and Environmental Chemicals*, Ekins, S. (Ed.), John Wiley & Sons, Hoboken, NJ, 2007, pp. 725–749.

40. Richard, A. M.; Yang, C.; Judson, R. S. Toxicity data informatics: supporting a new paradigm for toxicity prediction, *Toxicology Mechanisms and Methods*, **2008**, *18*, 103–118.

41. Mortensen, H. M.; Euling, S. Y. Integrating mechanistic and polymorphism data to characterize human genetic susceptibility for environmental chemical risk assessment in the 21st century, *Toxicol. Appl. Pharmacol.*, **2011**. DOI:10.1016/j.taap.2011.01.015.

42. Chiu, W. A.; Euling, S. Y.; Scott, C. S.; Subramaniam, R. P. Approaches to advancing quantitative human health risk assessment of environmental chemicals in the post-genomic era, *Toxicol. Appl. Pharmacol.*, **2010**. DOI:10.1016/j.taap.2010.03.019.

43. Zvinvashe, E.; Murk, A. J.; Rietjens, I. M. Promises and pitfalls of quantitative structure–activity approaches for predicting metabolism and toxicity, *Chem. Res. Toxicol.*, **2008**, *21*, 2229–2236.

44. Davis, A. P.; King, B. L.; Mockus, S.; Murphy, C. G.; Saraceni-Richards, C.; Rosentsein, M.; Wiegers, T.; Mattingly, C. J. The comparative toxicogenomics database: update 2011, *Nucleic Acids Res.*, **2011**, *39*, D1067–D1072.

45. European Commission. REACH in Brief, 2007.

46. REACH Implementation Project. REACH Proposal Process Description; June 2004.

47. Tucker, J. L. Green chemistry, a pharmaceutical perspective, *Org. Proc. Res. Dev.,* **2006**, *10*, 315–319.

INDEX

Green Chemistry and Engineering: A Pathway to Sustainability,
Anne E. Marteel-Parrish and Martin A. Abraham.
© 2014 American Institute of Chemical Engineers, Inc. Published 2014 by John Wiley & Sons, Inc.